S0-ANT-787

COMPUTER MODELLING OF MICROPOROUS MATERIALS

COMPUTER MODELLING OF MICROPOROUS MATERIALS

Edited by

C.R.A. Catlow

Royal Institution of Great Britain
21 Albermarle Street
London W1S 4BS, UK

R.A. van Santen

Schuit Institute of Catalysis
Laboratory of Inorganic Chemistry and Catalysis
Technical University of Eindhoven
5600 MB Eindhoven, The Netherlands

B. Smit

Department of Chemistry
University of Amsterdam
Nieuwe Achtergracht 166
1018 WV Amsterdam
The Netherlands

2004

Amsterdam – Boston – Heidelberg – London – New York – Oxford
Paris – San Diego – San Francisco – Singapore – Sydney – Tokyo

ELSEVIER B.V.	ELSEVIER Inc.	ELSEVIER Ltd.	**ELSEVIER Ltd.**
Sara Burgerhartstraat 25	525 B Street, Suite 1900	The Boulevard	**84 Theobalds Road**
P.O. Box 211	San Diego	Langford Lane, Kidlington	**London**
1000 AE Amsterdam	CA 92101-4495	Oxford OX5 1GB	**WC1X 8RR**
The Netherlands	USA	UK	**UK**

First edition 2004

Library of Congress Cataloging in Publication Data
A catalog record is available from the Library of Congress.

British Library Cataloguing in Publication Data
A catalogue record is available from the British Library.

ISBN: 0-12-164137-6

♾ The paper used in this publication meets the requirements of ANSI/NISO Z39.48-1992 (Permanence of Paper).
Printed in Great Britain.

Preface

Modelling methods are now well established in physical, biomedical and engineering sciences; and are widely used in assisting the interpretation of experimental data and increasingly in a predictive mode. Applications to inorganic materials are widespread, and indeed, such methods now play a major role in modelling structures, properties and reactivities of these materials.

This book focuses on the use of modelling techniques in the science of microporous materials whose complexity and extensive range of applications both stimulates and requires modelling methods to solve key problems relating to their structural chemistry, synthesis and use in catalysis, separation technologies and ion exchange. The book is mainly concerned with modelling at the microscopic level — the level of atoms and molecules — and aims to give a survey of the state-of-the-art of the application of both interatomic potential-based and quantum mechanical methods in the field.

The authors are grateful to many scientific colleagues for their contributions to the themes of the book. We would also like to thank Mrs Jean Conisbee for her assistance in the preparation of the manuscript.

Richard Catlow
Berend Smit
Rutger van Santen

Foreword and Introduction

Microporous materials, including both zeolites and aluminophosphates are amongst the most fascinating classes of material, with wide ranging important applications in catalysis, gas separation and ion exchange. The breadth of the field has, moreover, been extended in the last ten years by the discovery of the versatile and exciting range of mesoporous materials.

Computational methods have a long and successful history of application in solid state and materials science, where they are indeed established tools in modelling structural and dynamic properties of the bulk and surfaces of solids; and where they are playing an increasingly important role in understanding reactivity. Their application to zeolite sciences developed strongly in the 1980s, with initial successes in modelling structure and sorption, and with an emerging capability in quantum mechanical methods. The field was reviewed over ten years ago [1], since when there have been major developments in techniques and of course in the power of the available hardware, which have promoted a whole range of new applications to real complex problems in the science of microporous materials. This book aims to summarise and illustrate the current capabilities of atomistic computer modelling methods in this growing field.

Atomistic simulation methods can be divided into two very broad categories. The first rests on the use of interatomic potentials (force fields). Here no attempt is made to solve the Schrodinger equation; rather, we use functions (normally analytical) which express the energy of the system as a function of nuclear coordinates. These may then be implemented in minimisation methods to calculate structures and energies; in Monte Carlo simulations to calculate ensemble averages; or molecular dynamics simulations to model dynamical processes (such as molecular diffusion) explicitly. The early chapters of the book describe the application of these methods to modelling structures, and molecular sorption and diffusion in microporous materials. The second class of methods does solve the Schrodinger equation at some level of approximation. Such methods are essential for modelling processes that depend explicitly on bond breaking or making, which include, of course, catalytic reactions. Both Hartree Fock (HF) and Density Functional Theory (DFT) approaches have been used in modelling zeolites, although, as will be apparent from the work discussed in the book, DFT methods have predominated in recent applications.

The book therefore opens with an update on the field of static lattice techniques — a field which enjoyed a number of successes during the 1980s in modelling both framework structures and extra-framework cation distributions. The chapter highlights recent developments in predictive structural modelling and the new and exciting field of simulations of zeolite surfaces.

The next three chapters focus on the modelling of sorbed molecules in zeolites. Chapter 2 describes the state-of-the-art of Monte Carlo methods in simulating sorption isotherms. Molecular dynamics simulations of sorbate diffusion are reviewed in Chapter 3, while Chapter 4 focuses on the growing applications of dynamical Monte Carlo methods to molecular transport in microporous solids.

Probably the biggest development in the last ten years has been in the application of quantum mechanical methods, the theme of Chapters 5–7. Different techniques and applications are reviewed, including both periodic and cluster methods, with the main emphasis being on techniques based on density functional theory. Chapter 5 focuses on applications employing periodic methods. In Chapter 6, the emphasis is on catalysis effected by acid sites; while Chapter 7 describes applications to catalytic processes in which the active sites are metal ions.

Another significant feature of the field in recent years has been the use of modelling methods in understanding zeolite synthesis, in particular relating to the role of organic templates. These applications form the basis of Chapter 8.

Modelling methods are ultimately only of value if they solve real problems in real systems. The final chapter therefore presents a selection of applications where modelling methods have played a central role in solving problems in zeolite science.

The emphasis of the book is on microporous materials, especially zeolites, but applications to mesoporous materials are also reviewed. And while a comprehensive coverage is not possible in a book of this length, the key current techniques in atomistic modelling are surveyed. We hope that the book illustrates the power of these methods in solving problems in the science of microporous materials.

1. Catlow, C.R.A. (Ed.), *Modelling of Structure and Reactivity in Zeolites*. Academic Press Limited, London, 1992.

Contents

Computer Modelling of Microporous Materials
C.R.A. Catlow, R.A. van Santen and B. Smit (editors)

Chapter 1

Static lattice modelling and structure prediction of micro- and mesoporous materials

C.R.A. Catlow, R.G. Bell, and B. Slater

Davy Faraday Laboratory of the Royal Institution, 21 Albemarle Street, London W1S 4BS, UK

1. Introduction

Detailed structural models at the atomic level are an obvious prerequisite for a microscopic understanding of processes in solids. Computer modelling has become an increasingly standard technique in structural studies of complex materials, in particular micro- and mesoporous materials. Such methods may be used to refine approximate models and, more ambitiously, to predict new structures. They are an invaluable complement to experiment in studying local structures around defects and impurities, and they play a central role in the development of models for the surfaces of complex materials, including very recent studies of zeolites.

This chapter reviews the application of static lattice methods employing interatomic potentials, both to model long-range, local and surface structures of micro- and mesoporous systems, and to study energetics and stabilities. Such methods remain the most effective and economical approach for structure modelling. They are, moreover, complementary to quantum mechanical methods, which may explore and, if necessary, refine the structural models which they yield.

In the following section, we will summarise the, by now very standard, methodologies involved in these calculations. We will then describe their applications to modelling structures and energetics. Next, we consider the important, fast developing field of the prediction of new microporous structures, which we follow with a brief account of the development of models for mesoporous structures. We conclude with a survey of the role of static lattice methods in simulating the structures of the external surfaces of zeolites.

2. Methodology

Static lattice methods rest upon the calculation of an energy term — lattice, surface or defect energy — which is then minimised with respect to structural variables, i.e. atomic coordinates and cell dimensions. The methods, including their applications to microporous solids, have been reviewed several times in the older and more recent literatures [1–5]. Lattice energy calculations essentially rest on summation, first of the electrostatic energies arising from the interactions between the charges on the atoms; these are long range and must, in effect be summed to infinity in any accurate treatment. In contrast, the second, the short-range or non-Coulomb terms, comprising Pauli or overlap repulsion, and attractive forces due to covalence and dispersion may be safely truncated beyond a 'cut-off' which is typically 15–20 Å in contemporary calculations. Hence the lattice energy, E_{LAT}, is given by:

$$E_{\mathrm{LAT}} = E_{\mathrm{ELEC}} + E_{\mathrm{SR}}, \tag{1}$$

where

$$E_{\mathrm{ELEC}} = \frac{1}{2}\sum_{ij}^{N} \frac{q_i q_j}{r_{ij}}, \tag{2}$$

where r_{ij} is the separation between pairs of atoms i and j, and q_i is the charge of the ith ion. The sum, in principle, involves all atoms in the system, but for periodic solids it can in practice be made rapidly convergent by using the Ewald technique [6], which is based on a partial transformation into reciprocal space, and is employed in almost all modern lattice energy calculations. We note that the assignment of the charges q_i is a critical aspect of such calculations, as discussed below.

The short-range energy is given by:

$$E_{SR} = \frac{1}{2}\sum_{ij} V_{ij}(r_{ij}), \quad (r_{ij} < r_{cut}) \tag{3}$$

where V_{ij} is the short-range energy and where the summation is taken out to the cut-off radius r_{cut}.

Modelling of defects may use 3D periodic methods, which require the defect to be embedded in a supercell, which is periodically and infinitely repeated. Alternatively, we can embed an isolated defect in an infinite lattice, and calculate its interaction with the surrounding lattice out to a specific radius, with more distant interactions being treated using quasi-continuum methods — the basis of the 'Mott–Littleton' method, which has been widely and successfully used in calculations on defects in ionic and semi-ionic solids [1,3], and which has played a useful role in modelling impurities in zeolites.

In modelling surfaces, two approaches are commonly used.

(*i*) *The semi-infinite (2D) approach:* A slab is created from the bulk crystal with the desired face oriented to the surface normal. The system is 2D-periodic in the plane of the surface, but aperiodic parallel to the surface normal. The slab is then divided into two regions, one which represents the crystal bulk and one that is relaxed to mechanical equilibrium. The number of layers in the slab is chosen such that the electrostatic field in the lower section represents the Madelung field of the crystal bulk. Once this has been determined, the number of layers in the upper block is increased until the total energy per formula unit is converged.

(*ii*) *The 3D-periodic slab approach*, which lends itself to most simulation codes that are capable of describing cells with a wide variety of cell shapes using 3D-periodic boundary conditions. In essence, a slab of the desired thickness is created from the crystal bulk and chosen such that, as in case (i), the electrostatic potential is converged and equivalent to that of the crystal bulk. In this instance, the slab is placed in a three-dimensional periodic box where, for example, the cell width and length are concomitant with appropriate surface vectors, and the height of the box is chosen such that the 'vacuum gap' (i.e. the gap between the periodically repeated slabs) exceeds the maximum short-range cut off. In practice, it is often necessary to choose a vacuum gap in excess of the cut off, owing to the long-ranged Coulombic attraction between the periodic images. In the case that the gap is filled with, for example, a gas or fluid, the dimensions must be changed to be consistent with partial pressures and liquid densities.

Turning now to the minimisation methods used to obtain the minimum energy configuration of the unit-cell dimensions and the atoms in the crystal surface or around the defect, these are again based on quite standard iterative procedures. Gradient, particularly conjugate gradient methods, may be used, but most contemporary codes make use of information on second derivatives. Such methods involve constructing, inverting and updating (the inverse of) a matrix whose elements, $(\partial^2 E/\partial x_i \partial x_j)$, are the second derivative of the energy function with respect to atomic coordinates. The different methods are distinguished by the approximation used in the construction/update procedures. For further details, see Refs. [1–4]; the account by Watson et al. in Ref. [4] is particularly useful. We note that all standard minimisation methods can do no more than locate the nearest minimum to the starting configuration, and there is no guarantee that this is the 'global' minimum. There can, indeed, never be any guarantee that a global minimum has been located; but as discussed later, procedures are available to explore in a systematic manner the whole potential energy surface of the system: such methods are far more likely to identify the global minimum.

Interatomic potentials are the crucial input to static lattice calculations. They are essentially a representation of the energy of the system as a function of its nuclear coordinates. In practice, they normally comprise a set of charges (normally point entities) assigned to the atoms and parameterised analytical functions for the short-range interactions.

The nature of the potential model used depends on the character of the bonding in the system. For ionic solids, the *Born model* is the appropriate starting point. Here the solid is considered as a collection of ions, to which formal or partial charges may be assigned, interacting via short-range potentials which are commonly described using the 'Buckingham' potential:

$$V_{\mathrm{r}} = A\mathrm{e}^{-r/\rho} - Cr^{-6}, \tag{4}$$

where A, ρ and C are parameters characteristic of the interaction. This pair potential may be supplemented by simple three body terms of which the 'bond-bending' function, favoured for silica and silicate systems, takes the form:

$$V(\theta) = \frac{1}{2} K_{\mathrm{B}}(\theta - \theta_0)^2, \tag{5}$$

where θ is the angle subtended by an O–Si–O angle and θ_0 is the equilibrium value for the SiO_4 tetrahedron. Such functions crudely represent the angle dependence of the covalence in the tetrahedral units. Ionic polarisability may also be included, where the most widely used approach is the shell model originally formulated by Dick and Overhauser [7], which describes a polarisable ion in terms of a mass-less shell (representing the valence shell electrons) which is coupled by an harmonic spring to a core in which all the mass of the ion is concentrated. A dipole is created by the displacement of the shell relative to the core, and since short-range interactions act between the shells, the model includes the necessary coupling between polarisability and short-range repulsion.

In contrast, the conceptual starting point for constructing models of covalent systems is the chemical bond rather than the ion. Therefore, in 'molecular mechanics' potentials, simple analytical functions (e.g. bond harmonic or Morse) are used to model the interactions between bonded atoms; angle-dependent, torsional and non-bonded terms (including electrostatic and short range) are also included.

Having chosen the type of potential model, it is necessary to fix the variable parameters, for which there are two broad classes of procedure: *empirical methods* fit variable parameters to crystal properties (structural, elastic, dielectric, thermodynamic and lattice dynamical), while non-empirical methods calculate the interaction between a cluster or periodic array of atoms by a theoretical procedure (usually an ab initio method in recent studies); the resulting potential energy surface is then fitted to a potential function.

The field of interatomic potentials for silica and silicates is extensive, with several models proposed over the last 30 years. There has been a long debate over the nature of the bonding — ionic versus covalent — in silicas; although as argued in Ref. [8], the whole concept of 'ionicity scales' in solids is difficult: there are no unambiguous ways of partitioning charge between different atoms in solids, and there are no properties which can be used directly to establish an ionicity scale. Nevertheless, there is a general consensus that the bonding in silicas/silicates is intermediate in nature showing characteristics of both covalence and ionicity. Born model and molecular mechanics potentials have therefore been developed for these systems.

Of the Born model potentials, the simple *rigid ion* (i.e. no ion polarisability), pair potential models are the most widely usable as they can be readily implemented in dynamical as well as static lattice models. A highly successful parameterisation was developed by van Beest et al. [9] using ab initio calculations. The model uses partial charges on Si and O and has simple pairwise short-range potentials acting between S···O and O···O. The simplicity and flexibility of the model has led to its widespread and successful use. A successful *shell model* parameterisation was developed by Sanders et al. [10]; their model also included bond-bending terms of the type described above. The model was parameterised using empirical procedures, while shell model potentials based on ab initio calculations were derived by Purton et al. [11].

Several 'molecular mechanics' parameterisations are available. Perhaps the most widely used are those based on the 'cvff' models developed by BIOSYM Inc. (now Accelrys). In particular, the cff91_zeo potential [12] has enjoyed wide and successful usage.

Interatomic potential parameters are also available for Al···O and P···O interactions. For the former, the work of van Beest et al. [9] derived parameters that were consistent with their Si···O parameterisation. Shell model parameters for Al···O were reported by Catlow et al. [13] and were successfully incorporated into models for aluminosilicates including zeolites [14]. For aluminophosphates, Gale and Henson [15] developed an ionic shell model set. However, it would be desirable to develop different models in view of recent work of Corà et al. [16], which showed, using ab initio methods, that the bonding in these materials was molecular-ionic, i.e. aluminophosphates are best envisaged as comprising Al^{3+} and (covalently bonded) PO_4^{3-} ions.

2.1. Computer codes

Several general purpose codes are available for undertaking static lattice modelling. The GULP code written by Gale [17] provides a wide range of functionality for lattice and defect energy calculations, and can also be used to fit variable parameters in interatomic potential models to both empirical data and ab initio potential energy surfaces. The METADISE [18] and MARVIN [19] codes allow calculations on surfaces with 2D-periodicity boundary conditions (and 2D Ewald summations). Commercial software is available from Accelrys Inc., in particular the DISCOVER code [20] has extensive functionality for minimisation and dynamical simulations on both molecules and solids.

The output from all these codes may be interfaced with graphical software permitting the display of the structures generated, the power and importance of which is evident in several chapters in the book.

3. Applications

We now review three main areas of application: the first is the straightforward application of lattice energy calculations to modelling structures and stabilities of solids; next, we consider the rapidly developing field of predicting new structures of microporous materials; and thirdly we summarise the new field of modelling zeolite surfaces.

3.1. Structures and stabilities

This field, which was developed in the 1980s and 90s, is now mature, and has been reviewed previously [1–4]. Several good illustrations are given in Chapter 9. Early work established the viability of using lattice energy minimisation methods in modelling cation distributions [21] and framework structures of zeolites. There were notable successes in modelling the monoclinic distortion of silicalite [22]. And as discussed in Chapter 9, the methods were successfully used in assisting the solution of the structures of the zeolite Nu 87 [23].

In addition to modelling crystal structures, several successful studies have been reported of local structures, including the detailed investigation of FeZSM-5 [24], where models were obtained of the local structure around framework Fe^{3+} (replacing Si) which compared well with experimental data employing the EXAFS technique.

Applications of these now routine methods continue to be of value. Three developments over the last 10 years deserve, however, special mention. The first is the success of calculations of *energetics* as well as of structures. It has been well known for many years that microporous materials are all metastable with respect to dense structures; in the case of high silica zeolites, calorimetric data have established that the enthalpy difference between the microporous structures and quartz is in the range 10–20 kJ/mol [25,26]. Henson et al. [27] reported a detailed comparison of experimental and calculated energetics of a range of microporous structures; the calculations all refer to pure silica systems, and the experimental to high silica materials. The comparison, which is summarised in Fig. 1, shows excellent quantitative agreement between calculation and experiment. The same study also examined, in detail, the comparison between calculated and experimental structures for a range of high silica materials, and found good agreement, with those calculations employing the shell model potentials of Sanders et al. [10] performing particularly well.

Another significant development concerns the study of Si/Al distributions in the clinoptilolite/heulandite group of zeolites, where

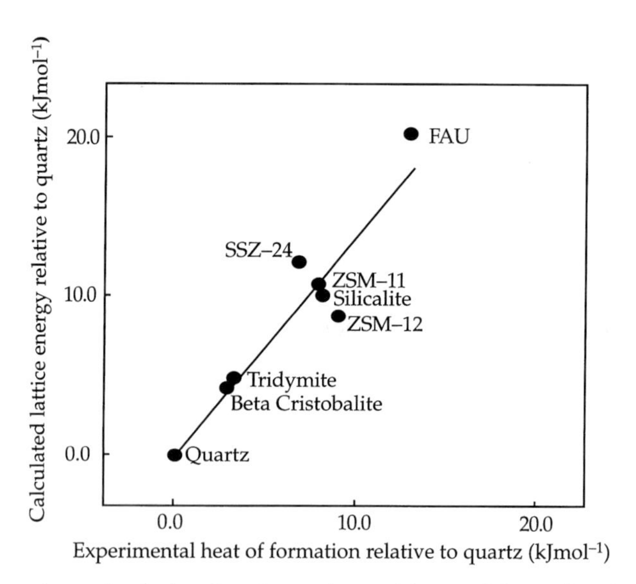

Fig. 1. Comparison of calculated and experimental heats of formation for high silica microporous materials (after Ref. [27]).

work of Ruiz-Salvador et al. [28] has combined a simple Monte Carlo procedure with lattice energy minimisation procedures to make successful predictions of Al distributions in these important natural zeolites. Channon et al. [29] also used lattice energy calculations to explore the Al distribution and cation locations in these materials. The success of their work suggests that these methods may be used increasingly routinely for modelling Si/Al distributions — a long-standing problem in zeolite science.

Thirdly, we should draw attention to the role of 'simulated annealing' methods in predicting zeolite structures. These methods use MD and MC techniques to explore configurational space for the system simulated (employing usually a simple readily computable energy function or a function based on simple geometric criteria). This stage of the calculation identifies plausible candidate structures, which are then refined by full lattice energy minimisation methods. More details are given in Chapter 9, and a successful example of the use of such methods was the impressive solution of the structure of a new AlPO material UI07 [30], where the simulated annealing methods generated a structure which successfully solved the high-resolution powder diffraction data for this material. We should note that in the first stage of the procedure, MD/MC simulations can be replaced by 'evolutionary' or genetic algorithm techniques, which allow candidate structures to evolve by exchange of features and by imitation. The viability of these methods in modelling zeolite structures has recently been demonstrated by Woodley et al. [31].

3.2. Hypothetical zeolites and lattice energy minimisation

There have been many attempts to predict new microporous structures, most of which have rested on the fact that the very definition of these materials is based on geometry, rather than on precise chemical composition, occurrence or function. In order to be considered as a zeolite, or zeolite-type material (zeo-type), a mineral or synthetic material must possess a three-dimensional four-connected inorganic framework [32], i.e. a framework consisting of tetrahedra which are all corner-sharing. There is an additional criterion that the framework should enclose pores or cavities which are able to accommodate sorbed molecules or exchangeable cations, which leads to the exclusion of denser phases. Topologically, the zeolite frameworks may thus be thought of as four-connected nets, where each vertex is connected to its four closest neighbours. So far 145 zeolite framework types are known [33], either from the structures of natural minerals or from synthetically produced inorganic materials. In enumerating microporous structures, a number of fruitful approaches have been developed. Some have involved the decomposition of existing structures into their various structural subunits, and then recombining these in such ways as to generate novel frameworks [34–42]. Methods which involve combinatorial, or systematic, searches of phase space have also been successfully deployed [43–45]. Recently, an approach based on mathematical tiling theory has also been reported [46]. It was established that there are exactly 9, 117 and 926 topological types of four-connected uninodal (i.e. containing one topologically distinct type of vertex), binodal and trinodal networks, respectively, derived from simple tilings (tilings with vertex figures which are tetrahedra), and at least 145 additional uninodal networks derived from quasi-simple tilings (the vertex figures of which are derived from tetrahedra, but contain double edges). In principle, the tiling approach offers a complete solution to the problem of framework enumeration, although the number of possible nets is infinite.

Potentially therefore we may be able to generate an unlimited number of possible zeolitic frameworks. Of these, only a portion is likely to be of interest as having desirable properties, with an even smaller fraction being amenable to synthesis in any given composition. It is this last problem, the feasibility of hypothetical frameworks, which is the key question in any analysis of such structures. The answer is not a simple one, since the factors which govern the synthesis of such materials are not fully understood. As discussed earlier, zeolites are metastable materials. Aside from this thermodynamic constraint, the precise identity of the phase or phases formed during hydrothermal synthesis is said to be under 'kinetic control', although there is increasing sophistication in targeting certain types of framework using various templating methods, fluoride media and other synthesis parameters [47]. Additionally, certain structural motifs are more likely to be formed within certain compositions, e.g. double 4-rings in germinates, 3-rings in beryllium-containing compounds. A full characterisation of any hypothetical zeolite must therefore include an analysis of framework topology and of the types of building unit present, as well as some estimate of the thermodynamic stability of the framework. Using an appropriate potential model, lattice energy minimisation can, as shown above, provide a very good measure of this stability as well as optimising structures to a high degree of accuracy.

In the method adopted by Foster and co-workers [48], networks derived from tiling theory were first transformed into 'virtual zeolites' of composition SiO_2 by placing silicon atoms at the vertices of the nets, and bridging oxygens at the midpoints of connecting edges. The structures were then refined using the geometry-based DLS procedure [49], before final optimisation by lattice energy minimisation. Among the 150 or so uninodal structures examined, all 18 known uninodal zeolite frameworks were found. Moreover, most of the unknown frameworks had been described by previous authors; in fact there is a considerable degree of overlap between the sets of uninodal structures generated by different methods. Most of the binodal and trinodal structures, however, are completely new. Using simulated lattice energy as an initial measure of feasibility, a number of more interesting structures are illustrated in Fig. 2. The challenge is now to synthesise these structures.

3.3. Modelling mesoporous structures

The existence of synthetic materials with ordered mesopores (channels with dimension in the range 20–100 Å) was first reported by scientists at Mobil in 1992 [50,51]. Since then a whole new field of material chemistry has developed based on such materials, in a host of compositions, and with a variety of potential applications. Compared to microporous zeolites, however, they present a problem for the computational

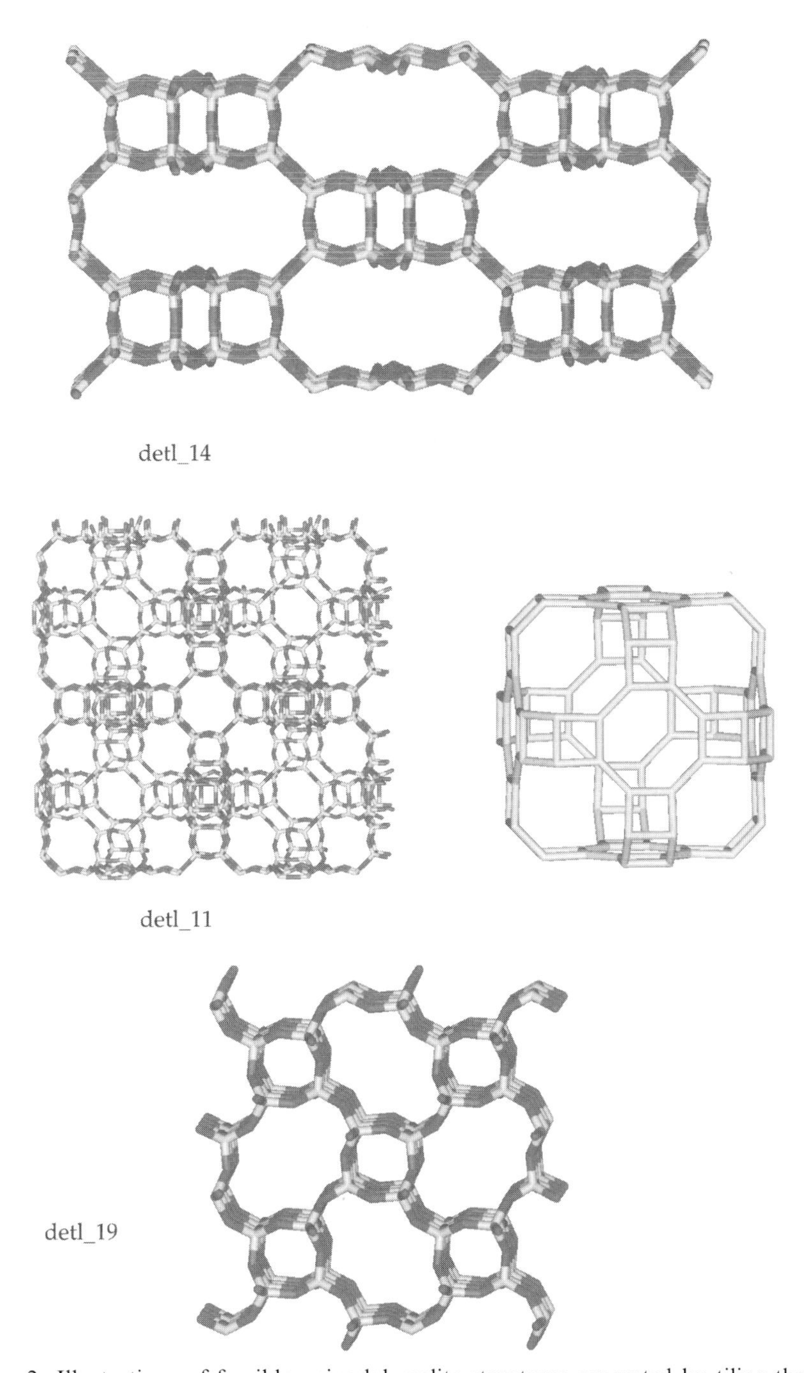

Fig. 2. Illustrations of feasible uninodal zeolite structures generated by tiling theory and modelled using lattice energy minimisation. (Continued on next page.)

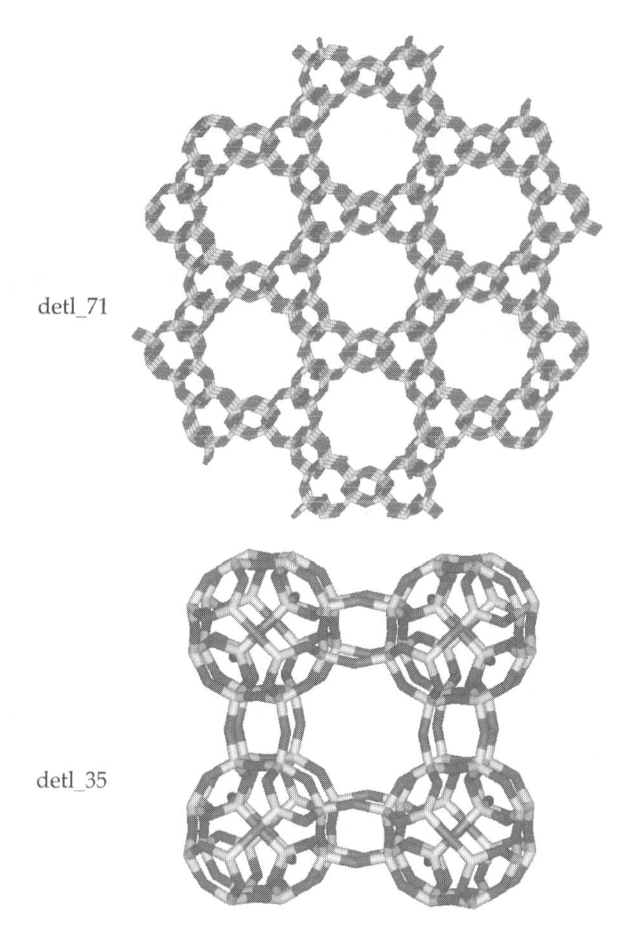

Fig. 2. (Continued)

chemist in that their short-range structure is poorly defined. The pores may be ordered and regular in size and shape, but the pore walls contain material which is crystallographically amorphous. A possible approach to modelling such structures involves taking a bulk amorphous structure obtained from high-temperature molecular dynamic simulations and then excising pores of a particular dimension from them. Periodic boundary conditions are then imposed, dangling bonds saturated with terminal OH groups, and the structure further 'annealed' using molecular dynamics prior to minimisation. Examples of such structures are shown in Fig. 3. These structures [52] have the silica composition and vary in the thickness of the pore walls. They were modelled using the *Discover* program [20] with the *cff91_zeo* [12] force field.

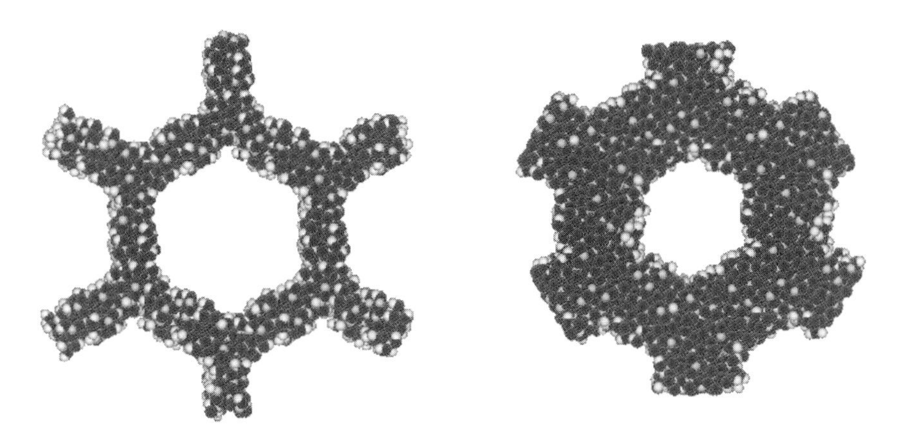

Fig. 3. Illustrations of two-model mesoporous silica structures with amorphous pore walls.

4. External zeolite surfaces

The earlier sections of this chapter have emphasised the utility of static lattice methods in predicting the structure and energetic properties of known and hypothetical structures. Central to the success of this method are the quality of the interatomic potentials, which are able to predict the structure and relative stability of synthetic siliceous aluminosilicate and aluminophosphate structures. We now review how simulation methodologies and force fields can be used to establish the structure and energies of zeolite external surfaces.

Surface science is currently a highly active area, where in particular experimental studies and computer simulation have enjoyed a fruitful, symbiotic relationship. Our understanding of, for example, elementary steps in catalysis has been revolutionised by the rapid increase in computer power coupled with fundamental theoretical developments. Whilst the surface structures of metals, metal oxides and minerals have been widely explored and characterised using AFM and other techniques, few investigators have attempted to use these techniques to probe zeolite surface structure. A similar trend is observed in theoretical literature, where there have been very few attempts to use simulation methods to predict the surface structures of simple and complex zeolites. However, it is increasingly clear that interatomic potential-based methods are capable of predicting the surface structures of these materials and hence of providing the necessary structural information to allow us to begin to understand transport, selectivity and catalysis at the interface between zeolites and other solids, liquids or gases.

In modelling the surfaces of microporous and other materials, we seek to answer a number of questions such as:

(i) What is the surface structure on the atomic scale?
(ii) What is the chemical integrity of the surface?
(iii) Does the surface geometry resemble that of the bulk?

Atomistic simulation methods can provide the answers to these questions. Consider, for example, the large body of work concentrating on inorganic solids and minerals [53]. In zeotypes, technical complications arise because one is not generally dealing with a low symmetry 'infinite' framework material, with substantial internal void space. Hence there are many chemically distinct planes that can be cleaved or expressed. In the three-dimensional network of bonds within the zeolite structure, it is not possible to cleave the crystal without breaking an Si–O or Al–O, semi-ionic/semi-covalent bond, in contrast to, for example, calcite ($CaCO_3$), which consists of sub-lattices of Ca^{2+} ions and CO_3^{2-} ions. Consequently, when we consider what surface terminations can be expressed on a given growth plane, we ignore the possibility of cleaving through CO_3^{2-} ions. In contrast, given that the framework material must have a finite and presumably ordered surface structure, one has to consider how many bonds are broken when the surface is created, which is expected to be proportional to the work done. The act of breaking bonds creates under-coordinated sites, and hence we cannot pre-judge what the terminating structure will be because as well as considering the number of bonds that can be broken, the strength of bonds varies considerably. Hence it is necessary to evaluate the bond strength of the material under investigation, where computer simulation is invaluable.

Furthermore, the phenomenon of reconstruction, well known in materials such as Si (for example, the Takayangi 7×7 reconstruction on the (111) plane) must also be considered. Another factor that adds to the computational expense and complexity of the atomistic calculation is the number of atoms in the typical zeolite unit cell, which for natural zeolites is 54 for EDI and 576 for LTA (framework atoms only). This point is emphasised by Fig. 4, which shows a comparison of the possible cleavage planes on a $CaCO_3$ $(10\bar{1}4)$ surface, compared to the ERI material.

Clearly, a number of these potential cuts can be eliminated on the grounds of symmetry, but we have to be able to discriminate between the possible terminations using a cost function of some sort. In earlier sections, the utility of interatomic potential methods to describe the

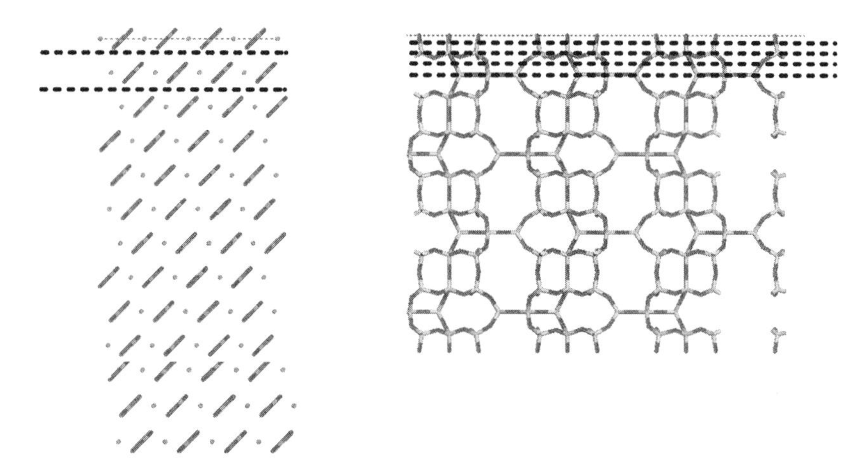

Fig. 4. (Left) The calcite (101.4) surface, where one termination is expressed. The surface is shown in cross-section, the upper black line signifies the surface mesh, whilst the dashed blue lines indicate possible cleavage planes. (Right) The erionite (001) surface is shown. The possible cleavage planes are signified by blue lines. Silicon atoms are shown in yellow and oxygen atoms in red.

lattice properties of materials has been emphasised and, as noted, the same basic approach can be used to describe surface properties. And in modelling surfaces, we recall that we can use both 2D- and 3D-periodic methods.

To model the stability of surfaces, we can proceed from the Gibbs equation for surface energy, which describes the work done in separating a crystal block:

$$\gamma = (E_{\text{Surface}} - nE_{\text{Bulk}})/A, \tag{6}$$

where n is the number of layers, E_{Surface} is the total energy of the slab, E_{Bulk} is the lattice energy per unit cell and A is the surface area. The value of γ is usually low for low-index faces, and is of the order of 0–2 J/m^2 for purely siliceous materials such as quartz [54], with similar values for relaxed purely siliceous zeolite surfaces [55]. Low-energy surfaces are expected to be stable and to be morphologically prevalent, whilst high-energy faces, which are by definition relatively unstable, are expected to occupy the lowest fraction of the expressed crystal surface area. It should be noted that a negative surface energy indicates that energy can be gained by spontaneous cleavage along a given crystal plane, which may be manifested by cracking of the surface. The surface energy can be used to predict the morphology, assuming that the morphological importance is inversely proportional to the surface energy. Using a Wulff plot, a prediction of the crystal morphology can be viewed and compared with experimental samples. This type of approach has proved to be particularly appropriate for ionic minerals, where growth is thought to be nucleation rather than diffusion controlled and driven by strong Coulomb forces. A potent use of this method, and validation of its efficiency, is in the modelling of the effect of impurities upon morphology of crystallites, a recent successful example being the work of Fleming et al. [56].

Aside from insights into growth rate, atomistic simulation methods have lent themselves to evaluation of the reaction enthalpy of water with the zeolite surface. This reaction is fundamental to the growth of zeolites, since under hydrothermal conditions, the usual synthetic natural environment, water is of course able to react with evolving or 'stable' terminating structures. This reaction can be considered by using a Born–Haber cycle. The principle is relatively simple and elegant, and has been described in work by Parker et al. [53] and also in work on quartz by de Leeuw [54]. A particularly lucid account of this methodology is given by Fleming et al. [56]. Recent work by Mistry et al. [57] has shown that contrary to popular belief, not all zeolite surfaces are hydroxyl terminated; indeed, the high-index faces of some zeolites reconstruct to self-passivate the growing surface. The resultant crystal is therefore endowed with hydrophilic character on low-index surfaces where the surface is coated with protons, and hydrophobic on the high-index faces where the sites contain a large number of dangling bonds, where atoms are uncoordinated. This phenomenon is consistent with chemical intuition, where one expects that hydrophilic surfaces formed in the presence of water are morphologically important, whilst hydrophobic surfaces are less stable and therefore less evident.

4.1. Surface structure

A key deliverable from the atomistic computer simulation approach is the structure of the microporous surfaces. Experimental studies using AFM and HRTEM have brought Angstrom resolution to the crystal surface and have allowed a unique insight into the surface structures that are characterised by crenellated features. Atomistic computer prediction of zeolite surface structure, in combination with AFM and HRTEM measurements, provides the most reliable evidence of the true surface structure. Whilst structure is of itself important, the most revealing details originate from consideration of the evolution of the structure. The fact that the regular crennelated structures occur repeatedly indicates that the structure is fundamental to the crystal growth, and that the growth structures are controlled by basic thermodynamic or kinetic factors; that is, the surface structures certainly do not arise from the random condensation of monomeric species on the surface, giving a continuum of surface structures. Several questions are prompted by these observations. Firstly, what dictates which structures are expressed? Secondly, do they signify any relation between the nature of the species in the solution and the structures evidenced at the surface? Thirdly, are there any unique properties that are manifested due to the expression of surface geometry, which may structurally (due to strong relaxation effects) or chemically unrelated to the bulk (due to expression of, for instance, terminal or geminal hydroxyl groups).

An important example of the insight obtainable from computer simulation is the most stable plane of zeolite Y, the (111) surface. In Fig. 5a, the unit cell of zeolite Y is shown orientated parallel (111) to illustrate the possible cleavage planes across the cell. Note that because the structure has a framework nature, there is no reason to presuppose that the surface should be planar. However, one expects that the minimum surface area should be exposed, for the simple reason that this minimises the density of under-coordinated bonds. In Fig. 5b and c, two possible terminations are shown that correspond exactly with those reported by Terasaki and co-workers [58,59].

It is clear that the difference between the two structures is a double 6-ring unit (D6R), which may suggest that the D6R is required to assemble in solution before reacting with the crystal surface. To answer whether the structures that are observed are long-lived signatures of crystal growth, one can use computer simulation methods to investigate the reaction of potential growth units with the growing surface. Atomistic computer simulation results suggest that a third intermediate structure, formed by adding an S6R to structure 5b is never observed because the D6R unit preforms in solution. This assertion is further supported by detailed work of Agger et al. [60], who showed that the pattern of nucleation observed by AFM can only be explained by a D6R-mediated mechanism (when the material is synthesised under hydrothermal conditions).

A similar finding is obtained for zeolite beta C, a purely siliceous material recently reported by Liu et al. [61]. This material is built from 4-, 5- and 6-rings, and crucially the material contains a double

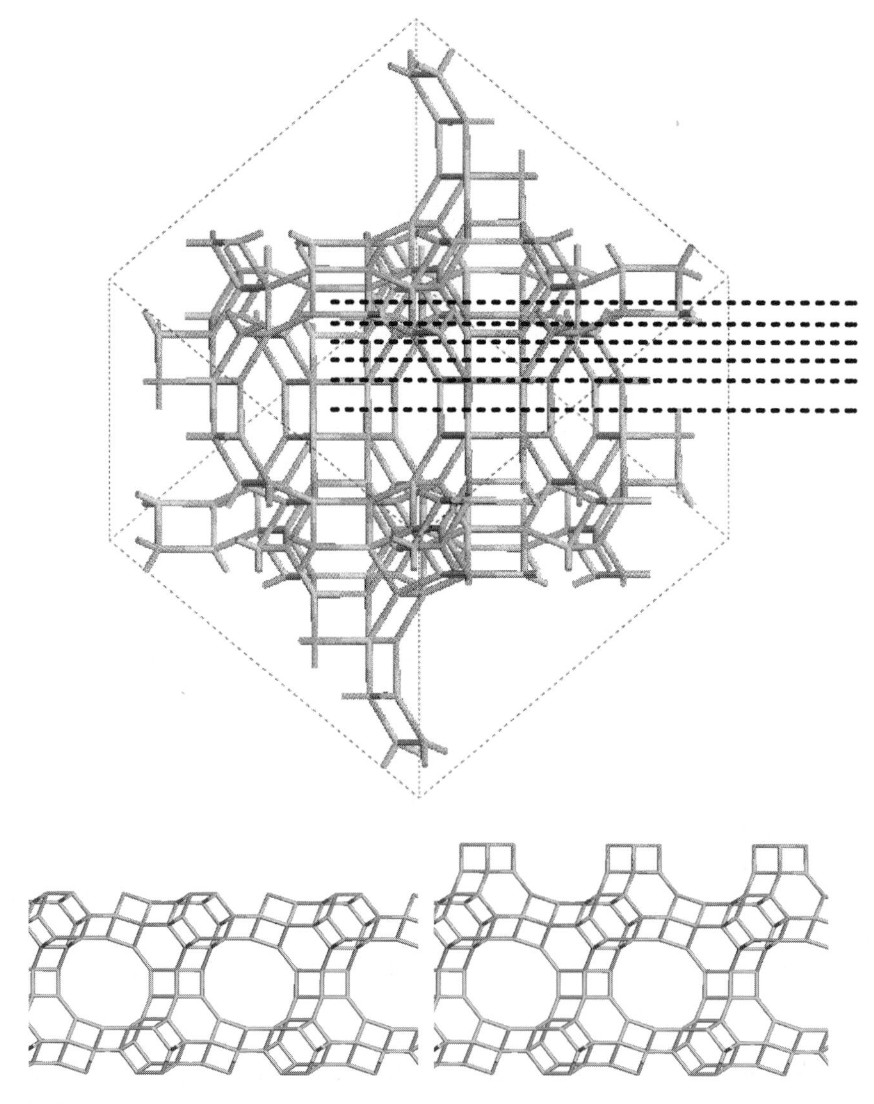

Fig. 5. The unit cell of Faujasite is shown in the upper figure (a). Only the silicon atoms (yellow) and aluminium (atoms) are shown. The blue dashed lines indicate some of the possible cleavage planes parallel to the (111) plane. In the lower figures, on the left (b) the 6-ring terminated structure is shown, whilst on the right-hand figure (c), the double 6-ring terminated structure is shown. The surface is shown in cross-section and the grey area indicates the lower bulk-like region of the crystal.

4-ring parallel to the (100) plane. The (100) face is dominant in the morphology, and Ohsuna and Terasaki reported HREM images [61] indicating extremely clear surface structures. Simulations of beta C [62]

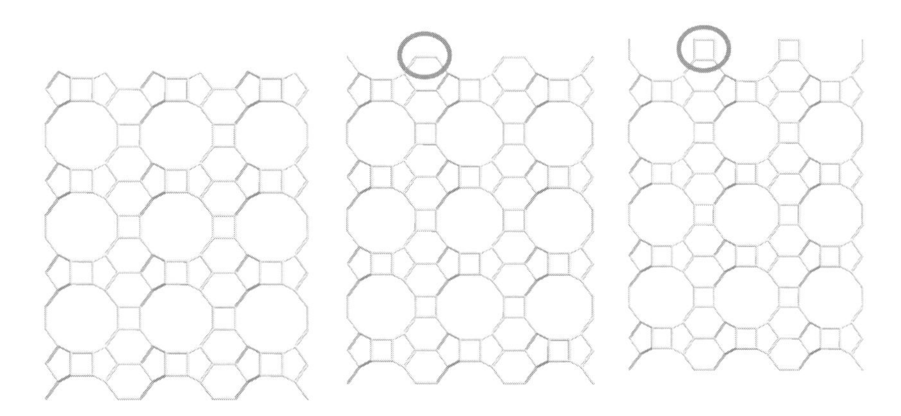

Fig. 6. The (100) face of zeolite beta C in cross-section. Only the silicon atoms are shown. From left to right, the surface is grown by stepwise addition condensation of a single 4-ring to give structure 6b and further addition to give structure 6c. Alternatively, addition of double 4-ring to structure 6a could result in a single-step growth mechanism giving rise to structure 6c. The dark rings highlight the potential growth units.

using the MARVIN code revealed that the three terminations of the (10) surface shown in Fig. 6 have identical surface energy.

Terminations 6a and 6c were observed experimentally, whilst termination 6b could not be identified on the single crystal — a result which prompted an investigation using ab initio methods, of which species are likely to be present in the mother liquor. In recent work [62], we described how a double 4-ring was found to be a stable entity, as was a single 4-ring. The condensation energetics linking these prototypical growth fragments to the growing surface was considered using planewave-based, periodic density functional theory, and a Born–Haber cycle to compute the gas-phase condensation of the growth units with the growth surface (6a). It was found that the reaction of a 4-ring was slightly endothermic, whilst addition of a further 4-ring was exothermic. From this result, we concluded that the reaction was either thermodynamically unfavourable, in which case it may not occur, or the reaction of a second 4-ring proceeded quickly, and hence the intermediate phase was kinetically unfavourable. Conversely, addition of a double four-membered ring was found to be favourable under reaction conditions. In this way, we were able to propose an explanation of the absence of one of the possible terminating structures, which clearly has strong implications for our understanding of the role of oligomeric species or secondary building units in controlling crystal growth mechanisms and the growth rate of zeolitic materials. Given, for example, the work of Loiseau et al. [63] and Kirschhok et al. [64], there is an increasing body of evidence which points to the organisation of material in solution to form secondary building units and subsequent deposition onto the crystal surface. It seems contrary to expectation to suppose that similar mechanisms may not be at work in dictating the formation of, for example, zeolite Y. Recent calculations on zeolite L [65] also support this interpretation. More crucially, work on natural zeolites where charge ordering is often a feature, suggests that ordering may well take place in solution [57].

The extent of surface relaxation at zeolite surfaces is very small, where generally, computation suggests that only the terminating atoms undergo any form of relaxation that significantly affects the chemical or geometric properties of the material [66]. This observation is supported by both AFM [67–69] and HRTEM [70–72] work, where the observed surface structure geometry is in essence identical to that of the bulk. This result is again consistent with intuition, where one expects that in low-density materials where the forces between atoms are dominated by chemical bonds, and the higher-coordination shells contain a relatively small number of atoms, relaxation will be dominated by the first coordination shell. Additionally, it is known that the Si–O–Si angle is flexible, allowing strain induced from cleaving the crystal to be dissipated without causing long-range deformation of the structure, in marked contrast, for example, to the case with ionic oxides, where the surface relaxation is often dramatic, arising from the need to balance long-ranged Coulomb forces between layers. Regarding surface reconstruction, unlike many other materials, evidence is scant, again consistent with the notion that the directionality of covalent bonds leads to a rigid framework that is stable, resulting in little drive to form new surface structures. These highly directional bonds preclude facile rearrangement under thermal agitation and again because of the large distances between atoms, the formation of, for example, charge density waves that could drive a surface-phase transition is presumably less probable than in denser, more ionic materials. However, we note that it is often easy to prepare zeolite phases upon existing zeolites, for example FAU and EMT, where overlayers of EMT are easily induced upon FAU [73]. The stacking faults almost certainly arise from growth units deviating from perfect stacking regimes, forcing overgrowth of a new phase because of the misalignment of the growth units with the surface sub-structure. It is important to distinguish these stacking faults from thermally induced phase transformations.

For many materials, it turns out that very few terminating structures are thermodynamically stable, the consequence of which has general chemical implications: firstly, only particular surface structures are observed and often the surface consists of cage-like structures, which may or may not have reactive properties distinct from those of the crystal interior. The second point is that because of the well-defined structure of the crystal, it also has well-defined acidity. This in turn dictates the surface reactivity, and hence one can begin to probe the complex surface chemistry of microporous materials, such as pore-mouth catalysis, using simulation methods.

To summarise this section, the examples presented above show that classical simulation methods provide a rigorous and reliable guide to zeolite surface stability. The structural complexity of these materials is such that only atomistic methods are appropriate to discriminate between the multifarious terminating structures with the required degree of accuracy. Moreover, this method allows us to address fundamental steps in the crystal-growth processes and to predict surface morphologies. These studies are only a start. They may even be used to model the influence of the surface on sorption and reactivity.

5. Summary

Static lattice methods employing interatomic potentials are simple, cheap and often very effective ways of modelling the structures and energetics of microporous materials and their surfaces. Moreover, when combined with other approaches — simulated annealing, genetic algorithm optimisation methods or topological approaches — the methods may have real predictive content. And where this class of simulation is applicable, it should always be used first, with quantum mechanical methods being, when appropriate, used to refine and extend predictions of the interatomic potential-based simulations.

References

1. Catlow, C.R.A. and Mackrodt, W.C. (Eds.), *Lecture Notes in Physics*. Springer-Verlag, Berlin, 1982, Vol. 166.
2. Catlow, C.R.A. (Ed.), *Modelling of Structure and Reactivity in Zeolites*. Academic Press Limited, London, 1992.
3. Catlow, C.R.A. and Stoneham, A.M. (Eds.), *Mott-Littleton 50th Anniversary Special Issue, J. Chem. Soc., Faraday Trans. 2*, **85**(5) (1989).

4. Watson, G., Tschaufeser, P., Wall, A., Jackson, R.A. and Parker, S.C., In: Catlow, C.R.A. (Ed.), *Computer Modelling in Inorganic Crystallography*. Academic Press, 1997, Chapter 3.
5. Catlow, C.R.A. and Price, G.D., *Nature*, **347**, 243 (1990).
6. Tosi, M.P., *Solid State Phys.*, **16**, 1 (1964).
7. Dick, B.G. and Overhauser, A.W., *Phys. Rev.*, **112**, 90 (1958).
8. Catlow, C.R.A. and Stoneham, A.M., *J. Phys. C*, **16**(22), 4321 (1983).
9. van Beest, B.W.H., Kramer, G.J. and van Santen, R.A., *Phys. Rev. Lett.*, **60**, 1955 (1990).
10. Sanders, M.J., Leslie, M. and Catlow, C.R.A., *J. Chem. Soc., Chem. Commun.*, 1271 (1984).
11. Purton, J., Jones, R., Catlow, C.R.A. and Leslie, M., *Phys. Chem. Miner.*, **19**(6), 392 (1993).
12. Hill, J.R. and Sauer, J., *J. Phys. Chem.*, **98**, 1238 (1994).
13. Catlow, C.R.A., James, R., Mackrodt, W.C. and Stewart, R.F., *Phys. Rev. B: Condens. Matter*, **25**(2), 1006–1026 (1982).
14. Jackson, R.A. and Catlow, C.R.A., *Mol. Simul.*, **1**, 207 (1988).
15. Gale, J.D. and Henson, N.J., *J. Chem. Soc., Faraday Trans.*, **90**, 3175 (1994).
16. Corà, F. and Catlow, C.R.A., *J. Phys. Chem. B*, **105**(42), 10278 (2001).
17. Gale, J.D., *J. Chem. Soc., Faraday Trans.*, **93**, 629 (1997).
18. Watson, G.W., Kelsey, E.T., de Leeuw, N.H., Harris, D.J. and Parker, S.C., *J. Chem. Soc., Faraday Trans.*, **92**, 433 (1996).
19. Gay, D.H. and Rohl, A.L., *J. Chem. Soc., Faraday Trans.*, **91**, 925 (1995); Gale, J.D. and Rohl, A.L., *Mol. Simul.*, **29**, 291 (1995).
20. Insight II v 400, Discover 9.50. Accelrys Inc., San Diego, CA.
21. Sanders, M.J., Catlow, C.R.A. and Smith, J.V., *J. Phys. Chem.*, **88**(13), 2796 (1984).
22. Bell, R.G., Jackson, R.A. and Catlow, C.R.A., *J. Chem. Soc., Chem. Commun.*, 782 (1990).
23. Shannon, M.D., Casci, J.L., Cox, P.A. and Andrews, S.J., *Nature*, **353**, 417 (1991).
24. Lewis, D.W., Carr, S., Sankar, G. and Catlow, C.R.A., *J. Phys. Chem.*, **99**, 2377 (1995).
25. Petrovic, I., Navrotsky, A., Davis, M.E. and Zones, S.I., *Chem. Mater.*, **5**, 1805 (1993).
26. Piccione, P.M., Laberty, C., Yang, S.Y., Camblor, M.A., Navrotsky, A. and Davis, M.E., *J. Phys. Chem. B*, **104**, 10001 (2000).
27. Henson, N.J., Cheetham, A.K. and Gale, J.D., *Chem. Mater.*, **6**, 1647 (1994).
28. Ruiz-Salvador, A.R., Lewis, D.W., Rubayo-Soneira, J., Rodriguez-Fuentes, G., Sierra, L.R. and Catlow, C.R.A., *J. Phys. Chem. B*, **102**, 8417 (1998).
29. Channon, Y.M., Catlow, C.R.A., Jackson, R.A. and Owens, S.L., *Microporous Mesoporous Mater.*, **24**, 153 (1998).
30. Akporiaye, D.E., Fjellvag, H., Halvorsen, E.N., Hustveit, J., Karlsson, A. and Lillerud, K.P., *J. Phys. Chem.*, **100**(41), 16641 (1996).
31. Woodley, S.M., Gale, J.D., Battle, P.D., Catlow, C.R.A., *J. Chem. Soc. Chem. Commun.*, in press, 2003.
32. McCusker, L.B., Liebau, F. and Engelhardt, G., *Pure Appl. Chem.*, **73**, 381 (2001).
33. Meier, W.M., Olson, D.H., Baerlocher, C., *Atlas of Zeolite Structure Types*. Fifth Elsevier, Amsterdam, 2001 (updates on http://www.iza-structure.org/databases).

34. Smith, J.V., *Chem. Rev.*, **88**, 149 (1988) and references therein.
35. Alberti, A., *Am. Miner.*, **64**, 1188 (1979).
36. Sato, M., *Z. Kristallogr.*, **161**, 187 (1982).
37. Sherman, J.D. and Bennett, J.M., *ACS Adv. Chem. Ser.*, **121**, 52 (1973).
38. Barrer, R.M. and Villiger, H., *Z. Kristallogr.*, **128**, 352 (1969).
39. O'Keeffe, M. and Brese, N.E., *Acta Crystallogr.*, **A48**, 663 (1992).
40. O'Keeffe, M., *Acta Crystallogr.*, **A48**, 670 (1992).
41. O'Keeffe, M., *Acta Crystallogr.*, **A51**, 916 (1995).
42. Akporiaye, D.E. and Price, G.D., *Zeolites*, **9**, 23 (1989).
43. Boisen, M.B., Gibbs, G.V., O'Keeffe, M. and Bartelmehs, K.L., *Microporous Mesoporous Mater.*, **29**, 219 (1999).
44. Draznieks, C.M., Newsam, J.M., Gorman, A.M., Freeman, C.M. and Férey, G., *Angew Chem. Int. Ed.*, **39**, 2270 (2000).
45. Treacy, M.M.J., Randall, K.H., Rao, S., Perry, J.A. and Chadi, D.J., *Z. Kristallogr.*, **212**, 768 (1997).
46. Delgado Friedrichs, O., Dress, A.W.M., Huson, D.H., Klinowski, J. and Mackay, A.L., *Nature*, **400**, 644 (1999).
47. Cundy, C.S. and Cox, P.A., *Chem. Rev.*, **103**, 663 (2003).
48. Foster, M.D., Delgado-Friedrichs, O., Bell, R.G., Almeida Paz, F.A., Klinowski, J., *Angew. Chem. Int. Ed.*, **42**, 3896 (2003).
49. Meier, W.M. and Villiger, H., *Z. Kristallogr.*, **128**, 352 (1969).
50. Kresge, C.T., Leonowicz, M.E., Roth, W.J., Vartuli, J.C. and Beck, J.S., *Nature*, **359**, 710 (1992).
51. Beck, J.S., Vartuli, J.C., Roth, W.J., Leonowicz, M.E., Kresge, C.T., Schmitt, K.D., Chu, C.T.-W., Olsen, D.H., Sheppard, E.W., McCullen, S.B., Higgins, J.B., Schlenker, J.L., *J. Am. Chem. Soc.*, **114**, 10834 (1992).
52. Bell, R.G., *Proc. Twelfth Intl. Zeolite Conf.* MRS, Warrendale, 1999, p. 839.
53. Parker, S.C., de Leeuw, N.H. and Redfern, S.E., *Faraday Discuss.*, **114**, 381 (1999).
54. de Leeuw, N.H., Higgins, F.M. and Parker, S.C., *J. Phys. Chem. B*, **103**(8), 1207 (1999).
55. Whitmore, L., Slater, B. and Catlow, C.R.A., *Phys. Chem. Chem. Phys.*, **2**(23), 5354 (2000).
56. Fleming, S.D., et al., *J. Phys. Chem. B*, **105**(22), 5099 (2001).
57. Mistry, M., Slater, B., Catlow, C.R.A., manuscript in preparation, 2003.
58. Terasaki, O., et al., *Ultramicroscopy*, **39**(1–4), 238 (1991).
59. Terasaki, O., et al., *Chem. Mater.*, **5**(4), 452 (1993).
60. Agger, J.R., Anderson, M.W., Crystal growth of zeolite Y studied by computer modelling and atomic force microscopy, In: *Impact of Zeolites and Other Porous Materials on the New Technologies at the Beginning of the New Millennium, Parts a and b*. 2002, p. 93.
61. Liu, Z., et al., *J. Am. Chem. Soc.*, **123**(22), 5370 (2001).
62. Slater, B., et al., *Angewandte Chemie-International Edition*, **41**(7), 1235 (2002).
63. Loiseau, T., et al., *J. Am. Chem. Soc.*, **123**(50), 12744 (2001).
64. Kirschhock, C.E.A., et al., *J. Phys. Chem. B*, **106**(19), 4897 (2002).
65. Slater, B., manuscript in preparation, 2003.
66. Slater, B., et al., *Curr. Opin. Solid State Mater. Sci.*, **5**(5), 417 (2001).
67. Agger, J.R., et al., *J. Am. Chem. Soc.*, **120**(41), 10754 (1998).
68. Agger, J.R., Hanif, N. and Anderson, M.W., *Angew Chem. Int. Ed.*, **40**(21), 4065 (2001).

69. Agger, J.R., et al., *J. Am. Chem. Soc.*, **125**(3), 830 (2003).
70. Terasaki, O., et al., *Curr. Opin. Solid State Mater. Sci.*, **2**(1), 94 (1997).
71. Terasaki, O., et al., *Supramol. Sci.*, **5**(3–4), 189 (1998).
72. Terasaki, O., *J. Electron Microsc.*, **43**(6), 337 (1994).
73. Alfredsson, V., et al., *Angewandte Chemie International Edition in English*, **32**(8), 1210 (1993).

Computer Modelling of Microporous Materials
C.R.A. Catlow, R.A. van Santen and B. Smit (editors)

Chapter 2

Adsorption phenomena in microporous materials

B. Smit*

Department of Chemical Engineering, Universiteit van Amsterdam, Nieuwe Achtergracht 166, 1018 WV Amsterdam, The Netherlands

1. Introduction

In this chapter, the first of three examining the application of simulation technique to the study of sorbed molecules in zeolites, we focus on the use of Monte Carlo simulations to study the adsorption in zeolites. We concentrate on those systems for which the conventional molecular simulation techniques, molecular dynamics, and Monte Carlo, are not sufficiently efficient. In particular, to simulate the adsorption of long-chain hydrocarbons novel Monte Carlo techniques have been developed. Here we discuss configurational-bias Monte Carlo (CBMC) which has been developed to compute the thermodynamic properties. The use of these methods is illustrated with some examples of technological importance.

The fact that the sorption behavior of molecules depends on the details of the structure of a microporous material is the basis of many applications of these materials. Therefore, it is important to have some elementary knowledge on the number of molecules that are adsorbed at a given condition. In fact, many monographs and review articles have been written on these adsorption phenomena [1–3]. Yet, our understanding of the sorption behavior is far from complete. Most experiments yield important macroscopic data, for example, heats of adsorption or adsorption isotherms, from which one can only indirectly extract molecular information on these adsorbed molecules.

*E-mail: b.smit@science.uva.nl

Compared to pure component adsorption our knowledge on competitive adsorption in mixtures is very poor. Yet, most applications involves mixtures. As a consequence most of the experimental data on these applications have been analyzed with incomplete data on the number of molecules that are adsorbed. In addition, even if one would have all pure component adsorption data available, the number of mixtures one could form with these pure components is simply too large to handle. Therefore, the probability is very small that the literature gives an answer to a question related to the number of molecules of a particular component that are adsorbed at a given pressure and temperature in a given microporous material. It is therefore important to have reliable theoretical methods that allow us to approximate the sorption behavior.

In this review we will illustrate the importance of detailed knowledge of the sorption behavior to understand better the properties of the system. The monograph of Ruthven [1] contains an excellent summary of the experimental techniques to measure adsorption isotherms and theoretical methods to analyze these experimental data. Over the last few years molecular simulation techniques have become an attractive alternative to study the sorption in microporous materials. In this work we focus on the applications of these simulation techniques upon. Therefore, it is important to emphasize that although in the examples the sorption behavior has been studied using molecular simulations, this is however, not essential. Similar results could have been obtained from experiments, but for these types of systems only simulation results are available. Some details on the simulation techniques that are used to study the adsorption of molecules in microporous materials are discussed in the next section. Additional information on the computational aspects of adsorption of molecules in zeolites are given in the review by Fuchs and Cheetham [4] and on diffusion aspects in a review by Demontis and Suffritti [5].

2. Molecular simulations

Several molecular simulation techniques have been used to study the adsorption in zeolites. The earlier studies used Molecular Mechanics to study the conformation or docking of molecules. From a computational point of view such simulations are relatively simple since they only involve the conformation of the molecule with the lowest energy. From a Statistical Thermodynamics point of view such a conformation corresponds to the equilibrium distribution at $T = 0$ K, where entropy effects do not play a role. All applications of zeolites, however, are at elevated temperatures. Simulations at these conditions require the use of molecular dynamics or Monte Carlo techniques. For such simulations one needs to sample many million configurations, which does require much more CPU time.

Because of the CPU requirement most of the systems that have been studied by Monte Carlo techniques and molecular dynamics concern the adsorption of noble gases or methane. Only a few studies of ethane or propane have been published. Only very recently the computers have become sufficiently powerful to perform molecular dynamics simulations of long-chain alkanes [6,7]. The reason why only small molecules have been studied becomes clear from the work of June et al. [8], in which molecular dynamics was used to investigate the diffusion of butane and hexane in the zeolite silicalite. June et al. showed that the diffusion of butane from one channel of the zeolite into another channel is very slow compared to diffusion of bulk butane. As a consequence many hours of computer time were required to obtain reliable results. In addition, the diffusion decreases significantly with increasing chain length.

The above example illustrates the fundamental problem of molecular dynamics. In a molecular dynamics simulation the approach is to mimic the behavior of the molecules as good as possible. If successful, all properties will be like in nature, including the diffusion. If the molecules diffuse slowly this will be reflected in very long simulation times and in the case of long-chain alkanes these simulation times can only recently be reached.

In principle, one can circumvent this intrinsically slow dynamics by using a Monte Carlo technique. In a Monte Carlo simulation one does not have to follow the 'natural path' and one can, for example, perform a move in which it is attempted to displace a molecule to a random position in the zeolite. If such a move is accepted, it corresponds to a very large jump in phase space. Again, utilization of such type of 'unnatural' Monte Carlo moves turned out to be limited to small molecules as is shown in the next section.

2.1. Monte Carlo simulation of adsorption

It may not be obvious why we need efficient Monte Carlo methods to simulate chain molecules. In general, a molecular dynamics approach is much easier to generalize to complex molecules. An example of an experiment that is 'impossible' to simulate using molecular dynamics is the computation of an adsorption isotherm.

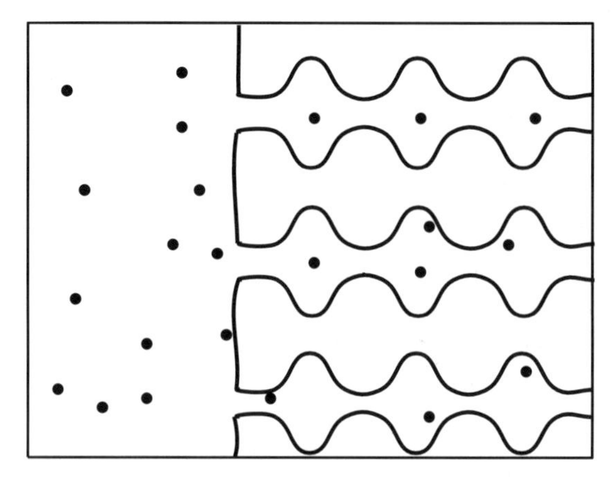

Fig. 1. A zeolite in direct contact with a gas, equilibrium is obtained via the diffusion of molecules from the gas phase into the zeolite.

Experimentally, one measures, for example, the weight increase of a zeolite sample as a function of the external pressure. In a simulation we can mimic this experimental setup (see Fig. 1); one needs a reservoir that is in open contact with a zeolite. For long-chain hydrocarbons the equilibration in the laboratory may take hours or even several days. It would be very impractical to simulate this experiment with a simulation. Even if one would have the patience to wait several million years before our computer experiment is equilibrated, one has to worry about the zeolite surface and one has to simulate a large reservoir of uninteresting molecules. It is therefore much more convenient to perform a grand-canonical Monte Carlo simulation (see Fig. 2). In such a simulation one imposes the temperature and chemical potential and computes the average number of particles in the (periodically) repeated zeolite crystal. It is important to note that in such a simulation the number of particles is not fixed but varies during the simulation. In such a simulation one therefore has to perform Monte Carlo moves which attempt to add or to remove particles.

For small absorbents such as methane or the noble gases, convention grand-canonical Monte Carlo simulations can be applied to calculate the adsorption isotherms in the various zeolites [9–15]. An example of an adsorption isotherm of methane in the zeolite silicalite is shown in Fig. 3. These calculations are based on the model of Goodbody et al. [11]. The agreement with the experimental data is very good, which shows that for these well-characterized systems simulations can give data that are comparable with experiments.

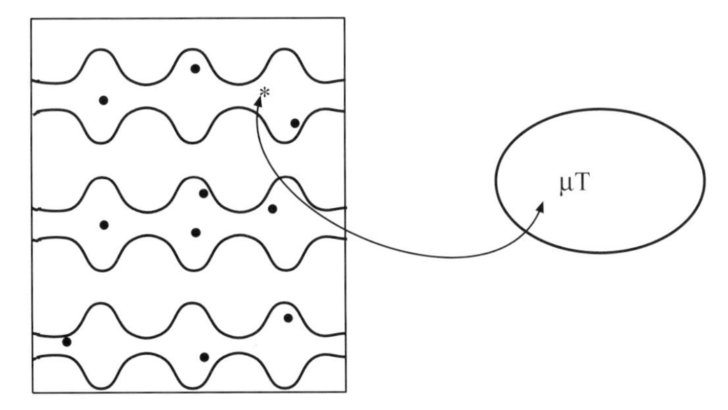

Fig. 2. Grand-canonical Monte Carlo; a zeolite in indirect contact with a reservoir, which imposes the chemical potential and temperature by exchanging particles and energy.

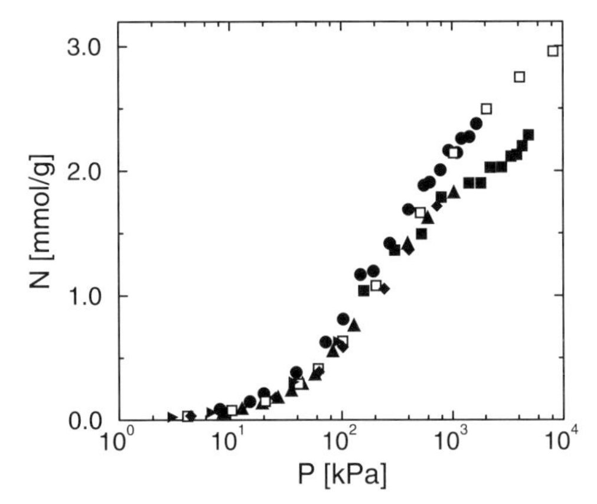

Fig. 3. Adsorption isotherms of methane in silicalite, showing the amount of methane adsorbed as a function of the external pressure. The black symbols are experimental data (see Ref. [16] for details). The open squares are the results of grand-canonical simulations using the model of Ref. [11].

In these simulations an attempt to insert a molecule is performed by generating a random position in the zeolite. If this position overlaps with one of the zeolite atoms the probability is high that such an attempt is rejected. The success of such a simulation depends on the number of successful attempts to insert a particle. To apply such a simulation for a long-chain alkane, one has to be able to insert such a molecule in a zeolite. In such a simulation one can observe that out of the 1000 attempts to move a methane molecule to a random position in the zeolite 999 attempts will be rejected because the methane molecule overlaps with a zeolite atom. If we were to perform a similar move with an ethane molecule, we would need 1000×1000 attempts to have one that was successful. Clearly, this random insertion scheme will break down for any but the smallest alkanes.

2.2. Monte Carlo simulations of chain molecules

2.2.1. Configurational-bias Monte Carlo

To make Monte Carlo simulations of long-chain molecules possible the configurational-bias Monte Carlo (CBMC) technique was developed [17]. The principle idea of the CBMC technique is to grow a molecule atom by atom instead of attempting to insert the entire molecule at random. Figure 4 shows one of the steps in this algorithm. Important to note is that this growing procedure introduces a bias, such that only the most favorable configurations are being generated. If one were to use the ordinary Metropolis acceptance rule, such a bias in the configurations of the molecules would lead to an incorrect distribution of configurations. This bias can be removed exactly by adjusting the acceptance rules [17].

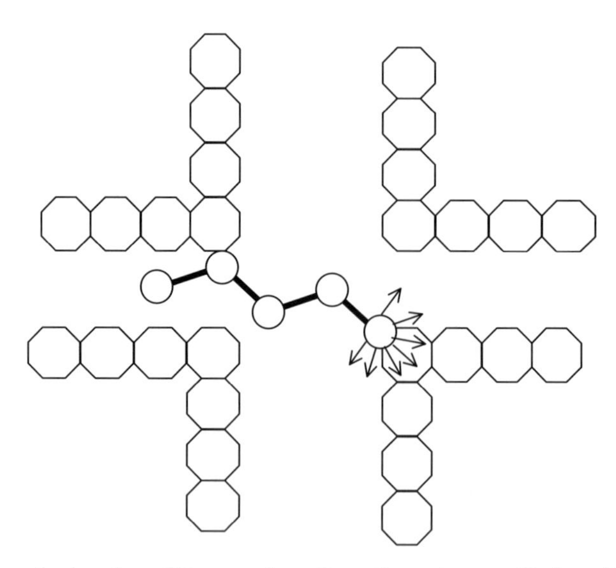

Fig. 4. Schematic drawing of the growing of an alkane in a zeolite in a CBMC move. The octagons represent the atoms of the zeolite and the circles represent the atoms of the alkane. Four atoms have been inserted successfully, and an attempt is made to insert the fifth.

It is not the purpose of this review to give an extensive discussion on the implementation of this algorithm for the adsorption of linear and branched alkanes in zeolites; details can be found in Refs. [18,19]. Smit and Siepmann estimated that for the adsorption of dodecane in silicalite a CBMC simulation can be up to 10–20 orders of magnitude (!) more efficient than the conventional techniques [20].

2.2.2. Free-energy calculation

In the CBMC algorithm the Rosenbluth scheme is used to generate new conformations of the hydrocarbons. This method can also be used to compute the free energy of chain molecule in a zeolite. At infinite dilution this free energy is related to the Henry coefficient. In this scheme a molecule is grown atom by atom using the algorithm of Rosenbluth and Rosenbluth [21]. During the growing of an atom several trial positions are probed; of each of these positions the energy is calculated, and the one with the lowest energy is selected with the highest probability according to:

$$p_i(j) = \frac{\exp[-\beta u_i(j)]}{\sum_{l=1}^{k} \exp[-\beta u_i(l)]} = \frac{\exp[-\beta u_i(j)]}{w(i)},$$

where $u_i(l)$ is the energy of atom i at trial position l. When the entire chain is grown, the normalized Rosenbluth factor of the molecule in configuration Γ can be computed:

$$\mathcal{W}(\Gamma) = \prod_{i=1}^{l} w(i)/k.$$

In Ref. [17] it is shown that the average Rosenbluth factor is related to the chemical potential of the molecule:

$$\langle \exp(-\beta u) \rangle = \mathcal{C}\langle \mathcal{W} \rangle,$$

where $\mathcal{C}$ is a constant defining the reference chemical potential (see Ref. [17] for more details).

These free-energy calculations can be used to compute the Henry coefficient. If the external pressures of interest are sufficiently low, a good estimate of the adsorption isotherm can be obtained from the

Henry coefficient K_H. Under these conditions, the number of adsorbed molecules per unit volume (ρ_a) is proportional to the Henry coefficient and external pressure P:

$$\rho_a = K_H P.$$

The Henry coefficient is directly related to the excess chemical potential of the adsorbed molecules. To see this, consider the ensemble average of the average density in a porous medium. In the grand-canonical ensemble, this ensemble average is given by

$$\left\langle \frac{N}{V} \right\rangle = \frac{1}{Q} \sum_{N=0}^{\infty} \frac{q(T)^N V^N \exp(\beta \mu N)}{N!} \int \mathrm{d}s^N \exp[-\beta \mathcal{U}(s^N)] N/V$$

$$= \frac{q(T) \exp(\beta\mu)}{Q} \sum_{N=0}^{\infty} (q(T)V)^{N-1} \exp[\beta\mu(N-1)]/(N-1)!$$

$$\times \int \mathrm{d}s^{N-1} \exp\left[-\beta\mathcal{U}(s^{N-1})\right] \int \mathrm{d}s_N \exp[-\beta\mathcal{U}(s_N)]$$

$$= q(T) \exp(\beta\mu) \langle \exp(-\beta \Delta \mathcal{U}^+) \rangle,$$

where $\Delta\mathcal{U}^+$ is defined as the energy of a test particle and $q(T)$ is the kinetic contribution to the molecular partition function. In the limit $P \to 0$, the reservoir can be considered an ideal gas

$$\beta\mu = \ln\left(\frac{\beta P}{q(T)}\right).$$

Substitution of this equation gives

$$\exp(\beta\mu^{\text{ex}}) = \langle \exp(-\beta\Delta\mathcal{U}^+) \rangle = \frac{\langle N/V \rangle}{\beta P}.$$

This gives, for the Henry coefficient,

$$K_H = \beta \exp(-\beta\mu^{\text{ex}}).$$

2.3. Intermolecular potentials

In the previous section simulation techniques are discussed that allow us to compute adsorption isotherms. The input of such a simulation is the intermolecular potentials.

Most simulations start with the assumption of Kisilev and co-workers [22] that the zeolite crystal is rigid. The atomic positions can be taken from the X-ray diffraction. For most structures the atomic data are published on the Web [23]. From a computational point of view the use of a rigid zeolite is very attractive. Since the zeolite atoms do not participate in the simulation, the total number of atoms for which the force has to be computed is reduced significantly. In addition, the potential energy at a given point inside the zeolite can be calculated a priori. If this is done for points on a grid, the potential energy at an arbitrary point can be estimated from interpolation during the simulations [24,25]. With such an interpolation scheme a gain in cpu-time of one to two orders of magnitude can be gained.

In some studies the importance of a flexible zeolite structure is emphasized [26,27]. It can be expected that framework flexibility can be of importance for the modeling of the diffusion of the molecules, since a flexible framework may reduce the diffusion barriers. Since these barriers correspond to positions in which the molecules have a relatively high energy and therefore do not contribute much to the equilibrium properties, it can be expected that the assumption of a rigid zeolite lattice is less severe for these properties. Important to note is that the adsorption of molecules may induce structural transitions of the zeolite lattice [28].

Some zeolites can be synthesized in the all-silica form. In practice, however, the none all-silica zeolites are very important. For example, zeolites are made catalytically active by substitution of trivalent aluminum for tetravalent silicon into the framework. This introduces chemical disorder which has to be taken into account in the simulations.

If we assume an all-silica structure and consider the adsorption of nonpolar molecules, for example alkanes, it is reasonable to assume that the alkane–zeolite interactions are dominated by dispersive forces, which are described with a Lennard-Jones potential

$$U(r_{ij}) = \begin{cases} 4\varepsilon_{ij}[(\sigma_{ij}/r_{ij})^{12} - (\sigma_{ij}/r_{ij})^{6}] & r_{ij} \leq R_c \\ 0 & r_{ij} > R_c \end{cases},$$

where r_{ij} is the distance between atoms i and j, ε is the energy parameter, σ is the size parameter, and R_c is the cut-off radius of the potential. The contribution of the atoms beyond the cut-off to the total energy is estimated using the usual tail corrections [29]. Since the size as well as the polarizability of the Si-atoms are much smaller than those of the O-atoms of the zeolite, one can assume that the contributions of these Si-atoms can be accounted for by using effective interactions with the O-atoms.

In many studies the adsorbed molecules are modeled as united atoms, for example, in case of an alkane the CH_4, CH_3, and CH_2 groups are considered a single interaction center. Despite the simplification these models do very well in reproducing the thermodynamic data of liquid hydrocarbons [30]. Also here one has to keep in mind that such a model has its limitations. For example, it is well known that a united atom model of an alkane cannot reproduce the experimental crystal structure. Details on the parameters of the various models can be found in Ref. [19].

With the above assumption the zeolite–alkane interactions are reduced to finding the optimal Lennard-Jones parameters between the oxygen of the zeolite and the united atoms of the alkane. Because of this assumption it is very difficult to use, for example, quantum chemical calculations to systematically develop methods to compute interaction parameters. The assumption that are being made are very specific for zeolites and therefore difficult to transfer methods that have been developed to generate interaction parameters to these systems. Therefore, most models are obtained by fitting to some experimental data [16].

For zeolites that contain aluminum two additional aspects have to be addressed. One aspect is the position of the aluminum atoms in the zeolite framework and the counterions to compensate for the charge deficit. The question where the aluminum positions are in a certain zeolite has been addressed by many researchers, but is far from being solved. It is beyond the scope of this review to give a detailed discussion on the various methods that are employed to determine the location of the Al atoms. Three approaches appear in the literature. To assume that the net positive charge is distributed over all T-sites, i.e. no distinction between Si and Al atom is made but both are considered as a single T-site. For some zeolites and for some specific Si/Al, the position of the Al atoms can be obtained from the crystal structure. Finally, theoretical methods have been developed to assign the Al atoms using semiempirical rules (see Ref. [31] for more details). The exact location of the Al atoms has important consequences for the preferred location of the cations. It is reasonable to assume that, unless the temperature is very high, the cations prefer to be close to the Al atoms. Therefore, important information on the location of the Al atoms is also contained in the location of the cations. Mellot-Draznieks et al. [32] studied the cation distribution in NaX faujasite using a model in which the charge was uniformly distributed of the T-sites [33] with a method in which an explicit distinction between the Al and Si sites was made. The conclusion of this study was that the uniform distribution gave a reasonable prediction of the location of the cations, but to obtain a correct location of the sodium ions in the supercages a more detailed model was required.

The presence of the counterions implies for adsorption studies that additional intermolecular potentials have to be introduced to take into account the interactions of the adsorbed molecules with these ions and the ions with the zeolite framework [34].

3. Adsorption isotherms

From a practical point of view it is important to have information on the number of molecules adsorbed in the pores of the zeolite as a function of the gas pressure. Here we illustrate how molecular simulation can give us some molecular insights in some special features of these adsorption isotherms.

3.1. Pure components

Most of the simulation studies investigate the energetics, siting, or diffusion of the adsorbed molecules and only a few results on the simulation of isotherms have been reported. An overview on the pure component adsorption that has been studied using molecular simulation has been compiled by Fuchs and Cheetham [4]. The early work on the simulation of adsorption isotherms was focused on small molecules such as noble gases or methane (see Fig. 3 for a typical example) [9,10,12,14,35–37] or mixtures of these gases [13,15,36].

At low pressures the adsorption can be computed from the Henry coefficient. For example, Maginn et al. [38] and Smit and Siepmann [20,25] used the approach described in the previous section to compute the Henry coefficients of linear alkanes adsorbed in the zeolite silicalite. Since the Henry coefficient is calculated at infinite dilution, there are no intramolecular alkane–alkane interactions. In Fig. 5 the Henry coefficients of the *n*-alkanes in silicalite as calculated by Smit

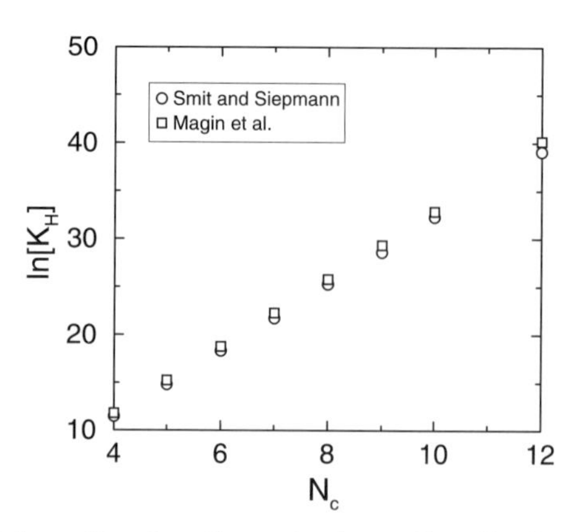

Fig. 5. Henry coefficients K_H of n-alkanes in the zeolite silicalite as a function of the number of carbon atoms N_c, as calculated by Maginn et al. [38] and Smit and Siepmann [25].

and Siepmann are compared with those of Maginn et al. If we take into account that the models considered by Maginn et al. and Smit and Siepmann are slightly different, the results of these two independent studies are in very good agreement.

Knowledge on the adsorption of pure components in zeolites is not only of practical importance but also of scientific interest since steps or kinks in the adsorption curve may signal transitions occurring in the pores of the zeolite. A typical example of this behavior is the adsorption of methane in $AlPO_4$-5 (AFI). Experimentally, one can find two steps in the adsorption isotherm at $T = 77$ K [39]. One step at a loading of approximately four molecules per unit cell and another step at a loading of six molecules per unit cell. These steps are also found via molecular simulations [40,41]. Simulations predict that these steps should disappear if the temperature is raised above $T = 100$ K, suggesting a phase transition to occur in the pores of the zeolite.

The adsorption isotherms of branched alkanes in silicalite also show a step for a given number of molecules per unit cell (see Fig. 6). Such a step is not observed for the linear isomers. For these branched molecules the steps are explained in terms of a preferential siting of these molecules at the intersections of the linear and zig-zag channels. Figure 6 shows that for isobutane a plateau is formed for four molecules per unit cell. There are four intersections per unit cell and the branched alkanes first adsorb at these sites. Once all intersections are occupied, the next atom has to adsorb in between two intersections. Since these sites are less favorable for the bulky branched molecules, this requires much higher pressure before these other sites are occupied.

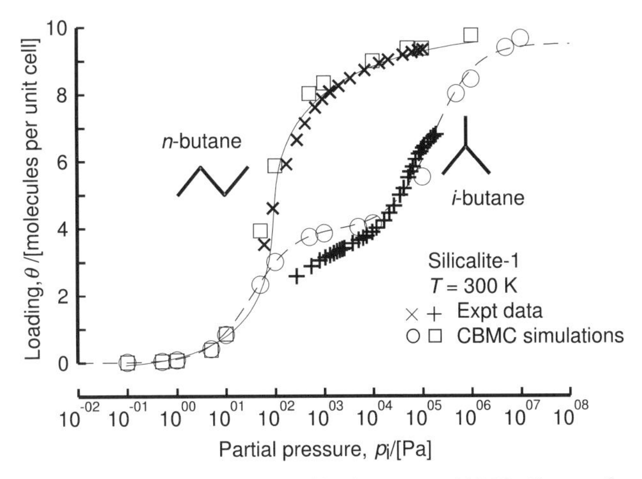

Fig. 6. Sorption isotherms for normal and isobutane at 300 K. Comparison of CBMC simulations with experiments. The data are taken from Ref. [42].

Figure 6 also shows a comparison of simulated and experimental adsorption isotherms of linear and branched alkanes in the zeolite silicalite. The simulations give a nearly quantitative description of the experimental adsorption isotherms. Also for other alkanes in silicalite a similar agreement has been obtained [19]; the simulations reproduce all qualitative features found in the experiments.

The good agreement of the simulated adsorption isotherms of the linear and branched hydrocarbons with the experimental ones is an encouraging result. Experimental adsorption isotherms are not readily available for a given zeolite at a given condition. These results show that one can get a reasonable estimate from a molecular simulation. However, it is important to point out that most simulations have been performed for silicalite for which the potentials have been developed as well. Unfortunately, there are not many experimental adsorption isotherms of other all-silica zeolites. It is therefore not known how accurate these simulations extrapolate to other zeolites.

At this point it is important to mention that these simulations use a rigid zeolite lattice. To see the limitation of this assumption, let us consider the transition, which is observed in the adsorption of benzene or xylene isomers in silicalite [43–45]. Olsen et al. [43] observed a step in the adsorption isotherm for *p*-xylene at 70°C, a plateau at a loading of four molecules per unit cell with a saturation at six molecules per unit cell (see Fig. 7). van Koningsveld showed that four molecules per unit cell in a structural transition of the zeolite framework from the *ortho* to the *para* structure occur [44].

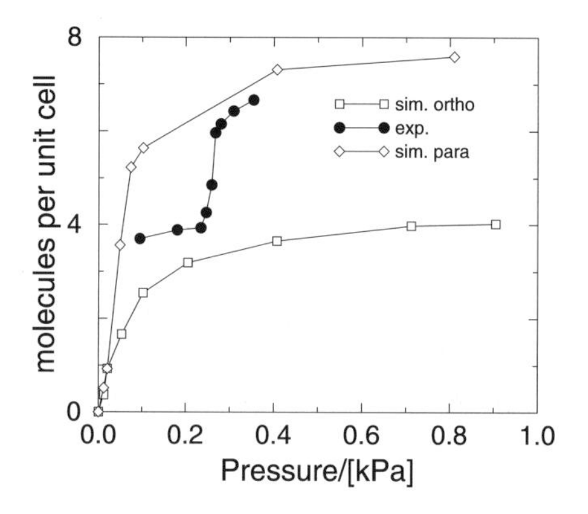

Fig. 7. Comparison of the simulated (open symbols) and experimental adsorption isotherms of *p*-xylene in silicalite. The simulation use the *para* and *ortho* structure of silicalite, and the simulation results are taken from Ref. [46]. The experimental data are taken from Ref. [43].

From a molecular simulation point of view this is a very challenging system to study. Most simulation studies use a rigid zeolite structure. For molecules that do not have a tight fit in the zeolite framework, this appears to be a good assumption. In the case of aromatics in silicalite the fit is very tight and can even induce a phase transition. The differences between the *ortho* and the *para* structure of silicalite are relatively small, yet these small differences result in very different adsorption behavior. Snurr et al. [46] have computed the adsorption isotherms of *p*-xylene in both the *ortho* and the *para* structure of silicalite (see Fig. 7). For both the *ortho* and the *para* structure a simple Langmuir isotherm is observed. The maximum loading for the *ortho* and the *para* structure was four and eight molecules per unit cell, respectively. Comparison with the experimental data shows that the jump in the adsorption isotherm is consistent with a change in the structure.

A similar behavior phase transition was observed for the adsorption of benzene in silicalite [47]. For this system, however, the agreement between experiments and the simulations of Snurr et al. was good at high temperatures but less satisfactory at low temperatures. This discrepancy motivated Clark and Snurr [48] to study the adsorption of benzene in silicalite in detail. Their study showed that the adsorption isotherms of benzene are very sensitive for small changes in the structure of the zeolite. Also these calculations were performed with a rigid zeolite and one would expect that the zeolite structure would 'respond' to the presence of these molecules. Clark and Snurr point out that this requires simulation with a model of a zeolite with accurate flexible lattice potentials.

3.2. Mixtures

Most experimental techniques to determine adsorption isotherms are based on measuring the weight increase of the zeolite due to the adsorption of molecules. For a pure component this directly relates to the number of adsorbed molecules, but for a mixture additional experiments are required to analyze the composition of the adsorbed molecules. As a consequence far less experimental data on mixtures are available.

From a practical point of view the separation of xylene isomers using zeolites is an important system. Lachet et al. [49,50] used molecular simulations to study the effect of cations on the adsorption selectivity. The simulations showed a reversal of the selectivity if Na^+ is exchanged by K^+. The differences in selectivity are related to a combination of differences in size and location of the cations which results in a completely different adsorption behavior.

For the mixtures of small hydrocarbons, adsorption isotherms have been obtained by Dunne et al. [51] and Abdul-Rehman [52]. These mixture isotherms can be reproduced using molecular simulations [53,54]. For these small molecules the observed adsorption behavior is consistent with the theoretical calculations of Talbot [55], in which it is shown that for a mixture of molecules, because of entropic reasons, the smallest component at sufficiently high pressures absorbs better than the bigger components. For mixtures of linear and branched alkanes the situation is more complex. For example, Vlugt et al. [19] have shown that for a mixture of *n*-hexane and 3-methyl pentane in silicalite *n*-hexane is preferentially adsorbed at sufficiently high pressures (see Fig. 8). More complex mixture of linear, mono- and di-branched hydrocarbons have been studied by Calero et al. [56]. These simulations show that, due to configurational entropy effects, in a mixture of mono- and di-branched isomers at sufficiently high pressure the mono-branched isomers are preferentially adsorbed in silicalite.

4. Applications

In this section we illustrate the use of adsorption data to explain some experimental observations in zeolites. The interesting aspect is that at

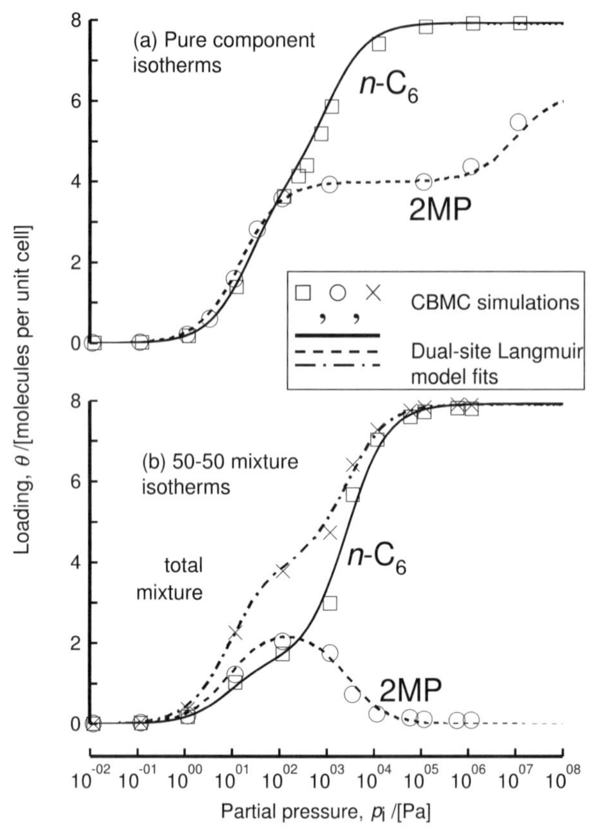

Fig. 8. Configurational-bias Monte Carlo simulations of the adsorption isotherms of *n*-hexane and 2-methyl pentane at $T = 300$ K [19]. The lines are fits to the dual-site Langmuir isotherm model.

first sight it may not be obvious that the explanation is related to adsorption phenomena.

4.1. Permeation through membranes

A description of the permeation through membranes requires the knowledge of the fluxes as a function of the concentration of the molecules. Krishna and Wesselingh [57] have shown that the fluxes can be described using the Maxwell–Stefan equation, which relates the fluxes to the diffusion coefficients and the gradient of the chemical potential. This relation tells us that we need to know the chemical potential as a function of the loading of the molecules in a zeolite. To be more precise one needs to know the loading of each individual component. For pure components there are not many experimental adsorption isotherms, for mixtures it is even worse. Therefore, most estimations of the permeation are based on experimental Henry coefficients of the pure components. These Henry coefficients show the 'expected' temperature dependence, and therefore any anomalous behavior is often attributed to a dependence of the diffusion coefficients on temperature or pressure.

To compute the Maxwell–Stefan diffusion coefficient or Darken-corrected diffusion coefficient, from the experimentally measured fluxes (or Fick diffusion coefficient), one has to convert a gradient in the concentration to a gradient in the chemical potential. To be able to make this conversion one has to know the adsorption isotherms. For a normal Langmuir isotherm an increase of the chemical potential (or pressure) results in an increase in the concentration inside the pores of the zeolite. However, if one approaches maximum loading an increase in the pressure hardly increases the loading, which results in a large Darken correction. In the previous section, we have seen that the adsorption isotherm of benzene in silicalite (at 30°C) shows a step at four molecules per unit cell. Such a step has consequences for the diffusion since a large thermodynamic (Darken) correction can also be expected at the plateau of the adsorption isotherm. This results in a nonmonotonic dependence of the Fick diffusion coefficient as a function of the loading [58]. The Maxwell–Stefan diffusion coefficient is nearly independent of the loading. The practical importance of this result is that if the adsorption isotherm of the system is known, a much better estimation of the (Fick) diffusion coefficient as a function of loading can be made.

For some of the pure components the adsorption isotherms have been determined experimentally. For mixtures, however, far less is known. It would therefore be interesting to investigate how well one can approximate the mixture isotherm using pure component data. In Fig. 8 the pure component isotherms and a mixture isotherm for a mixture of linear and branded isomers are shown. At low pressures the adsorption isotherm is simply the sum of the pure component isotherms. At elevated pressures, however, one observed that the branched alkane is expelled from the zeolite. Also here such a nonmonotonic dependence of the adsorption of the components as a function of the pressure has its consequences for the diffusion. For the permeation through a membrane of the mixture of these components, thermodynamic contributions to the diffusivity results in an enhancement of the selectivity by a factor of 20 compared to what one can predict on the basis of the pure components. In fact, these observations indicate that such

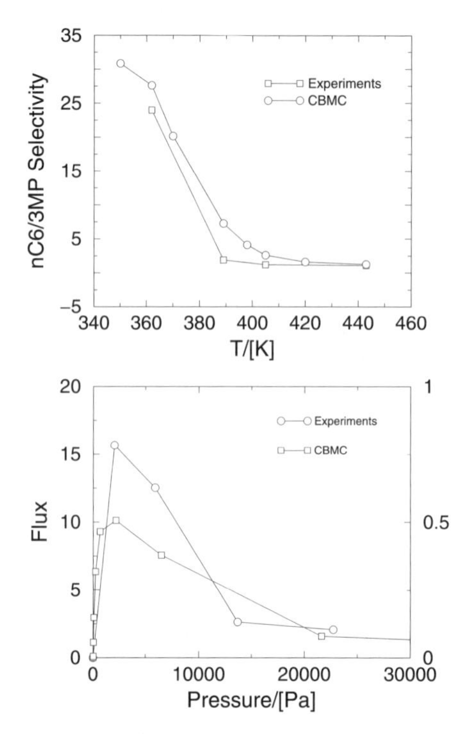

Fig. 9. Effect of temperature and pressure on selectivity or fluxes of the membrane. In the top figure the experimental data are taken from Ref. [60] and in the bottom figure from Ref. [62] and the CBMC simulations from Ref. [61].

mixture effects can be used for a novel concept to separate mixtures of hydrocarbons [59].

In Fig. 9 the temperature and pressure dependence of the selectivity of a zeolite membrane is shown. Funke et al. [60] found a sharp decrease of the selectivity when the temperature was increased above 380 K. The comparison with the simulation results, as obtained from the adsorption isotherms [61], show that this temperature dependence can be explained in terms of differences in adsorption. A similar explanation exists for the effect of pressure as observed by Gump et al. [62]. The maximum in the selectivity can be related to loading [61], which can be obtained from molecular simulations.

4.2. Compensation effect in zeolite catalysis

Haag [63] was among the first to realize the importance of understanding the adsorption behavior for the interpretation of catalytic data. A famous example is the 'compensation effect'. The reaction rate constants of the cracking of *n*-paraffins as a function of carbon number show a higher activation energy which is compensated by an increase of the exponential factor. However, detailed calculations on the reaction mechanism do not support an increase of the activation energy for longer carbon chains. Haag showed that the kinetic data were analyzed using the gas-phase concentration and did not take into account differences in adsorption, i.e. at a given pressure the number of adsorbed molecules in the pores depends on the carbon number. Haag showed that if the kinetic data are corrected for these differences in number of adsorbed molecules, he obtained a constant activation energy.

The ideas of Haag were used by Maesen and co-workers [64,65] to explain shape selectivity of hydroconversion reactions in zeolites. Maesen and co-workers computed the free energy of various reaction intermediates in the pores of the zeolite. It is argued that those intermediates with the lowest free energy are preferentially formed in the pores of the zeolite. Whereas for the large-pore zeolites the zeolite structure has little influence on the thermodynamics, for the small pore zeolite pronounced effects are observed. An important aspect is that some of these reaction intermediates are favored because they have a structure that is commensurate with the zeolite. Some intermediates form inside the pores of a zeolite but are too bulky to diffuse out of the zeolite. Yet, the products that originate from such intermediates can be observed in the product distribution.

Figure 10 shows the contribution of the free energy for various reaction intermediates of a hydroconversion reaction of *n*-decane. In a large-pore zeolite (FAU) all reaction intermediates can form and the zeolite contributes little to the relative free energies of formation. In a small-pore zeolite (TON), however, comparison of the various free energies of formation shows that in TON the formation of the large tri- and di-branched intermediates are suppressed. The zeolite MFI and MEL are very similar, yet there is a marked difference in the free energy of formation of 2,4-dimethyloctane and 4,4-dimethyloctane. Schenk et al. argue that these differences explain the differences in the experimental product distribution.

In the approach of Maesen and co-workers it is assumed that the shape selectivity is determined by the 'stable' reaction intermediates. In fact, they assume that the Polanyi–Bronsted principle holds. This implies that for a given reaction where there are competing paths to various stable reaction intermediates, if a particular reaction intermediate is favored, the zeolite lowers its free energy of formation. The

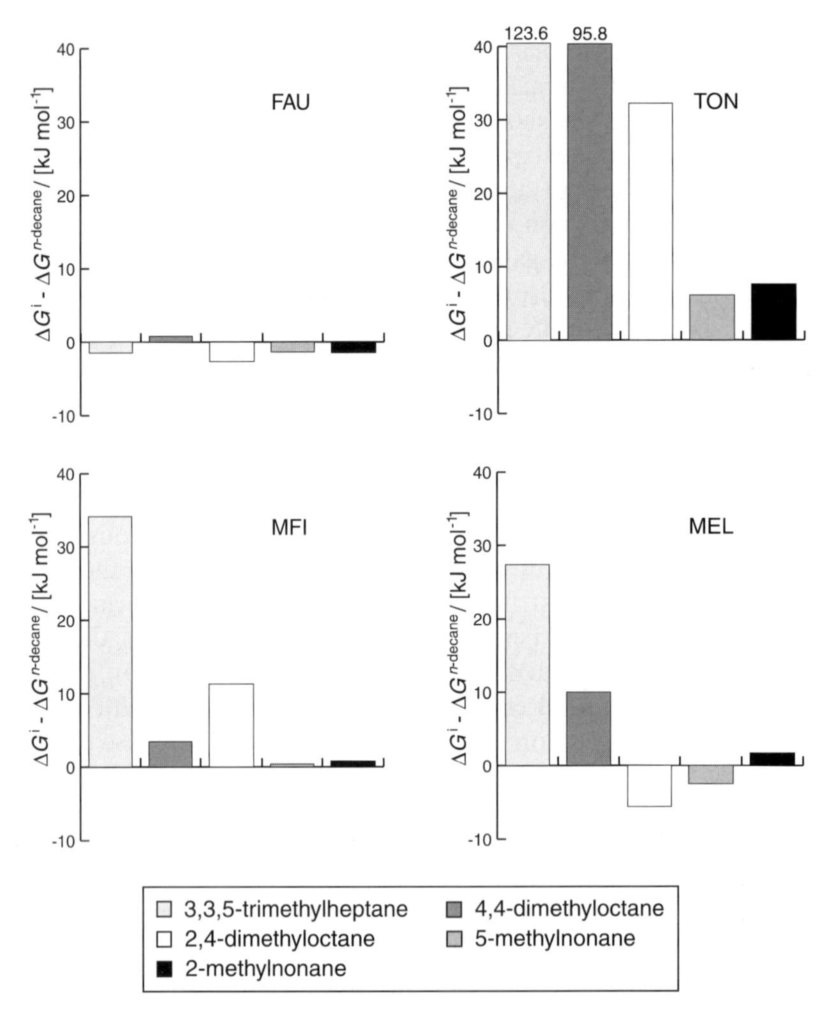

Fig. 10. The Gibbs free energy of formation of hydrocarbons relative to decane in the zeolites FAU, TON, MFI, and MEL as obtained from CBMC simulations [65].

Polanyi–Bronsted principle implies that a similar shift of the free energy of the transition state associated with this particular path can be expected. This approach will therefore fail, if in a particular competing reaction path the free energy of the corresponding transition state does not follow the same trend as its product. The specific effects of confinement on the transition state has been studied by Macedonia and Maginn [66]. In this work the free energy of a transition state in a zeolite is computed assuming that this free energy is dominated by 'classical' interactions, i.e. this free energy is dominated by steric effects rather than electronic effects.

5. Concluding remarks

In this review the focus has been on the use of modern simulation techniques to compute adsorption isotherms. It is shown that for hydrocarbons in zeolite good results can be obtained. However, most of the force fields used in these studies have been developed for the zeolite silicalite. For this zeolite ample experimental data is available. It remains to be seen whether the results extrapolate equally well to other zeolites.

Linear and branched paraffins fit loosely in the channels of silicalite. Small errors in the potential related to the parameters of the potential, the positions of the zeolite atoms, or the assumption of a rigid zeolite lattice can be compensated by the use of effective potentials. For tight-fitting molecules, such as the aromatics, such an effective potential is far less successful. Hence, for such systems it is essential to further investigate the role of lattice vibrations on the adsorption. This, however, requires accurate potentials for the zeolite–zeolite interactions.

Despite the fact that the simulations do not give a perfect prediction of the experimental adsorption isotherms and that in applications the zeolites are often far from perfect crystals, these simulation methods do allow us to obtain a reasonable estimate whether for a given application for a given zeolite whose components are adsorbed. Knowledge on the adsorption is often necessary to interpret experimental data for many applications. Here, we have used the permeation through zeolite membrane and shape selectivity as a typical example in which a detailed understanding of the sorption behavior is essential to correctly interpret the experimental results.

Acknowledgements

The author gratefully acknowledges grants from the Netherlands Organization for Scientific Research (NWO-CW).

References

1. Ruthven, D.M., *Principles of Adsorption and Adsorption Processes*. John Wiley, New York, 1984.
2. Stach, H., Lohse, U., Thamm, H. and Schirmer, W., *Zeolites*, **6**, 74 (1986).
3. Ruthven, D.M., *Ind. Eng. Chem. Res.*, **39**, 2127 (2000).
4. Fuchs, A.H. and Cheetham, A.K., *J. Phys. Chem. B*, **105**, 7375 (2001).
5. Demontis, P. and Suffritti, G.B., *Chem. Rev.*, **97**, 2845 (1997).
6. Runnebaum, R.C. and Maginn, E.J., *J. Phys. Chem. B*, **101**, 6394 (1997).

7. Webb III, E.B., Grest, G.S. and Mondello, M., *J. Phys. Chem. B*, **103**, 4949 (1999).
8. June, R.L., Bell, A.T. and Theodorou, D.N., *J. Phys. Chem.*, **96**, 1051 (1992).
9. Soto, J.L. and Myers, A.L., *Mol. Phys.*, **42**, 971 (1981).
10. Wood, G.B. and Rowlinson, J.S., *J. Chem. Soc., Faraday Trans.*, **285**, 765 (1989).
11. Goodbody, S.J., Watanabe, K., MacGowan, D., Walton, J.P.R.B. and Quirke, N., *J. Chem. Soc., Faraday Trans.*, **87**, 1951 (1991).
12. Snurr, R.Q., June, R.L., Bell, A.T. and Theodorou, D.N., *Mol. Sim.*, **8**, 73 (1991).
13. Karavias, F. and Myers, A.L., *Langmuir*, **7**, 3118 (1991).
14. van Tassel, P.R., Davis, H.T. and McCormick, A.N., *J. Chem. Phys.*, **98**, 8919 (1993).
15. Maddox, M.W. and Rowlinson, J.S., *J. Chem. Soc., Faraday Trans.*, **89**, 3619 (1993).
16. Smit, B., *J. Phys. Chem.*, **99**, 5597 (1995).
17. Frenkel, D. and Smit, B., *Understanding Molecular Simulations: From Algorithms to Applications*. Academic Press, San Diego, 2nd ed., 2002.
18. Vlugt, T.J.H., Martin, M.G., Smit, B., Siepmann, J.I. and Krishna, R., *Mol. Phys.*, **94**, 727 (1998).
19. Vlugt, T.J.H., Krishna, R. and Smit, B., *J. Phys. Chem. B*, **103**, 1102 (1999).
20. Smit, B. and Siepmann, J.I., *Science*, **264**, 1118 (1994).
21. Rosenbluth, M.N. and Rosenbluth, A.W., *J. Chem. Phys.*, **23**, 356 (1955).
22. Bezus, A.G., Kiselev, A.V., Lopatkin, A.A. and Du, P.Q., *J. Chem. Soc., Faraday Trans. II*, **74**, 367 (1978).
23. http://WWW.IZA-structure.org/.
24. June, R.L., Bell, A.T. and Theodorou, D.N., *J. Phys. Chem.*, **94**, 1508 (1990).
25. Smit, B. and Siepmann, J.I., *J. Phys. Chem.*, **98**, 8442 (1994).
26. Titiloye, J.O., Parker, S.C., Stone, F.S. and Catlow, C.R.A., *J. Phys. Chem.*, **95**, 4038 (1991).
27. Demontis, P., Suffritti, G.B., Fois, E.S. and Quartieri, S., *J. Phys. Chem.*, **96**, 1482 (1992).
28. van Koningsveld, H., Tuinstra, F., Jansen, J.C. and van Bekkum, H., *Acta Crystallogr. B*, **45**, 423 (1989).
29. Allen, M.P. and Tildesley, D.J., *Computer Simulation of Liquids*. Clarendon Press, Oxford, 1987.
30. Smit, B., Karaborni, S. and Siepmann, J.I., *J. Chem. Phys.*, **102**, 2126 (1995); erratum: *J. Chem. Phys.* **109**, 352 (1998).
31. Jaramillo, E. and Auerbach, S.M., *J. Phys. Chem. B*, **103**, 9589 (1999).
32. Mellot-Draznieks, C., Buttefey, S., Boutin, A. and Fuchs, A.H., *Chem. Commun.*, 2200 (2001).
33. Buttefey, S., Boutin, A., Mellot-Draznieks, C. and Fuchs, A.H., *J. Phys. Chem. B*, **105**, 9569 (2001).
34. Macedonia, M.D., Moore, D.D., Maginn, E.J. and Olken, M.M., *Langmuir*, **16**, 3823 (2000).
35. Maginn, E.J., Bell, A.T. and Theodorou, D.N., *J. Phys. Chem.*, **97**, 4173 (1993).
36. van Tassel, P.R., Davis, H.T. and McCormick, A.V., *Langmuir*, **10**, 1257 (1994).
37. Jameson, C.J., Jameson, A.K., Baello, B.I. and Lim, H.-M., *J. Phys. Chem.*, **100**, 5965 (1994).

38. Maginn, E.J., Bell, A.T. and Theodorou, D.N., *J. Phys. Chem.*, **99**, 2057 (1995).
39. Martin, C., Tosi-Pellenq, N., Patarin, J. and Coulomb, J.P., *Langmuir*, **14**, 1774 (1998).
40. Lachet, V., Boutin, A., Pellenq, R.J.M., Nicholson, D. and Fuchs, A.H., *J. Phys. Chem.*, **100**, 9006 (1996).
41. Maris, T., Vlugt, T.J.H. and Smit, B., *J. Phys. Chem. B*, **102**, 7183 (1998).
42. Vlugt, T.J.H., Zhu, W., Kapteijn, F., Moulijn, J.A., Smit, B. and Krishna, R., *J. Am. Chem. Soc.*, **120**, 5599 (1998).
43. Olsen, D.H., Kokotailo, G.T., Lawton, S.L. and Meier, W.M., *J. Phys. Chem.*, **85**, 2238 (1981).
44. van Koningsveld, H., van Bekkum, H. and Jansen, J.C., *Acta Crystallogr. B*, **43**, 127 (1987).
45. Guo, C.J., Talu, O. and Hayhurst, D.T., *AIChE J.*, **35**, 573 (1989).
46. Snurr, R.Q., Bell, A.T. and Theodorou, D.N., *J. Phys. Chem.*, **97**, 13742 (1993).
47. Talu, O., Guo, C.-J. and Hayhurst, D.T., *J. Phys. Chem.*, **93**, 7294 (1989).
48. Clark, L.A. and Snurr, R.Q., *Chem. Phys. Lett.*, **308**, 155 (1999).
49. Lachet, V., Boutin, A., Tavitian, B. and Fuchs, A.H., *J. Phys. Chem. B*, **102**, 9224 (1998).
50. Lachet, V., Buttefey, S., Boutin, A. and Fuchs, A.H., *Phys. Chem. Chem. Phys.*, **3**, 80 (2001).
51. Dunne, J.A., Rao, M., Sircar, S., Gorte, R.J. and Myers, A.L., *Langmuir*, **13**, 4333 (1997).
52. Abdul-Rehman, H.B., Hasanain, M.A. and Loughlin, K.F., *Ind. Eng. Chem. Res.*, **29**, 1525 (1990).
53. Macedonia, M.D. and Maginn, E.J., *Fluid Phase Equilibria*, **160**, 19 (1999).
54. Du, Z., Vlugt, T.J.H., Smit, B. and Manos, G., *AIChE J.*, **44**, 1756 (1998).
55. Talbot, J., *AIChE J.*, **43**, 2471 (1997).
56. Calero, S., Smit, B. and Krishna, R., *Phys. Chem. Chem. Phys.*, **3**, 4390 (2001).
57. Krishna, R. and Wesselingh, J.A., *Chem. Eng. Sci.*, **52**, 861 (1997).
58. Shah, D.B., Guo, C.J. and Hayhurst, D.T., *J. Chem. Soc., Faraday Trans.*, **91**, 1143 (1995).
59. Krishna, R. and Smit, B., *Chem. Inv.*, **31**, 27 (2001).
60. Funke, H.H., Argo, A.M., Falconer, J.L. and Noble, R.D., *Ind. Eng. Chem. Res.*, **36**, 137 (1997).
61. Calero, S., Smit, B. and Krishna, R., *J. Catal.*, **202**, 395 (2001).
62. Gump, C.J., Noble, R.D. and Falconer, J.L., *Ind. Eng. Chem. Res.*, **38**, 2775 (1999).
63. Haag, W.O., In: Weitkamp, J., Karge, H.G., Pfeifer, H. and Holderich, W. (Eds.), *Zeolites and Related Microporous Materials: State of the Art 1994, Studies in Surface Science and Catalysis*. Elsevier, Amsterdam, 1994, Vol. 84, pp. 1375–1394.
64. Maesen, Th.L.M., Schenk, M., Vlugt, T.J.H., de Jonge, J.P. and Smit, B., *J. Catal.*, **188**, 403 (1999).
65. Schenk, M., Smit, B., Vlugt, T.J.H. and Maesen, T.L.M., *Angew. Chem. Int. Ed. Engl.*, **40**, 736 (2001).
66. Macedonia, M.D. and Maginn, E.J., *AIChE J.*, **46**, 2544 (2000).

Computer Modelling of Microporous Materials
C.R.A. Catlow, R.A. van Santen and B. Smit (editors)

Chapter 3

Dynamics of sorbed molecules in zeolites*

Scott M. Auerbach**

Department of Chemistry and Department of Chemical Engineering, University of Massachusetts, Amherst, MA 01003, USA

Fabien Jousse*** and Daniel P. Vercauteren

Computational Chemical Physics Group, Institute for Studies in Interface Science, Facultés Universitaires Notre-Dame de la Paix, Rue de Bruxelles 61, B-5000 Namur, Belgium

1. Introduction

This chapter continues the exploration of the properties of sorbed molecules in zeolites by examining recent efforts to model their dynamics with either atomistic methods or lattice models. We discuss the assumptions underlying modern atomistic and lattice approaches, and detail the techniques and applications of modeling both rapid dynamics and activated diffusion. We summarize the major findings discovered over the last several years, and enumerate future needs for the frontier of modeling dynamics in zeolites.

With a rich variety of interesting properties and industrial applications [1–3], and with over 100 zeolite framework topologies [4–6] synthetically available — each with its own range of compositions — zeolites offer size, shape, and electrostatically selective adsorption [7], diffusion [8,9], and reaction [7] up to remarkably high temperatures.

---

*The majority of this chapter has previously been published in: J. Kärger, S. Vasenkov, and S. M. Auerbach, "Diffusion in Zeolites" in *Handbook of Zeolite Science and Technology*, edited by S. M. Auerbach, K. A. Carrado and P. K. Dutta, pp. 341–422, Marcel Dekker, Inc., New York, 2003. Reprinted by courtesy of Marcel Dekker, Inc.

**E-mail: auerbach@chem.umass.edu
***E-mail: fjousse@scf.fundp.ac.be

The impressive selectivities produced by these materials result from strong guest–zeolite interactions; however, these same interactions can severely retard the eventual permeation of desired products from zeolites. This has led to growing interest in modeling the transport of molecules in zeolites, to seek an optimal balance between high selectivity and high flux by identifying the fundamental interaction parameters that determine these key properties. In this review, we describe recent efforts using atomistic methods and lattice models to simulate the dynamics of sorbed molecules in zeolites.

Practical applications of zeolites are typically run under steady-state conditions, making the relevant transport coefficient the Fickian diffusivity or other related permeability coefficient. However, modeling such steady-state transport through zeolites with atomistic models is challenging, prompting many researchers instead to simulate self-diffusion, which is the stochastic motion of tagged particles at equilibrium. Although self-diffusivities for molecular liquids over a wide temperature range typically fall in the range of 10^{-9}–10^{-8} $m^2 s^{-1}$, self-diffusivities for molecules in zeolites cover a much larger range, from 10^{-19} $m^2 s^{-1}$ for benzene in Ca–Y [10] to 10^{-8} $m^2 s^{-1}$ for methane in silicalite-1 [11]. Such a wide range offers the possibility that diffusion in zeolites, probed by both experiment and simulation, can provide an important characterization tool complementary to diffraction, NMR, IR, etc., because diffusive trajectories of molecules in zeolites sample all relevant regions of the guest–zeolite potential energy surface. Below we assess the accuracy with which modern dynamics simulations can predict self-diffusivities of molecules in zeolites, and discuss the insights gained from such simulations regarding guest–zeolite structure.

The wide range of diffusional timescales encountered by molecules in zeolites presents unique challenges to the modeler, requiring that various simulation tools, each with its own range of applicability, be brought to bear on modeling dynamics in zeolites. In particular, when transport is relatively rapid, the molecular dynamics technique can be used to simulate both the temperature and loading dependencies of self-diffusion [12,13]. On the other hand, when molecular motion is relatively slow because free-energy barriers separating sorption sites are large compared to thermal energies, transition-state theory and related methods must be used to simulate the temperature dependence of site-to-site jump rate constants. In this regime, kinetic Monte Carlo and mean-field theory (MFT) can then be used to model the loading dependence of activated diffusion in zeolites [14,15]. In this review we describe the techniques and applications of these methods, focusing on how the interplay between guest–zeolite adhesion and guest–guest cohesion controls diffusion in zeolites.

The goal of most diffusion simulations is to predict the temperature and loading dependencies of self-diffusion in various zeolites with different framework topologies, and over a range of Si:Al ratios. One generally expects self-diffusivities to exhibit an Arrhenius temperature dependence, with the apparent activation energy controlled by migration through bottlenecks such as narrow channels or cage windows. In addition, one typically observes that self-diffusivities decrease linearly with loading as site blocking decreases the number of successful jump attempts. While these ideas provide useful rules of thumb, we see below that guest–zeolite systems provide many fascinating examples that break these long-honored rules. We also find below that with modern tools of theory and simulation, researchers have produced remarkably useful insights and accurate predictions regarding the dynamics of sorbed molecules in zeolites.

2. Atomistic dynamics in zeolites

The goals of simulating molecular dynamics in zeolites with atomistic detail are two-fold: to predict the transport coefficients of adsorbed molecules, and to elucidate the mechanisms of intracrystalline diffusion. Below we discuss the basic assumptions and force fields underlying such simulations, as well as the dynamics methods used to model both rapid and activated motion through zeolites.

2.1. Basic model and force fields

2.1.1. Zeolite model

Ordered models. Modeling the dynamics of sorbates in zeolites requires an adequate representation of the zeolite sorbent. Zeolites are crystalline materials, which tremendously simplifies the modeler's task as compared to the task of modeling amorphous or disordered microporous materials such as silica gels or activated carbons. Zeolite framework structures are well known from many crystallographic studies and easily accessible from reference material such as Meier and Olson's Atlas of Zeolite Structure Types [4], commercial [5], or internet databases [6]. Moreover, the typical size of a zeolite crystallite is 1–100 μm, that is, much larger than the length scale probed by atomistic molecular dynamics simulations. Size effects therefore can often be neglected except for single-file systems [16], and an adequate modeling of the sorbent is obtained with only a few unit cells included in the simulation cell, with periodic boundary conditions to represent the crystallite's extent.

However, a zeolite structure presents some heterogeneities at the atomistic scale: the arrangement of Si and Al atoms in the structure (or Al and P for $AlPO_4$'s) usually does not present any long-ranged ordering; and in the general case, extra-framework cations also occupy crystallographic positions without full occupancy or long-range ordering. The simplest way to tackle this problem is to ignore it completely; indeed, a good 80% of all molecular dynamics (MD) studies of guest dynamics in zeolites published since 1997 concern aluminum-free, cation-free, defect-free all-silica zeolite analogs rather than zeolites. These structures sometimes exist, such as silicalite-1, silicalite-2, and ZDDAY, the respective analogs of ZSM-5 (structure MFI), ZSM-11 (structure MEL), and Na–Y (structure FAU). However, the siliceous analogs sometimes do not exist but in the modeler's view, such as LTL, the analog of the cation-containing zeolite L. Nevertheless, these models can be very useful for studying the influence of zeolite structure or topology on an adsorbate's dynamics, irrespective of the cations [17], or to determine exactly, by comparison, the cations' influence [18,19]. Furthermore, some zeolites of industrial interest such as ZSM-5 present high Si:Al ratios, so that their protonated forms have very few protons per unit cell. Heink et al. have shown, for example, that the Si:Al ratio of ZSM-5 has very little influence on hydrocarbon diffusivity [20]. In these cases, it is safe to assume that studying diffusion in a completely siliceous zeolite analog will display most characteristics of the diffusion in the protonated form. This assumption simplifies several factors of the simulation and of the subsequent analysis: fewer parameters for the guest–zeolite interaction potential are needed, the system does not present any heterogeneity, and electrostatic interactions can be neglected when using adequate van der Waals interaction parameters, therefore decreasing the computational cost of a force evaluation.

Charge distributions. There are many cases where such a simplified representation is inadequate: in particular, exchangeable cations create an intense local electric field (amounting to 3 $V\,Å^{-1}$ next to a Ca^{2+} cation in Na–A, according to induced IR measurements) [21] so that, unless the cation is inaccessible to the sorbate, one cannot neglect its Coulombic interaction with an adsorbed molecule. The number of cations in the frame depends on the Si:Al ratio: each Al atom brings one negative charge to be compensated by the adequate number of mono- or multivalent cations. Hence the Si:Al ratio strongly influences the adsorptive properties of zeolites, so much that a change in the amount of Al brings a change in nomenclature: for example, FAU-type zeolites are denoted zeolite X for Si:Al < 1.5 and zeolite Y for a Si:Al > 1.5. Many groups have investigated the distribution of Al and Si atoms in zeolites, to determine whether there is any local arrangement of these atoms [22–27]. Since X-ray crystallography does not distinguish Si from Al, this is necessarily determined from indirect techniques such as Si or Al NMR. Löwenstein's rule forbids any Al–O–Al bonds, which brings perfect ordering for Si:Al = 1, such as in Na–A. In most other cases, no local ordering has been found in the studies mentioned above. An exception is zeolite EMT, where rich Si and Al phases have been found from crystallographic measurements, when synthesized using crown ethers as templates [28]. In zeolite L, aluminum atoms preferentially occupy T_1 rather than T_2 sites, as found out by neutron crystallography [29].

In the absence of local ordering, a common modeling procedure involves neglecting the local inhomogeneity of the Si:Al distribution, and replacing all Al or Si by an average tetrahedral atom T, which is exactly what is observed crystallographically. The Si:Al ratio then is reflected by the average charge of this T atom, the charges on framework oxygen atoms, and by the number of charge-compensating cations. This T-site model has been used in many recent modeling studies, and performs very well for reproducing adsorptive properties of zeolites [30,31]. Indeed, few studies of guest adsorption in zeolites consider explicit Al and Si atoms [32–34].

The most important inhomogeneity inside cation-containing zeolites comes from the cation distribution. Indeed, except for very special values of the Si:Al ratio, the possible cation sites are not completely or symmetrically filled, and crystallographic measurements give only average occupancies. A common procedure is to use a simplified model, with just the right Si:Al ratio that allows complete occupancy of the most probable cation sites and no cations in other sites. This has been used, e.g., by Santikary and Yashonath in their modeling of diffusion in zeolite Na–A: instead of Si:Al = 1, they used a model Na–A with Si:Al = 2, thus allowing complete occupancy of cation site I, which gives cubic symmetry of the framework [35]. Similarly, Auerbach and co-workers used a model zeolite Na–Y with Si:Al = 2 in a series of studies on benzene diffusion, so that the model would contain just the right number of Na cations to fill sites I′ and II, thereby giving tetrahedral symmetry [18,36,37]. In studying Na–X, which typically involves Si:Al = 1.2, they used Si:Al = 1 so that Na(III) would also be filled [18]. This type of procedure is generally used to level off inhomogeneities that complicate the analysis.

It is instructive to observe the effect of the Si:Al ratio of FAU-type zeotites on the behavior of benzene diffusion, as determined from modeling [18,36,38]. For very high Si:Al ratios no cations are accessible to sorbed benzene, which only feels a weak interaction with the framework, and hence diffuses over shallow energetic barriers. These reach only 10 $kJ\,mol^{-1}$ between the supercage sites and window sites, where benzene adsorbs in the plane of the 12 T-atom ring (12R) window separating two adjacent supercages [38]. As the Si:Al ratio decreases toward Na–Y, cation sites II begin to fill in as indicated in Fig. 1. These Na(II) cations at tetrahedral supercage positions create strong local adsorption sites for benzene (the S_{II} site), while the window site remains unchanged. As a consequence, the energetic barrier to diffusion increases to ca. 40 $kJ\,mol^{-1}$ [36]. The spread in measured activation energies for benzene in Na–Y, as shown in Fig. 1, reflects both intracage and cage-to-cage dynamics [39], because both NMR relaxation data (intracage) and diffusion data (cage-to-cage) are shown. When the Si:Al ratio further decreases toward Na–X, the windows are occupied by strongly adsorbing site III cations. As a consequence, the window site is replaced by a strong S_{III} site where benzene is facially coordinated to the site III cation, so that transport is controlled by smaller energy barriers reaching only about 15 $kJ\,mol^{-1}$ [18]. Figure 1 (top and middle) schematically presents this behavior, while on the bottom part we compare the expected behavior of the activation energy (full line) as a function of Si:Al ratio to the available experimental observations (points). The correlation between simulation and experiments is qualitatively reasonable considering the spread of experimental data. Figure 1 shows the success of using a particular Si:Al ratio to simplify the computation, and furthermore shows that adding cations in the structure does not necessarily result in an increase of the diffusion activation energy.

Despite the success of treating disordered charge distributions as being ordered, Chen et al. have suggested that electrostatic traps created by disordered Al and cation distributions can significantly diminish self-diffusivities from their values for corresponding ordered systems [48]. In addition, when modeling the dynamics of exchangeable cations [49] or molecules in acidic zeolites [34], it may be important to develop more sophisticated zeolite models which completely sample Al and Si heterogeneity, as well as the possible cation distributions. For example, Newsam and co-workers proposed an iterative strategy allowing the placement of exchangeable cations inside a negatively

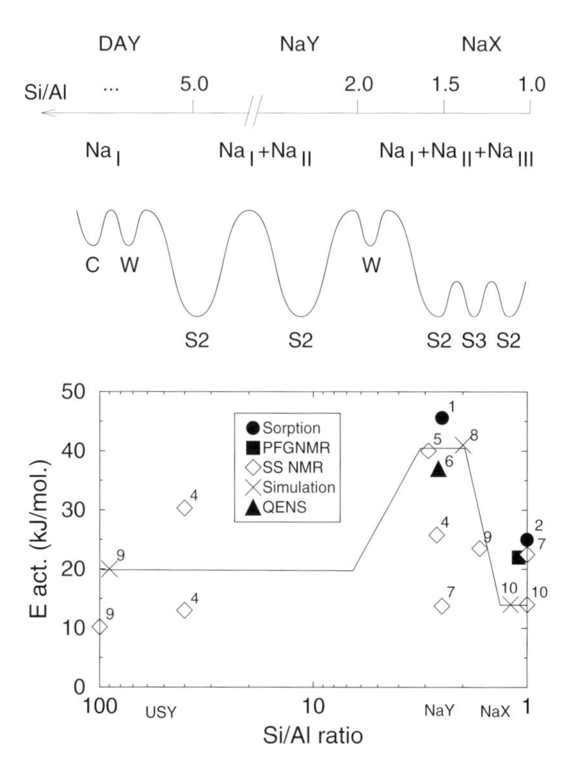

Fig. 1. Activation energies of benzene diffusion in FAU-type zeolites. The top part shows Si:Al ratios of FAU-type zeolites, with the corresponding occupied cation sites. The middle part represents schematic benzene adsorption sites, and the energy barriers between them arising from different cation distributions. C is a benzene supercage site far from a cation, W is a benzene window site far from a cation, S2 is a cage site close to an S_{II} cation, S3 is a window site close to an S_{III} cation. The bottom part gives diffusion activation energies for various Si:Al ratios. The solid line shows the overall trend from simulations, symbols are particular experiment or simulation results: 1, Forni et al. [40]; 2, Bülow et al. [41]; 3, Lorenz et al. [42]; 4, Sousa-Gonçalves et al. [43]; 5, Isfort et al. [44]; 6, Jobic et al. [45]; 7, Burmeister et al. [46]; 8, Auerbach et al. [36]; 9, Bull et al. [47]; and 10, Auerbach et al. [18].

charged framework [50], implemented within MSI's Cerius^2 modeling environment. In addition, we have constructed a model zeolite H–Y (Si:Al = 2.43) by randomly placing aluminum atoms in the frame, and distributing protons using the following three rules: (i) protons are linked to an oxygen close to an Al atom; (ii) no two hydroxyl groups can be linked to the same silicon atom; (iii) no proton can be closer than 4.0 Å from another [34]. Although these rules do not completely determine the proton positions, we found that several different proton distributions were broadly equivalent as far as sorption of benzene is concerned. It is clear from the above examples that the real issue in modeling the dynamics of sorbed molecules in zeolites comes from the interaction potentials, also known as force fields when computed from empirical functional forms. Before discussing these force fields in the context of dynamics, however, we examine a hot topic among scientist in the field: whether framework vibrations influence the dynamics of guest molecules in zeolites.

Framework flexibility. This question has long remained an open one, but many recent studies have made systematic comparisons between fixed and flexible lattice simulations, based on several examples: methane and light hydrocarbons in silicalite-l [51–55], methane in cation-free LTA [56], Lennard-Jones adsorbates in Na–A [35] and in Na–Y [57], benzene and propylene in MCM-22 [58], benzene in Na–Y [59–61], and methane in $AlPO_4$-5 [62]. In cation-free zeolites, these recent studies have found that diffusivities are virtually unchanged when including lattice vibrations. Fritzsche et al. [56] explained earlier discrepancies on methane in cation-free LTA zeolite by pointing out that inappropriate comparisons were made between rigid and flexible framework studies. In particular, the rigid studies used crystallographic coordinates for the framework atoms, while the force field used to represent the framework vibrations gave a larger mean window size than that in the rigid case, thereby resulting in larger diffusivities in the flexible framework. By comparing with a model-rigid LTA minimized using the same force field, they found almost no influence on the diffusion coefficient. Similarly, Demontis et al. have studied the diffusion of methane in silicalite-1, with rigid and flexible frameworks [53]. They conclude that the framework vibrations do not influence the diffusion coefficient, although they affect local dynamical properties such as the damping of the velocity autocorrelation function. Following these findings, numerous recent diffusion studies of guest hydrocarbons or Lennard-Jones adsorbates in cation-free zeolites keep the framework rigid [17,63–71].

There are, however, some counter examples in cation-free zeolites. In a recent MD study of benzene and propylene in MCM-22 zeolite, Sastre, Catlow, and Corma found differences between the diffusion coefficients calculated in the rigid and flexible framework cases [58]. Bouyermaouen and Bellemans also observe notable differences for *i*-butane diffusion in silicalite-1 [55]. Snurr, Bell, and Theodorou used TST to calculate benzene jump rates in a rigid model of silicalite-1 [72], finding diffusivities that are one to two orders of magnitude smaller than experimental values. Forester and Smith subsequently applied TST to benzene in flexible silicalite-1 [73], finding essentially quantitative agreement with experiment, thus demonstrating the importance of including framework flexibility when modeling tight-fitting guest–zeolite systems.

Strong framework flexibility effects might also be expected for molecules in cation-containing zeolites, where cation vibrations strongly couple to the adsorbate's motions, and where diffusion is mostly an activated process. However, where a comparison between flexible and fixed framework calculations has been performed, surprisingly little influence has been found. This has been shown by Santikary and Yashonath for the diffusion of Lennard-Jones adsorbates of varying size in Na–A. They found a notable difference on the adsorbate density distribution and external frequencies, but not on diffusion coefficients [35]. Mosell et al. found that the potential of mean force for the diffusion of benzene in Na–Y remains essentially unchanged when framework vibrations are included [59]. Jousse et al. also found that the site-to-site jump probabilities for benzene in Na–Y do not change when including framework flexibility, in spite of very strong coupling between benzene's external vibrations and the Na(II) cation [61]. The reasons behind this behavior remain unclear, and it is also doubtful whether these findings can be extended to other systems. Nevertheless, the direct examination of the influence of zeolite vibrations on guest dynamics suggests the following: a strong influence on local static and dynamical properties of the guest, such as low-frequency spectra, correlation functions, and density distributions; a strong influence on the activated diffusion of tight-fitting guest–zeolite systems; but a small influence on diffusion of smaller molecules such as unbranched alkanes.

The preceding discussion on framework flexibility, and its impact on molecular dynamics, has the merit of pointing out the two important aspects for modeling zeolites: structural and dynamical. On the structural side, the zeolite cation distribution, channel diameters, and window sizes must be well represented. On the dynamical side, for tight-fitting host–guest systems, the framework vibrations must allow for an accurate treatment of the activation energy for molecular jumps through flexing channels and/or windows. Existing zeolite framework force fields are numerous and take many different forms, but they are generally designed for only one of these purposes. It is beyond the scope of this chapter to review zeolite framework force fields [13], which are also discussed in Chapter 1; we simply wish to emphasize that one should be very cautious in choosing the appropriate force field designed for the properties to be studied.

2.1.2. Guest–zeolite force fields

The guest-framework force field is the most important ingredient for atomistic dynamical models of sorbed molecules in zeolites. Force fields for guest–zeolite interactions are at least as diverse as those for the zeolite framework: even more so, in fact, as most studies of guest molecules involve a reparameterization of potential energy functions to reproduce some typical thermodynamical property of the system, such as adsorption energies or adsorption isotherms. Since force fields are but an analytical approximation of the real potential energy surface, it is essential that the underlying physics is correctly captured by the analytical form. Every researcher working in the field has an opinion on what the correct form should be; therefore the following discussion must necessarily remain subjective, and we refer the reader to the original articles to sample different opinions.

Physical contributions to the interaction energy between host and guest are numerous: most important are the short-range dispersive and repulsive interactions, and the electrostatic multipolar and inductive interactions. Nicholson and co-workers developed precise potentials for the adsorption of rare gases in silicalite-1, including high-order dispersive terms [74], and have shown that all terms contribute significantly to the potential energy surface [75], the largest contributions coming from the two- and three-body dispersion terms. Cohen de Lara and co-workers developed and applied a potential function including inductive terms for the adsorption of diatomic homonuclear molecules in A-type zeolites [76,77]. Here also the induction term makes a large contribution to the total interaction energy. A general force field would have to account for all these different contributions, but most force fields completely neglect these terms for the sake of simplicity. Simplified expressions include only a dispersive–repulsive short-ranged potential, often represented by a Lennard-Jones 6–12 or a Buckingham 6-exp. potential, possibly combined with electrostatic interactions between partial charges on the zeolite and guest atoms, according to:

$$U_{\mathrm{ZG}} = \sum_I \sum_j \left\{ \frac{q_I q_j}{r_{Ij}} - \frac{A_{Ij}}{r_{Ij}^6} + \frac{B_{Ij}}{r_{Ij}^{12}} \right\}. \tag{1}$$

In general, the parameters A and B are determined by some type of combination rule from 'atomic' parameters, and adjusted to reproduce equilibrium properties such as adsorption energies or adsorption isotherms. It is unlikely, however, that such a potential is transferable between different guest molecules or zeolite structures. As such, the first step of any study utilizing such a simple force field on a new type of host or guest should be the computation of some reference experimental data, such as the heat of adsorption, and eventually the reparametrization of force-field terms. Indeed, general-purpose force fields such as CVFF do not generally give adequate results for adsorption in zeolites [78,79].

The simplification of the force field terms can proceed further: in all-silica zeolite analogs with small channels, the electric field does not vary much across the channel and as a consequence the Coulombic term in Eq. (1) can often be neglected. This is of course not true for cation-containing zeolites, where the cations create an intense and local electric field that generally gives rise to strong adsorption sites. Since evaluating electrostatic energies is so computationally demanding, neglecting such terms allows for much longer dynamics simulations. Another common simplification is to represent CH_2 and CH_3 groups in saturated hydrocarbons as united atoms with their own effective potentials. These are very frequently used to model hydrocarbons in all-silica zeolites [56,64,65,67,80]. There is, however, active debate in the literature whether such a simplified model can account for enough properties of adsorbed hydrocarbons [81–83].

The standard method for evaluating Coulombic energies in guest–zeolite systems is the Ewald method [84,85], which scales as $n \ln n$ with increasing number of atoms n. In 1987, Greengard and Rokhlin [86] presented the alternative 'Fast Multipole Method' (FMM) which only scales as n, and therefore offers the possibility of simulating larger systems. In general, FMM only competes with the Ewald method for systems with many thousand atoms [87], and therefore is of little use in zeolitic systems where the simulation cell can usually be reduced to a few hundreds or a few thousand atoms. However, in the special case where the zeolite lattice is kept rigid, most of the terms in FMM can be precomputed and stored; in this case we have shown that FMM becomes faster than Ewald summation for benzene in Na–Y [37].

This section would not be complete without mentioning the possibility of performing atomistic simulations in zeolites without force fields [88], using ab initio molecular dynamics (AIMD) [89,90]. Following the original work of Car and Parrinello, most such studies use density functional theory and planewave basis sets [91]. This technique has been applied recently to adsorbate dynamics in zeolites [92–100]. Beside the obvious interest of being free of systematic errors due to the force field, this technique also allows the direct study of zeolite catalytic activity [92–94]. However, AIMD remains so time consuming that a dynamical simulation of a zeolite unit cell with an adsorbed guest only reaches a few picoseconds at most. This timescale is too short to follow diffusion in zeolites, so that current simulations are mostly limited to studying vibrational behavior [92–97]. Similarly, catalytic activity is limited to reactions with activation energies on the order of thermal energies [92,94,98]. However, the potential of AIMD to simulate transport coefficients has been demonstrated for simpler systems [101,102], and will likely extend to guest–zeolite systems in the near future as computers and algorithms improve.

2.2. *Equilibrium molecular dynamics*

Since the first application of equilibrium MD to guest molecules adsorbed in zeolites in 1986 [103], the subject has attracted growing interest [13,15]. Indeed, MD simulations provide an invaluable tool for studying the dynamical behavior of adsorbed molecules over times ranging from picoseconds to nanoseconds, thus correlating atomistic interactions to experiments that probe molecular dynamics, including: solid-state NMR, pulsed field gradient NMR (PFG NMR), inelastic neutron spectroscopy (INS), quasi-elastic neutron scattering (QENS), IR, and Raman spectroscopy.

Molecular dynamics of guest molecules in zeolites is conceptually no different from MD simulations of any other nanosized system. Classical MD involves numerically integrating classical equations of motion for a many-body system. For example, when using Cartesian coordinates, one can integrate Newton's second law: $\mathbf{F}_i = m_i\mathbf{a}_i$ where m_i is the mass of the ith particle, $\mathbf{a}_i = \mathrm{d}^2\mathbf{r}_i/\mathrm{d}t^2$ is its acceleration, and $\mathbf{F}_i = -\nabla\mathbf{r}_i V$ is the force on particle i. The crucial inputs to MD are the initial positions and velocities of all particles, as well as the system potential energy function $V(\mathbf{r}_1, \mathbf{r}_2, \ldots, \mathbf{r}_n)$. The output of MD is the dynamical trajectory $[\mathbf{r}_i(t), \mathbf{v}_i(t)]$ for each particle. All modern techniques arising in the field can be applied to the simulation of zeolites, including multiple timescale techniques, thermostats, and constraints. The interested reader is referred to textbooks on the method [85,104], and to modern reviews [105,106]. In this section we shall describe only those aspects of MD that are especially pertinent to molecules in zeolites. A comprehensive review on MD of guest molecules in zeolites was published in 1997 by Demontis and Suffritti [13]. Because the review by Demontis and Suffritti discusses virtually all applications of the method up to 1996, we will limit our examples to the most recent MD studies.

2.2.1. Parameters and ensembles

Parameters. Equilibrium MD is generally composed of two stages: an equilibration run, allowing the system to relax to equilibrium, and a production run, during which data are gathered for later analysis. Typically, the equilibration is initiated from some initial configuration of the adsorbate (randomly chosen or from an energy minimum) with initial velocities assigned from a Maxwell–Boltzmann distribution. The duration required to reach equilibrium depends on the relaxation time of the system: in general larger systems presenting strong correlations, e.g. at high loading, require much longer equilibration times than do smaller systems. For example, Gergidis and Theodorou have used equilibration times ranging from 0.5 to 2 ns for low to high loading of mixtures of methane and *n*-butane in silicalite-1 [70]. Other groups, however, used much shorter equilibration runs: Clark et al. [64] or Schuring et al. [67] used equilibration runs of 50–125 ps for long alkanes in silicalite-1, while Sastre and co-workers, who used a much more complex and computationally demanding force field with a flexible framework, limited the equilibration runs to 25 ps [58,82,83]. Schrimpf et al. have directly studied the relaxation of adsorbed xenon and one-center methane in a model Na–Y, using nonequilibrium molecular dynamics [57]. We have recently investigated the relaxation of benzene in Na–Y at infinite dilution [61]. Both these studies show that relaxation is influenced by framework vibrations, lateral interactions between guest molecules and coupling with the internal degrees of freedom. However, in all cases relaxation remains quite fast, decaying exponentially with a time constant of ca. 5 ps for benzene at 100 K [61], 11 ps for methane, and 25 ps for xenon at 300 K. The equilibration run is generally performed in the canonical ensemble to achieve a desired temperature [13], since the dynamics is not monitored, any method of temperature control can be used.

Equilibrium MD calculations are mostly performed to generate trajectories for studying adsorbate self-diffusion. Special care should be taken to ensure that the trajectories are indeed long enough to compute a statistically converged self-diffusion coefficient. We estimate that the current-limiting diffusivity, below which adsorbate motion is too slow for equilibrium MD, is around $D_{\mathrm{min}} \approx 5 \times 10^{-10}$ m^2 s^{-1}, obtained by supposing that a molecule travels over 10 unit cells of 10 Å during a 20 ns MD run. This value of D_{min} is higher than most measured diffusivities in cation-containing zeolites [8], explaining why so many MD studies focus on hydrocarbons in all-silica zeolite analogs. Even then, the simplifications discussed above are required in order to perform MD runs of several nanoseconds in a manageable time: simple Lennard-Jones force fields on united atom interaction centers without Coulombic interactions, bond constraints on C–C bonds allowing for longer time steps, and the use of fixed frameworks.

Ensembles. A flexible zeolite framework typically provides an excellent thermostat for the sorbate molecules. The framework temperature exhibits minimal variations around its average value, while the sorbate energy fluctuates in a way consistent with the canonical ensemble. This is valid either for a microcanonical (NVE) ensemble run, or a canonical (NVT) ensemble run involving mild coupling to an external thermostat. We caution that coupling the system too strongly to an external bath will almost surely contaminate the actual sorbate dynamics.

The problem is clearly more complex when the zeolite framework is kept rigid. Ideally, one should run the dynamics in the canonical ensemble, with just the right-coupling constant to reproduce the fluctuations arising from a flexible framework. When these fluctuations are unknown, however, it is not obvious whether a canonical or microcanonical run is better. In the NVE ensemble, the sorbate does not exchange energy with a bath, which may lead to incorrect energy statistics. This is particularly true at low loading, but may remain true for higher loadings as well. Indeed, in a direct study of the kinetic energy relaxation of Lennard-Jones particles in Na–Y, Schrimpf et al. found that the thermalization due to interactions with the framework is considerably faster than the thermalization due to mutual interactions between the adsorbates [57]. Therefore, it is probably better to run the dynamics in the NVT ensemble, with sufficiently weak coupling to an external thermostat to leave the dynamics uncontaminated. On the other hand, we have shown that for nonrigid benzene in Na–Y, there is very rapid energy redistribution from translational kinetic energy into benzene's internal vibrational degrees of freedom [61], which proceeds on a timescale comparable to the thermalization due to interactions with the flexible frame. This suggests that for sufficiently large, flexible guest molecules, the transport behavior can be adequately modeled in the NVE ensemble even at infinite dilution.

Although this section focuses on equilibrium MD, we note growing interest in performing nonequilibrium MD (NEMD) simulations on guest–zeolite systems. As an aside, we note that MD experts would classify thermostated MD, and any non-Newtonian MD for that matter, as NEMD [107,108]. We shall be much more restrictive and limit the nonequilibrium behavior to studies involving an explicit gradient along the system, resulting in a net flow of particles. This is especially interesting in zeolite science, because most applications of zeolites are run under nonequilibrium conditions, and also because of recent progress in the synthesis of continuous zeolite membranes [109,110]. In this case we seek the Fickian or 'transport' diffusivity, defined by Fick's law: $J = -D\nabla\theta$, where J is the net particle flux, D is the transport diffusivity, and $\nabla\theta$ is the local concentration gradient. These concepts are discussed more thoroughly in Section 3.3.2; here we only wish to discuss ensembles relevant to this NEMD.

A seminal study was reported in 1993 by Maginn, Bell, and Theodorou, reporting NEMD calculations of methane-transport diffusion through silicalite-1 [111]. They applied gradient-relaxation MD as well as color-field MD, simulating the equilibration of a macroscopic concentration gradient and the steady-state flow driven by an external field, respectively. They found that the color-field MD technique provides a more reliable method for simulating the linear response regime. Since then, NEMD methods in the grand canonical ensemble have been reported. Of particular interest is the 'dual control volume grand canonical molecular dynamics' (DCV-GCMD) method, presented by Heffelfinger and van Swol [112]. In this approach the system is divided into three parts, a central and two boundary regions. In the central region, regular molecular dynamics is performed, while in the boundary regions creation and annihilation of molecules are allowed to equilibrate the system with a given chemical potential, following the grand canonical Monte Carlo procedure. This or similar methods have been applied to the simulation of fluid-like behavior in slit pores of very small dimensions (down to a few σ) [113–118]. At the time of this writing, however, no such simulation has been applied to Fickian diffusion in structured zeolite pores, presumably because it would depend on details of zeolite crystallite surface structure. Nonetheless, this is likely to be an important area of future research.

2.2.2. Data analyses

Most equilibrium MD studies aim to determine the self-diffusion coefficient of the adsorbed molecules within the zeolite pores. The self-diffusivity is defined by Einstein's relation:

$$D_{\mathrm{s}} = \lim_{t\to\infty} \frac{1}{6t} \langle |\mathbf{r}(t) - \mathbf{r}(0)|^2 \rangle, \tag{2}$$

that is, it is proportional to the long-time limiting slope of the mean-square displacement (MSD). This expression assumes that for $t \to \infty$, the guest diffusion becomes Fickian, so that the MSD becomes linear with t. This is valid whenever the motions of the adsorbates are not too strongly correlated. An extreme case of correlation between molecular motions is single-file diffusion, where molecules diffusing in unidirectional narrow channels must necessarily diffuse all together or not at all. This type of behavior has recently been experimentally observed for the diffusion of tetrafluoromethane in $AlPO_4$-5 [119,120]. In that case, correlations extend to infinity and the behavior of the MSD as a function of t at long time depends on the boundaries of the model [121]: linear for open boundaries, plateau for closed boundaries, and $\sqrt{t}$ for an infinite system.

Although MD becomes inefficient for modeling-activated diffusion, MD can provide useful information about such transport when barriers are comparable to $k_B T$. In this case, MD can be used to define a coarse-grained model of diffusion [122,123]. This coarse graining requires two inputs: the lattice of sites on which diffusion takes place, and the kinetic law governing the motions between those sites. The analysis of MD trajectories as a jump diffusion process allows one to determine the adsorption sites, by monitoring the positions of maximum probability of the adsorbate during the dynamics [123], as well as the details of the kinetic law. It has generally been found that residence time distributions follow a simple exponential dependence, characteristic of random site-to-site jumps. In Fig. 2, we present such a residence time distribution for the example of benzene diffusing in zeolite LTL, clearly showing this signature. These observations support the usual

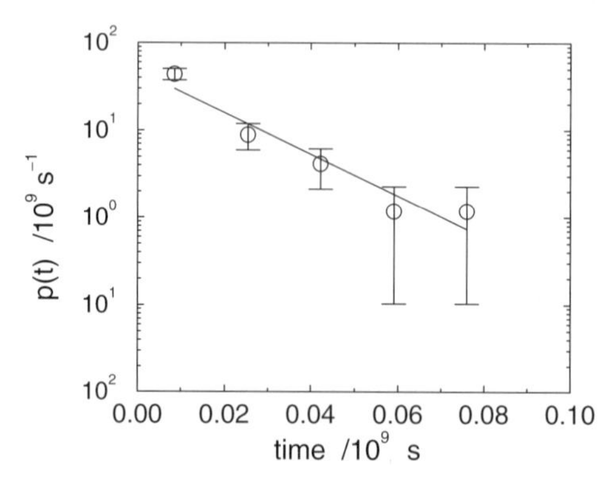

Fig. 2. Cage residence time distribution of benzene in zeolite LTL showing agreement with Poisson statistics, computed from a 1-ns molecular dynamics simulation at 800 K with a single benzene molecule in the simulation cell.

assumption of Poisson dynamics, central to many lattice models of guest diffusion in zeolites (see Section 3.1). However, one often finds correlations between jumps that complicate the coarse-grained representation of diffusion [123–125].

Jump diffusion analyses of MD are particularly useful for comparing with quasi-elastic neutron scattering (QENS) experiments. QENS experiments measure the scattering function $F(\mathbf{Q}, \omega)$, which is the space-time Fourier transform of the van Hove correlation function:

$$G(\mathbf{r}, t) = \frac{1}{N} \sum_{i=1}^{N} \left\langle \delta(\mathbf{r} + r_i(t) - \mathbf{r}_i(0)) \right\rangle \tag{3}$$

For an adsorbate containing hydrogen atoms, the largest part of the incoherent scattering comes from these atoms [126]. A model of their microscopic motions is required to determine the mobility of the adsorbed molecule [127]. MD simulations can be used to provide a direct analysis of the microscopic motions, and therefore to guide the interpretation of experiments. For example, Gaub et al. derived a simplified analytical formula for the van Hove correlation function of an adsorbate diffusing in a periodic zeolite structure [63]. Recently, Gergidis, Jobic, and Theodorou analyzed QENS experiments of mixtures of methane and butane in silicalite-1 using a jump diffusion model, with the distribution of jumps extracted from their MD simulations [80].

When $k_B T$ is comparable to or greater than barriers between sites, the self-diffusion coefficient can also be determined from the velocity autocorrelation function, according to:

$$D_s = \frac{1}{3} \int_0^{\infty} dt \left\langle \mathbf{v}(t) \cdot \mathbf{v}(0) \right\rangle, \tag{4}$$

where $\mathbf{v}(t)$ indicates the instantaneous velocity of the adsorbate's center-of-mass. This equation shows that the self-diffusion coefficient is proportional to the the zero-frequency component of the power spectrum $G(\omega)$ of the adsorbed molecule:

$$G(\omega) = \frac{1}{2\pi c} \int dt \frac{\left\langle \mathbf{v}(t) \cdot \mathbf{v}(0) \right\rangle}{\left\langle \mathbf{v}(0) \cdot \mathbf{v}(0) \right\rangle} e^{i\omega t}. \tag{5}$$

This spectrum, as well as spectra coming from other correlation functions, give particularly useful information about structure and dynamics, thereby providing additional ways to assess the validity of force fields used in dynamics simulations [79]. The interested reader is referred to classical textbooks on MD simulations for more details on obtaining these spectra [85,104]. Some recent applications include the computation of low-frequency IR and Raman spectra of cationic-exchanged EMT zeolites by Bougeard et al. [128], and our study of the external vibrations and rotations of benzene adsorbed in faujasite, with comparison to inelastic neutron scattering experiments [79].

2.2.3. Recent applications

Dynamics of hydrocarbons in silicalite-1 and 10R zeolites. Zeolite ZSM-5 is used in petroleum cracking, which explains the early interest in modeling the diffusion of alkanes in silicalite-1, the all-silica analog of ZSM-5 [51–53,122,129–131]. This early work has been reviewed by Demontis and Suffritti in 1997 [13], and therefore we only wish to outline recent studies.

As pointed out earlier, the relatively rapid diffusivity of alkanes in the channels of all-silica zeolites, at room temperature or above, makes these systems perfect candidates for MD simulations. In general, very good agreement is found between MD self-diffusivities and those of microscopic types of experiments, such as PFG NMR or QENS. Figure 3 gives an example of this agreement, for methane and butane in silicalite-1 at 300 K (MD data slightly spread for clarity). This good agreement, in spite of the crudeness of the potentials used, shows that the diffusivity of light alkanes in silicalite-1 depends on the force field properly representing the host–guest steric interactions, i.e. on the size and topology of the pores. Recognizing this, many recent studies focus on comparing diffusion coefficients for different alkanes in many different zeolite topologies, in an effort to rationalize different observed catalytic behaviors. Jousse et al. studied the diffusion of butene isomers at infinite dilution in 10R zeolites with various topologies: TON, MTT, MEL, MFI, FER, and HEU. They observed in all cases except for the structure TON, that *trans*-2-butene diffuses more rapidly than all other isomers [132]. Webb and Grest studied the diffusion of linear decanes and *n*-methylnonanes in seven 10R zeolites: AEL, EUO, FER, MEL, MFI, MTT, and TON [17]. For MEL, MTT, and MFI, they observe that the self-diffusion coefficient decreases monotonically as the branch position is moved toward the center (and the isomer becomes bulkier), while for the four other structures, D_s presents a minimum for another

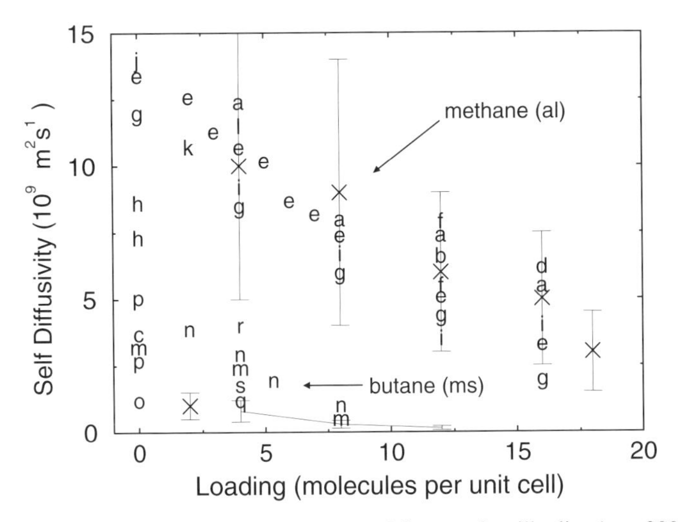

Fig. 3. Self-diffusion isotherms of methane and butane in silicalite-1 at 300 K, from PFG NMR, QENS, and MD simulations, showing good agreement with the $(1-\theta)$ loading dependence predicted by mean field theory. Crosses are NMR data from Caro et al. [11] for methane and Heink et al. [20] for butane, while the star shows QENS butane data from Jobic et al. [134]. In all cases, error bars represent an estimated 50% uncertainty. Letters are MD results (slightly spread for clarity): a–l for methane and m–s for butane, from the following references: (a) June et al. [129], (b) Demontis et al. [51], (c) Catlow et al. [52], (e) Goodbody et al. [131], (f) Demontis et al. [53], (g) Nicholas et al. [135], (h) Smirnov [54], (i) Jost et al. [71], (j) Ermoshin and Engel [136], (k) Schuring et al. [67], (l) Gergidis and Theodorou [70], (m) June et al. [122], (n) Hernández and Catlow [137], (o) Maginn et al. [138], (p) Bouyermaouen and Bellemans [55], (q) Goodbody et al. [131], (r) Gergidis and Theodorou [70], and (s) Schuring et al. [67].

branch position, suggesting that product shape selectivity might play some role in determining the zeolite selectivity. More recently, Webb et al. studied linear and branched alkanes in the range $n = 7$–30 in TON, EUO, and MFL [68]. Again they observe lattice effects for branched molecules, where D_s presents a minimum as a function of branch position dependent upon the structure. They note also some 'resonant diffusion effect' as a function of carbon number, noted earlier by Runnebaum and Maginn [133]: the diffusivity becomes a periodic function of carbon number, due to the preferential localization of molecules along one channel and their increased diffusion in this channel. Schuring et al. studied the diffusion of C_1 to C_{12} in MFI, MOR, FER, and TON for different loadings [67]. They also find some indication of a resonant diffusion mechanism as a function of chain length. Their study also indicates that the diffusion of branched alkanes is significantly slower than that of their linear counterparts, but only for structures with small pores where there is a tight fit between the adsorbates and the pores.

Another current direction of research concerns the diffusion of mixtures of adsorbates. Although the currently preferred atomistic simulation method applied to the adsorption of mixtures is grand canonical Monte Carlo [139–143], MD simulations are also used to determine how the dynamics of one component affects the diffusion of the other [70,71,80,144]. Sholl and Fichthorn investigated how a binary mixture of adsorbates diffuses in unidirectional pores [144], finding a dual mode of diffusion for certain mixtures, wherein one component undergoes normal unidirectional diffusion while the other performs single-file diffusion. Jost et al. studied the diffusion of mixtures of methane and xenon in silicalite-1 [71]. They find that the diffusivity of methane decreases strongly as the loading of Xe increases, while the diffusivity of Xe is nearly independent of the loading of methane, which they attribute to the larger mass and heat of adsorption of Xe. On the other hand, Gergidis and Theodorou in their study of mixtures of methane and *n*-butane in silicalite-1 [70], found that the diffusivity of both molecules decreases monotonically with increasing loading of the other. Both groups report good agreement with PFG NMR [71] and QENS experiments [80].

Single-file diffusion. Single-file diffusion designates the particular collective motion of particles diffusing along a one-dimensional channel, and unable to pass each other. As already mentioned, in that case the long-time motions of the particles are completely correlated, so that the limit of the MSD depends on the boundaries of the system. Exact treatments using lattice models show that the MSD has three limiting dependencies with time [121,145]: plateau for fixed boundaries, linear with t for periodic boundaries or open boundaries [16], and $\sqrt{t}$ for infinite pore length. Experimental evidence for the existence of single-file behavior in unidimensional zeolites [119,120,146,147] has prompted renewed interest in the subject during the last few years [16,65,66,148–152]. In particular, several molecular dynamics simulations of more or less realistic single-file systems have been performed, in order to determine whether the single-file $\sqrt{t}$ regime is not an artifact of the simple lattice model on which it is based [65,66,69,150,151]. Since the long-time motions of the particles in the MD simulations are necessarily correlated, great care must be taken to adequately consider the system boundaries. In particular, when using periodic boundary conditions, the system size along the channel axis must be sufficiently large to avoid the linear behavior due to the diffusion of the complete set of molecules.

Hahn and Kärger studied the diffusion of Lennard-Jones particles along a straight tube in three cases: (i) without external forces acting on the particles from the tube, (ii) with random forces, and (iii) with a periodic potential from the tube [151]. They find for the no-force case that the MSD is proportional to t, whereas for random forces and a periodic potential it is proportional to $\sqrt{t}$, in agreement with the random walk model. Keffer et al. performed MD simulations of Lennard-Jones methane and ethane in an atomistic model of $AlPO_4$-5 [150]. The methane molecules, which are able to pass each other, display undirectional but otherwise normal diffusion with the MSD linear with t; while ethane molecules, which have a smaller probability to pass each other, display single-file behavior with an MSD proportional to $\sqrt{t}$. For longer times, however, the nonzero probability to pass each other destroys the single-file behavior for ethane. Similar behavior was found by Tepper et al. [69]. Sholl and co-workers investigated the diffusion of Lennard-Jones particles in a model $AlPO_4$-5 [65,66,152], and found that diffusion along the pores can occur via concerted diffusion of weakly bound molecular clusters, composed of several adsorbates. These clusters can jump with much smaller activation energies than that of a single molecule. However, the MSD retains its single-file $\sqrt{t}$ signature because all the adsorbates in a file do not collapse to form a single supramolecular cluster.

These MD simulations of unidirectional and single-file systems confirm the lattice gas prediction, that the MSD is proportional to $\sqrt{t}$. They also show that whenever a certain crossing probability exists, this single-file behavior disappears at long times, to be replaced by normal diffusion. Similar 'anomalous' diffusion regimes, with the MSD proportional to t at long times and to t^{α} with $\alpha < 1$ at short times, have also been found in other systems that do not satisfy the single-file criteria, such as *n*-butane in silicalite-1 at high loadings [70]. Therefore, one should be very careful to define exactly the timescale of interest when working with single-file or other highly correlated systems.

2.3. *Reactive flux molecular dynamics and transition-state theory*

As discussed in Section 2.2.1, the smallest diffusivity that can be simulated by MD methods is well above most measured values in cation-containing zeolites [8], explaining why so many MD studies focus on hydrocarbons in all-silica zeolite analogs. This issue has been addressed by several groups within the last 10 years [153], by applying reactive flux molecular dynamics [154,155] (RFMD) and transition-state theory [156] (TST) to model the dynamics of rare events in zeolites. This subject has been reviewed very recently by Auerbach [15]; as a result, we give below only a brief outline of the theory.

2.3.1. Rare-event theory

The standard *ansatz* in TST is to replace the dynamically converged, net reactive flux from reactants to products with the instantaneous flux through the transition-state dividing surface. TST is inspired by the fact that, although a dynamical rate calculation is rigorously independent of the surface through which fluxes are computed [157], the duration of dynamics required to converge the net reactive flux is usually shortest when using the transition-state dividing surface. The TST approximation can be formulated for gas-phase or condensed-phase systems [154,155,158], using classical or quantum mechanics [159]. The rate coefficient for the jump from site i to site j can be expressed classically as [154,155]:

$$k_{i \to j} = k_{i \to j}^{\mathrm{TST}} \times f_{ij}, \tag{6}$$

where $k_{i \to j}^{\mathrm{TST}}$ is the TST rate constant, and f_{ij} is the dynamical correction factor also known as the classical transmission coefficient. The TST rate constant is given by:

$$k_{i \to j}^{\mathrm{TST}} = \frac{1}{2}\left(\frac{2k_{\mathrm{B}}T}{\pi m}\right)^{1/2} \frac{Q^{\ddagger}}{Q_i}, \tag{7}$$

where m is the reduced mass associated with the reaction coordinate, $Q^{\ddagger}$ is the configurational partition function on the dividing surface, and Q_i is the configurational partition function in the reactant state i. The last expression can be evaluated without recourse to dynamics, either by Monte Carlo simulation [160] or in the harmonic approximation by normal-mode analysis [161]. The dynamical correction factor is usually evaluated from short molecular dynamics simulations originating on the dividing surface. For classical systems, f_{ij} always takes a value between zero and one, and gives the temperature-dependent fraction of initial conditions on the dividing surface that initially point to products and eventually give rise to reaction.

When one has an educated guess regarding the reaction coordinate, but no knowledge of the transition state or the dividing surface, a reliable but computationally expensive solution is to calculate the free-energy surface along a prescribed path from one free energy minimum to another. The free-energy surface, $F(x_0)$, which is also known as the potential of mean force and as the reversible work surface, is given by:

$$F(x_0) = -k_B T \ln[L\langle\delta(x - x_0)\rangle] = -k_B T \ln Q(x_0), \tag{8}$$

where x is the assumed reaction coordinate, x_0 is the clamped value of x during the ensemble average over all other coordinates, the length L is a formal normalization constant that cancels when computing free-energy differences, and $Q(x_0)$ is the partition function associated with the free energy at x_0. In terms of the free-energy surface, the TST rate constant is given by:

$$k_{i\to j}^{\mathrm{TST}} = \frac{1}{2}\left(\frac{2k_B T}{\pi m}\right)^{1/2} \frac{\mathrm{e}^{-\beta F(x^{\ddagger})}}{\int_i \mathrm{d}x\, \mathrm{e}^{-\beta F(x)}}, \tag{9}$$

where the integral over x is restricted to the reactant region of configuration space. Computing TST rate constants is therefore equivalent to calculating free-energy differences. Numerous methods have been developed over the years for computing $\mathrm{e}^{-\beta F(x)}$, many of which fall under the name umbrella sampling or histogram window sampling [104,155,162].

While Eqs. (6)–(9) are standard expressions of rare-event theory, the exact way in which they are implemented depends strongly upon the actual system of interest. Indeed, if the transition-state dividing surface is precisely known (as for the case of an adatom), then $k_{i\to j}^{\mathrm{TST}}$ provides a good first approximation to the rate coefficient, and the dynamical correction factor accounts for the possibility that the particle does not thermalize in the state it has first reached, but instead goes on to a different final state. This process is called 'dynamical recrossing' if the final state is identical to the original state, and otherwise is called 'multisite jumping'. The importance of dynamical recrossing or multisite jumping depends on a number of factors, of which the height of the energy barriers and the mechanism of energy dissipation are essential.

For example, the minimum energy path for benzene to jump from a cation site to a window site in Na–Y is shown in Fig. 4, alongside the corresponding energy plot [36]. Despite benzene's anisotropy,

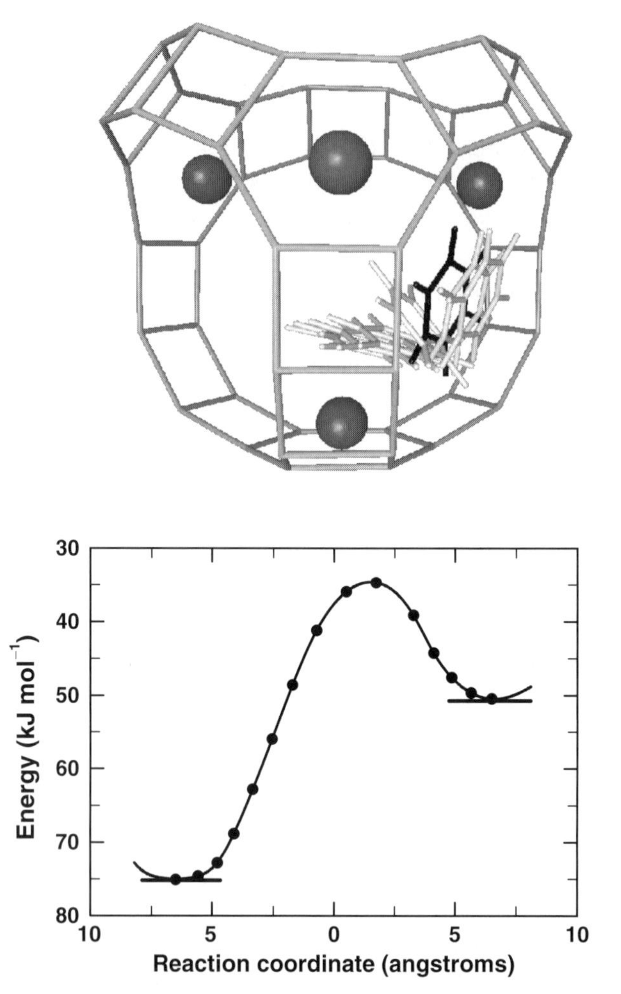

Fig. 4. Cation ↔ window path for benzene in Na–Y (transition state indicated in bold), with a calculated barrier of 41 kJ mol^{-1} [36].

a reasonable model for the cation ↔ window dividing surface turns out to be the plane perpendicular to the three-dimensional vector connecting the two sites. This simple approach yields dynamical correction factors mostly above 0.5 [37].

In a complex system with many degrees of freedom it might be difficult, or even impossible, to define rigorously the dividing surface between the states. In this case the transition-state approximation may fail, requiring the calculation of f_{ij}. Indeed, TST assumes that all trajectories initially crossing the dividing surface in the direction of the product state will eventually relax in this state. This statement will be qualitatively false if the supposed surface does not coincide with the actual dividing surface. In this case, the dynamical correction factor corrects TST for an inaccurately defined dividing surface, even when dynamical recrossings through the actual dividing surface are rare. The problem of locating complex dividing surfaces has recently been addressed using topology [163], statistics [164], and dynamics [165,166].

2.3.2. Recent applications

Siliceous zeolites. June, Bell, and Theodorou reported the first application of TST dynamically corrected with RFMD for a zeolite–guest system in 1991 [153], modeling the diffusion of Xe and 'spherical SF_6' in silicalite-1. This system is sufficiently weakly binding that reasonably converged MD simulations could be performed for comparison with the rare-event dynamics, showing excellent quantitative agreement in the diffusivities obtained. The dynamical correction factors obtained by June et al. show that recrossings can diminish rate coefficients by as much as a factor of ca. 3, and that multisite jumps along straight channels in silicalite-1 [124] contribute to the well-known diffusion anisotropy in MFI-type zeolites [167]. Jousse and co-workers reported a series of MD studies on butene isomers in all-silica channel zeolites MEL and TON [123,168]. Because the site-to-site energy barriers in these systems are comparable to the thermal energies studied in the MD simulations, rare-event dynamics need not apply. Nonetheless, Jousse and co-workers showed that even for these relatively low-barrier systems, the magnitudes and loading dependencies of the MD diffusivities could be well explained within a jump diffusion model, with residence times extracted from the MD simulations.

As discussed in Section 2.1.1, Snurr, Bell, and Theodorou applied harmonic transition-state theory (TST) to benzene diffusion in silicalite-1, assuming that benzene and silicalite-1 remain rigid, by using normal-mode analysis for the six remaining benzene degrees of freedom [72]. Their results underestimate experimental diffusivities by one to two orders of magnitude, probably more from assuming a rigid zeolite than from using harmonic TST. Forester and Smith subsequently applied TST to benzene in silicalite-1 using constrained reaction coordinate dynamics on both rigid and flexible lattices [73]. Lattice flexibility was found to have a very strong influence on the jump rates. Diffusivities obtained from the flexible framework simulations are in excellent agreement with experiment, overestimating the measured room temperature diffusivity (2.2×10^{-14} $m^2 s^{-1}$) by only about 50%. These studies suggest that including framework flexibility is very important for bulkier guest molecules, which may require framework distortions to move along zeolite channels or through windows separating zeolite cages.

Cation-containing zeolites. Mosell, Schrimpf, and Brickmann reported a series of TST and RFMD calculations on Xe in Na–Y [169,170] in 1996, and benzene and *p*-xylene in Na–Y [59,60] in 1997. They calculated the reversible work of dragging a guest species along the cage-to-cage [111] axis of Na–Y, and augmented this version of TST with dynamical corrections. In addition to computing the rate coefficient for cage-to-cage motion through Na–Y, Mosell et al. confirmed that benzene window sites are free-energy local minima, while *p*-xylene window sites are free-energy maxima, i.e. cage-to-cage transition states [59,60]. Mosell et al. also found relatively small dynamical correction factors, ranging from 0.08 to 0.39 for benzene and 0.24 to 0.47 for *p*-xylene.

At about the same time in 1997, Jousse and Auerbach reported TST and RFMD calculations of specific site-to-site rate coefficients for benzene in Na–Y [37], using Eq. (6) with jump-dependent dividing surfaces. As with Mosell et al., we found that benzene jumps to window sites could be defined for all temperatures studied. We were unable to use TST to model the window $\rightarrow$window jump because we could not visualize simply the anisotropy of the window $\rightarrow$window dividing surface. For jumps other than window $\rightarrow$window, we found dynamical correction factors mostly above 0.5, suggesting that these jump-dependent dividing surfaces coincide closely with the actual ones. Although the flavors of the two approaches for modeling benzene in Na–Y differed, the final results were remarkably similar considering that different force fields were used. In particular, Mosell et al. used MD to sample dividing-surface configurations, while we applied the Voter displacement-vector Monte Carlo method [160] for sampling dividing surfaces. The apparent activation energy for cage-to-cage motion in our study is 44 $\mathrm{kJ\,mol^{-1}}$, in very reasonable agreement with 49 $\mathrm{kJ\,mol^{-1}}$ obtained by Mosell et al.

Finite loadings. Tunca and Ford reported TST rate coefficients for Xe cage-to-cage jumps at high loadings in ZK-4 zeolite, the siliceous analog of Na–A (structure LTA) [171]. These calculations deserve several remarks. First, because this study treats multiple Xe atoms simultaneously, defining the reaction coordinate and dividing surface can become quite complex. Tunca and Ford addressed this problem by considering averaged cage sites, instead of specific intracage sorption sites, which is valid because their system involves relatively weak zeolite–guest interactions. They further assume a one-body reaction coordinate and dividing surface regardless of loading, which is tantamount to assuming that the window separating adjacent α-cages in ZK-4 can only hold one Xe at a time, and that cooperative many-Xe cage-to-cage motions are unlikely. Second, Tunca and Ford advocate separate calculations of $Q^{\ddagger}$ and Q_i for use in Eq. (7), as opposed to the conventional approach of calculating ratios of partition functions viz. free energies [160]. It is not yet obvious whether separating these calculations is worth the effort. Third, Tunca and Ford developed a recursive algorithm for building up $(N+1)$-body partition functions from N-body partition functions, using a 'test particle' method developed for modeling the thermodynamics of liquids. Although the approach of Tunca and Ford has a restricted regime of applicability, it nonetheless seems promising in its direct treatment of many-body diffusion effects.

Free-energy surfaces. Maginn, Bell, and Theodorou performed reversible work calculations with a TST flavor on long-chain alkanes in silicalite-1 [138], finding that diffusivities monotonically decrease with chain length until about n-C_8, after which diffusivities plateau and become nearly constant with chain length. Bigot and Peuch calculated free-energy surfaces for the penetration of n-hexane and isooctane into a model of H-mordenite zeolite with an organometallic specie, $Sn(CH_3)_3$, grafted to the pore edge [172]. Bigot and Peuch found that $Sn(CH_3)_3$ has little effect on the penetration barrier of n-hexane, but they predict that the organometallic increases the penetration barrier of isooctane by 60 kJ mol^{-1}. Sholl computed the free-energy surface associated with particle exchange of Ar, Xe, methane, and ethane in $AlPO_4$-5, a one-dimensional channel zeolite [173], suggesting timescales over which anomalous single-file diffusion is expected in such systems.

Jousse, Auerbach, and Vercauteren modeled benzene site-to-site jumps in H–Y zeolite (Si:Al = 2.43), using a force field that explicitly distinguishes Si and Al, as well as oxygens in Si–O–Si, Si–O–Al, and Si–OH–Al environments [34]. Such heterogeneity creates many distinct adsorption sites for benzene in H–Y. Multiple paths from site to site open as the temperature increases. To simplify the picture, we computed the free-energy surface for benzene motion along the [111] axis in H–Y, which produces cage-to-cage migration. Due to the multiplicity of possible cage-to-cage paths, the temperature dependence of the cage-to-cage rate constant as computed by umbrella sampling exhibits strong non-Arrhenius behavior. These calculations may help to explain intriguing NMR correlation times for benzene in H–Y, which also exhibit striking non-Arrhenius temperature dependencies [43].

Quantum dynamics. Of all the dynamics studies performed on zeolites, very few have explored the potentially quantum mechanical nature of nuclear motion in micropores [174–177]. Quantum modeling of proton transfer in zeolites [175,177,178] seems especially important because of its relevance in catalytic applications. Such modeling will become more prevalent in the near future, partially because of recent improvements in quantum dynamics approaches [177], but mostly because of novel electronic structure methods developed by Sauer and co-workers [179,180], which can accurately compute transition-state parameters for proton transfer in zeolites by embedding a quantum cluster in a corresponding classical force field.

To facilitate calculating quantum rates for proton transfer in zeolites, Fermann and Auerbach developed a novel semiclassical transition-state theory (SC-TST) for truncated parabolic barriers [177], based on the formulation of Hernandez and Miller [181]. Our SC-TST rate coefficient is stable to arbitrarily low temperatures as opposed to purely harmonic SC-TST, and has the form $k^{\text{SC-TST}} = k^{\text{TST}} \cdot \Gamma$ where the quantum transmission coefficient, Γ, depends on the zero-point corrected barrier and the barrier curvature. To parameterize this calculation, Fermann, Blanco, and Auerbach performed high-level cluster calculations [178] yielding an O(1)$\rightarrow$O(4) zero-point corrected barrier height of 86.1 kJ mol^{-1}, which becomes 97.1 kJ mol^{-1} when including long-range effects from the work of Sauer et al. [179]. Using this new approach, Fermann and Auerbach calculated rate coefficients and crossover temperatures for the O(1)$\rightarrow$O(4) jump in H–Y and D–Y zeolites, yielding crossover temperatures of 368 and 264 K, respectively. These results suggest that *tunneling dominates proton transfer* in H–Y up to and slightly above room temperature, and that true proton-transfer barriers are being underestimated by neglecting tunneling in the interpretation of experimental mobility data.

3. Lattice dynamics in zeolites

When modeling strongly-binding or tight-fitting guest–zeolite systems, theoretical methods specialized for rare-event dynamics such as TST and kinetic Monte Carlo (KMC) are required. These methods are

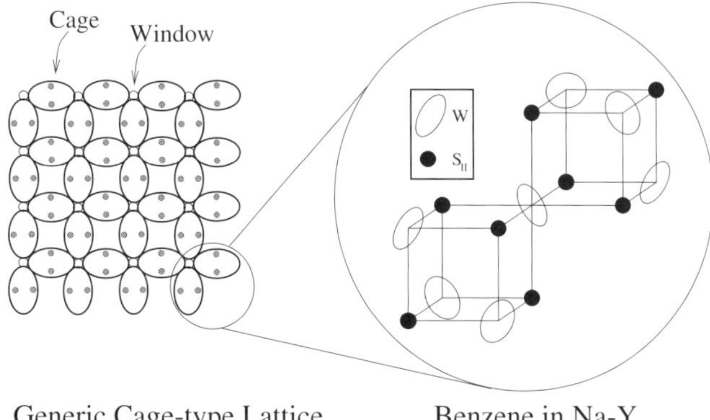

Fig. 5. Schematic lattice model for molecules in cage-type zeolites, showing cages, intracage sites, and window sites (left), as well as the specific lattice geometry for benzene in Na–Y zeolite (right).

applied by coarse-graining the molecular motions, keeping only their diffusive character. In zeolites, the well-defined cage and channel structure naturally orients this coarse-graining toward *lattice models*, which are the focus of this section.

The simplest such model was proposed by Ising in 1925 [182]. Many variants of the Ising model have since been applied to study activated surface diffusion [183]. Although, in principle, a lattice can be regarded simply as a numerical grid for computing configurational integrals required by statistical mechanics [184], the grid points can have important physical meaning for dynamics in zeolites, as shown schematically in Fig. 5. Applying lattice models to diffusion in zeolites rests on several (often-implicit) assumptions on the diffusion mechanism; here we recall those assumptions and analyze their validity for modeling dynamics of sorbed molecules in zeolites.

3.1. Fundamental assumptions

Temperature-independent lattice. Lattice models of transport in zeolites begin by assuming that diffusion proceeds by activated jumps over free-energy barriers between well-defined adsorption sites, i.e. site residence times are much longer than travel times between sites. These adsorption sites are positions of high probability, constructed either from energy minima, for example next to cations in cation-containing zeolites, and/or from high volume, for example channel intersections in silicalite-1. Silicalite-1 provides a particularly illustrative

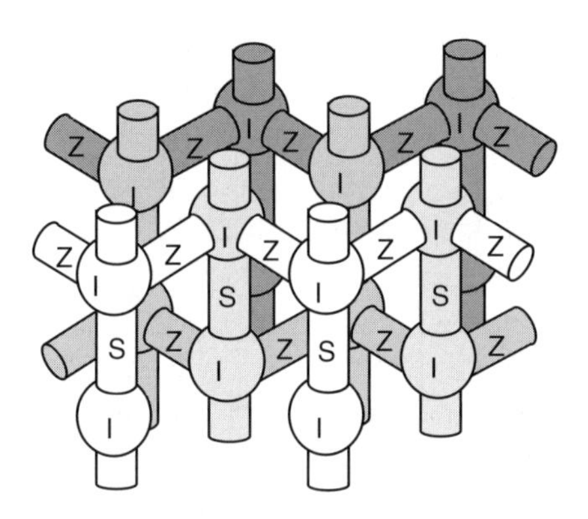

Fig. 6. Channel and site structure of silicalite-1 showing intersection sites (I), straight channel sites (S), and zig-zag channel sites (Z).

example [72]: its usual description in terms of adsorption sites involves two distinct channel sites, where the adsorbate is stabilized by favorable energy contacts with the walls of the 10R channels; and an intersection site at the crossing between the two-channel systems, where the large accessible volume compensates entropically for less-favorable contacts (see Fig. 6). Depending on the temperature, one or both types of sites can be populated simultaneously.

The silicalite-1 example points to the breakdown of the first assumption inherent in lattice models, namely, that adsorption and diffusion of guests in zeolites proceeds on a fixed lattice of sites, independent of external thermodynamic variables such as temperature. Clearly this is not the case. Indeed, when $k_B T$ becomes comparable with the activation energy for a jump from site i to site f, a new lattice that subsumes site i into site f may be more appropriate [124]. Alternatively, one may retain site i with modifications to the lattice model discussed below, taking into account so-called kinetic correlations that arise from the relatively short residence times in site i [123–125].

Poisson statistics. The second assumption inherent in most lattice models of diffusion, which is related to the first, is that subsequent jumps of a given molecule are uncorrelated from each other, i.e. a particular site-to-site jump has the same probability to occur at any time. This assumption results in a site residence time distribution that follows the exponential law associated with Poisson statistics [185]. In Fig. 2 we have seen that such a law can indeed result from the analysis of MD trajectories. As a result, lattice models can often be mapped onto master rate equations such as those in the chemical kinetics of first-order reactions [185,186]. This fact highlights the close connection between reaction and diffusion in zeolites, when modeled with lattice dynamics.

This second assumption is obtained from the following physical observations. Time-independent jump probabilities arise when the mechanism of activation involves simple energy transfer from the heat bath, which is usually *very* rapid compared to site-to-site jumping. Subsequent jumps are uncorrelated when center-of-mass velocity correlations of the jumping guest decay well before the next jump occurs. Both of these criteria are typically satisfied when free-energy minima are very deep compared to $k_B T$. However, when $k_B T$ becomes comparable to barriers separating sites, multisite jumps become important [187–191], requiring either the definition of a more coarse-grained lattice [124], the calculation of multisite jump rates [192], or a statistical model that estimates the kinetic correlations between subsequent jumps [123,125,193]. We have found that ignoring multisite jumps yields accurate results for diffusion through cage-type zeolites such as Na–Y [37], but that such an approximation can cause noticeable errors for transport through channel zeolites such as silicalite-1 (MFI) and silicalite-2 (MEL) [123,125].

Deviations from Poisson statistics would also arise if a molecule were most likely to jump in phase with a low-frequency zeolite framework vibration, such as a window breathing mode [194], or if a molecule were most likely to jump in concert with another guest molecule. An extreme case of this latter effect was predicted by Sholl and Fichthorn [65,152], wherein strong adsorbate–adsorbate interactions in single-file zeolites generated transport dominated by correlated cluster dynamics instead of single-molecule jumps. In this case, a consequence of Poisson statistics applied to diffusion in zeolites at finite loadings ceases to hold, namely, there no longer exists a time interval sufficiently short so that only one molecule can jump at a time.

Loading-independent lattice. The final assumption, which is typically invoked by lattice models of diffusion at finite loadings, is that the sites do not qualitatively change their nature with increasing adsorbate loading. This assumption holds when adsorption sites are separated by barriers such as windows between large cages [171], and also when host–guest interactions dominate guest–guest interactions. This loading-independent lattice model breaks down when the effective diameter of guest molecules significantly exceeds the distance between adjacent adsorption sites, as high loadings create unfavorable excluded-volume interactions between adjacent guests. This effect does not arise for benzene in Na–Y [195], which involves site-to-site distances and guest diameters both around 5 Å, but is predicted for Xe in Na–A by classical density functional theory calculations [196].

Despite these many caveats, lattice models have proven extremely useful for elucidating qualitatively and even semiquantitatively the following physical effects regarding: (i) host structure: pore topology [197–199], diffusion anisotropy [167,200], pore blockage [201], percolation [202], and open-system effects [16,200]; (ii) host–guest structure: site heterogeneity [203,204] and reactive systems [205]; and (iii) guest–guest structure: attractive interactions [168,198,199], phase transitions [206,207], concerted cluster dynamics [65,152], single-file diffusion [16,208], and diffusion of mixtures [144,209,210]. In what follows, we outline the theory and simulation methods used to address these issues.

3.2. Kinetic Monte Carlo

In Section 2.3 we outlined dynamical methods for computing site-to-site jump rate coefficients for molecules in zeolites. In order to make contact with measurements of transport through zeolites [8,9], we must relate these site-to-site rate coefficients with quantities such as the self-diffusivity and transport diffusivity, which arise from molecular translation; or we can model NMR correlation times, which are controlled by molecular rotation. At infinite dilution on an M-dimensional hypercubic lattice, i.e. 1-d, 2-d square, 3-d cubic, etc., both the self- and transport diffusivity are given by $D_0 = k_{\mathrm{hop}} a^2 = (1/2M) k a^2$, where k_{hop} is the fundamental rate coefficient for jumps between nearest neighbor sites, a is the distance between such sites, and $1/k$ is the mean site residence time [183]. This result neglects multisite hops, which have jump distances greater than a. An alternative formula exists that accounts for such jumps in terms of multisite jump rates.

Unfortunately, site lattices in zeolites are usually much more complicated than hypercubic, apparently defying such simple analytical formulas. To address this complexity, many researchers have applied kinetic Monte Carlo (KMC) to model diffusion in zeolites, parameterized either by ad hoc jump frequencies or by atomistically calculated jump rate coefficients. KMC models diffusion on a lattice as a random walk composed of uncorrelated, single-molecule jumps as discussed above, thereby providing a stochastic solution to the master equation associated with the lattice model. Although KMC models transport as sequences of uncorrelated events in the sense that jump times are extracted from Poisson distributions, KMC does account for spatial correlations at finite loadings. Indeed, when a molecule executes a jump at higher loadings, it leaves behind a vacancy that is likely to be occupied by a successive jump, thereby diminishing the diffusivity from the MFT estimate discussed in Section 3.3.

KMC is isomorphic to the more conventional Monte Carlo algorithms [85], except that in a KMC simulation random numbers are compared to ratios of rate coefficients, instead of ratios of Boltzmann factors. However, if the preexponential factors cancel in a ratio of rate coefficients, then a ratio of Boltzmann factors does arise, where the relevant energies are *activation energies*. KMC formally obeys detailed balance, meaning that all thermodynamic properties associated with the underlying lattice Hamiltonian can be simulated with KMC. In addition to modeling transport in zeolites, KMC has been used to model adsorption kinetics on surfaces [211], and even surface growth itself [212]. Further discussion of the technique is given in the following chapter.

3.2.1. Algorithms and ensembles

Algorithms. KMC can be implemented with either constant time-step or variable time-step algorithms. Variable time-step methods are efficient for sampling jumps with widely varying timescales, while fixed time-step methods are convenient for calculating ensemble averaged correlation functions. In the constant time-step technique, jumps are accepted or rejected based on the kinetic Metropolis prescription, in which a ratio of rate coefficients, k_{hop}/k_{ref}, is compared to a random number [39,213]. Here k_{ref} is a reference rate that controls the temporal resolution of the calculation according to $\Delta t_{bin} = 1/k_{ref}$. The probability to make a particular hop is proportional to k_{hop}/k_{ref}, which is independent of time, leading naturally to a Poisson distribution of jump times in the simulation. In the fixed time-step algorithm, all molecules in the simulation attempt a jump during the time Δt_{bin}. In order to accurately resolve the fastest molecular jumps, k_{ref} should be greater than or equal to the largest rate constant in the system, in analogy with choosing time steps for MD simulations. However, if there exists a large separation in timescales between the most rapid jumps, e.g. intracage motion, and the dynamics of interest, e.g. cage-to-cage migration, then one may vary k_{ref} to improve efficiency. The cost of this modification is detailed balance; indeed, tuning k_{ref} to the dynamics of interest is tantamount to simulating a system where all the rates larger than k_{ref} are replaced with k_{ref}.

A useful alternative for probing long-time dynamics in systems with widely varying jump times is variable time-step KMC. In the variable time-step technique, a hop is made every KMC step and the system clock is updated accordingly [201,214]. For a given configuration of random walkers, a process list of possible hops from occupied to empty sites is compiled for all molecules. A particular jump from site i to j is chosen from this list with a probability of $k_{i \to j}/k_{\text{tot}}$, where k_{tot} is the sum of all rate coefficients in the process list. In contrast to fixed time-step KMC, where *all* molecules *attempt* jumps during a KMC step, in variable time-step KMC a *single* molecule *executes* a jump every KMC step and the system clock is updated by an amount $\Delta t_n = -\ln(1-x)/k_{\text{tot}}$, where $x \in [0, 1)$ is a uniform random number and n labels the KMC step. This formula results directly from the Poisson distribution, suggesting that other formulas may be used in variable time-step KMC to model kinetic correlations [123]. In general, we suggest that simulations can be performed using the variable time-step method, with data analyses carried out by mapping the variable time-step KMC trajectories onto a fixed time-step grid [186] as discussed in Section 3.2.2.

Ensembles. Guest–zeolite systems at equilibrium are inherently multicomponent systems at constant temperature and pressure. Since guest molecules are continually adsorbing and desorbing from more-or-less fixed zeolite particles, a suitable ensemble would fix N_z = amount of zeolite, μ_G = chemical potential of guest, p = pressure, and T = temperature, keeping in mind that μ_G and p are related by the equation of state of the external fluid phase. However, constant-pressure simulations are very challenging for lattice models, since constant pressure implies volume fluctuations, which for lattices involve adding or deleting whole adsorption sites. As such, constant-volume simulations are much more natural for lattice dynamics. Since both the volume and amount of zeolite is virtually fixed during intracrystalline adsorption and diffusion of guests, we need to specify only one of these variables. In lattice simulations it is customary to specify the number of adsorption sites, N_{sites}, which plays the role of a unitless volume. We thus arrive at the natural ensemble for lattice dynamics in zeolites: the grand canonical ensemble, which fixes μ_G, N_{sites}, and T.

The overwhelming majority of KMC simulations applied to molecules in zeolites have been performed using the canonical ensemble, which fixes N_G = number of guest molecules, N_{sites}, and T. Although the adsorption–desorption equilibrium discussed above would seem to preclude using the canonical ensemble, fixing N_G is reasonable if zeolite particles are large enough to make the relative root-mean-square fluctuations in N_G rather small. Such closed-system simulations are usually performed with periodic boundary conditions, in analogy with atomistic simulations [85,104]. Defining the fractional loading, θ, by $\theta = N_G/N_{sites}$, typical KMC calculations produce the self-diffusion constant D_s as a function of T at fixed θ for Arrhenius analysis, or as a function of θ at fixed T, a so-called diffusion isotherm.

There has recently been renewed interest in grand canonical KMC simulations for three principal reasons: (i) to relax periodic boundary constraints to explore single-file diffusion with lattice dynamics [16], (ii) to study nonequilibrium permeation through zeolite membranes [200], and (iii) in general to explore the interplay between adsorption and diffusion in zeolites [140,215,216]. Grand canonical KMC requires that the lattice contain at least one edge that can exchange particles with an external phase. In contrast to grand canonical MC used to model adsorption, where particle insertions and deletions can occur anywhere in the system, grand canonical KMC must involve insertions and deletions only at the edges in contact with external phases, as shown in Fig. 8.

The additional kinetic ingredients required by grand canonical KMC are the rates of adsorption to and desorption from the zeolite [217]. Because desorption generally proceeds with activation energies close to the heat of adsorption, desorption rates are reasonably simple to estimate. However, adsorption rates are less well known, because they depend on details of zeolite crystallite surface structure. Although qualitative insights on rates of penetration into microporous solids are beginning to emerge [218,219], zeolite-specific models have yet to take hold [172]. Calculating precise adsorption rates may not be crucial for parameterizing qualitatively reliable simulations, because adsorption rates are typically much larger than other rates in the problem. For sufficiently simple lattice models, adsorption and desorption rates can be balanced to produce the desired loading according to the adsorption isotherm [200]. If one assumes that the external phase is an ideal fluid, then insertion frequencies are proportional to pressure p. As such, equilibrium grand canonical KMC produces the self-diffusion coefficient as a function of p and T. Alternatively, for nonequilibrium systems involving different insertion frequencies on either site of the membrane, arising from a pressure (chemical potential) gradient across the membrane, grand canonical KMC produces the Fickian or transport diffusion coefficient, D, as a function of T and the local loading in the membrane.

3.2.2. Data analyses

Computing ensemble averages and correlation functions is extremely straightforward using fixed time-step KMC. Ensemble averages take the form:

$$\langle A \rangle = \frac{1}{T_{\mathrm{K}}} \sum_{n=1}^{N_{\mathrm{F}}} \Delta t_{\mathrm{bin}} A(n) = \frac{1}{N_{\mathrm{F}}} \sum_{n=1}^{N_{\mathrm{F}}} A(n), \tag{10}$$

where $T_{\mathrm{K}} = N_{\mathrm{F}} \Delta t_{\mathrm{bin}}$ is the total KMC time elapsed in a fixed time-step simulation, N_{F} is the number of steps in the fixed time-step simulation, and $A(n)$ is some dynamical variable evolving during the KMC trajectory. In addition, correlation functions are obtained according to:

$$C(t) = \langle A(0)B(t) \rangle = \frac{1}{N_{\mathrm{F}}} \sum_{n=1}^{N_{\mathrm{F}}-m} A(n)B(n+m), \tag{11}$$

where $t = m\Delta t_{\mathrm{bin}}$. Ensemble averages from variable time-step KMC take the form:

$$\langle A \rangle = \frac{1}{T_{\mathrm{K}}} \sum_{n=1}^{N_{\mathrm{V}}} \Delta t_n A(n), \tag{12}$$

where N_{V} is the number of steps in a variable time-step simulation, and Δt_n are the variable time steps. We note that while the physical time T_{K} may not vary between fixed and variable time-step simulations, the number of steps, N_{F} and N_{V} respectively, will generally be different because each fixed time-step involves a full system sweep, while each variable time-step effects a single-molecule jump.

Computing correlation functions with variable time steps is not as straightforward as that for ensemble averages. We first choose a time bin width, Δt_{bin}, which must be adjusted to encompass the dynamics we wish to study. For example, Δt_{bin} should be a fraction of the estimated cage residence time when modeling diffusion in cage-type zeolites [203]. For the time $t = n\Delta t_{\mathrm{bin}}$, the correlation function $C(t)$ is given by:

$$C(t) = \langle A(0)B(t) \rangle = \frac{1}{Q_n} {\sum_{lm}}' A(l)B(m), \tag{13}$$

where the sum is restricted to those pairs (l, m) for which $t_m - t_l$ falls into the nth time bin, characterized by $n = \text{int}[(t_m - t_l)/\Delta t_{\text{bin}}]$, and Q_n is the number of such pairs.

Probably the most important quantity that is calculated from a KMC simulation is the MSD, which was discussed in Section 2.2.2. The long-time limit required to relate the MSD to diffusion indicates the use of time bins that contaminate short-time dynamics, as can arise for both fixed and variable time-step KMC. The MSD is calculated with variable time-step KMC by replacing $A(l)B(m)$ in Eq. (13) with $|\mathbf{r}(l) - \mathbf{r}(m)|^2$. Great care must be taken to ensure that KMC simulations are run long enough to approach the long-time limit implicit in the Einstein's relation. Indeed, one can obtain linear MSDs and still *not* sample truly diffusive motion through zeolites [39].

To make direct contact with NMR probes of dynamics [220] such as NMR relaxation [43,221], exchange-induced sidebands NMR [10], and multidimensional exchange NMR [222], one can use KMC to calculate the orientational correlation function (OCF) given by $C(t) = \langle P_2(\cos \beta_t) \rangle$, where $P_2(x) = (1/2)(3x^2 - 1)$ is the second-degree Legendre polynomial, and β_t is the angle between a molecular axis at time 0 and t. In practice, an efficient way to evaluate OCFs using KMC is to precompute and store a matrix of values of $P_2[\cos \beta(ij)]$, where i and j label different sites in the lattice. In analogy with Eq. (13), the KMC-calculated OCF thus becomes:

$$C(t) = \frac{1}{Q_n} \sum_{lm}{}' P_2[\cos \beta(i_l j_m)], \tag{14}$$

where i_l is the site occupied at time t_l, and likewise for j_m. Calculating OCFs at long times with KMC is typically more computationally intensive than that for MSDs, because Monte Carlo algorithms are generally inefficient at converging exponentially small numbers arising from sign cancellation. This is known as the 'sign problem' which is especially dire for real-time quantum Monte Carlo and many-fermion quantum Monte Carlo. In actual applications, Auerbach and co-workers have used brute-force computer power to converge OCFs [39,186], which severely limits the system sizes one can study.

3.2.3. Models of finite loading

The great challenge in performing KMC simulations at finite loadings is that the rate coefficients $\{k_{i \to j}\}$ should depend upon the local configuration of molecules because of guest–guest interactions. That is, in compiling the process list of allowed jumps and associated rate constants on the fly of a KMC simulation, TST or related calculations should be performed to account for the effect of specific guest configurations on the jump rate coefficients. To date, this 'ab initio many-body KMC' approach has not been employed because of its daunting computational expense. Instead, researchers either ignore how guest–guest interactions modify rate coefficients for site-to-site jumps; or they use many-body MD at elevated temperatures when guest–guest interactions cannot be ignored [165,166].

A popular approach for modeling many-body diffusion in zeolites with KMC is thus the 'site-blocking model', where guest–guest interactions are ignored, except for exclusion of multiple-site occupancy. This model accounts for entropic effects of finite loadings, but not energetic effects. Calculating the process list and available rate coefficients becomes particularly simple; one simply sums the available processes using rates calculated at infinite dilution [223]. This model is attractive to researchers in zeolite science [224], because blocking of cage windows and channels by large, aromatic molecules that form in zeolites, i.e. 'coking', is a problem that zeolite scientists need to understand and eventually control.

The site-blocking model ignores guest–guest interactions that operate over medium to long length scales, which modify jump activation energies for site-to-site rate coefficients depending upon specific configurations of neighboring adsorbates. By incorporating these additional interactions, diffusion models reveal the competition between guest–zeolite adhesion and guest–guest cohesion [168,225,226]. Qualitatively speaking, the diffusivity is generally expected to increase initially with increasing loading when repulsive guest–guest interactions decrease barriers between sites, and to decrease otherwise. At very high loadings, site blocking lowers the self-diffusivity regardless of the guest–guest interactions.

To develop a quantitative model for the effects of guest–guest attractions, Saravanan et al. proposed a 'parabolic jump model', which relates binding energy shifts to transition-state energy shifts [203,227]. This method was implemented for lattice gas systems whose thermodynamics is governed by the following Hamiltonian:

$$H(\vec{n}) = \sum_{i=1}^{M} n_i f_i + \frac{1}{2} \sum_{i,j=1}^{M} n_i J_{ij} n_j, \tag{15}$$

where M is the number of sites in the lattice, $\vec{n} = (n_1, n_2, \ldots, n_M)$ are site occupation numbers listing a configuration of the system, and $f_i = \varepsilon_i - Ts_i$ is the free energy for binding in site i. In Eq. (15), J_{ij} is the nearest neighbor interaction between sites i and j, i.e. $J_{ij} = 0$ if sites i and j are not nearest neighbors.

Saravanan et al. assumed that the minimum energy hopping path connecting adjacent sorption sites is characterized by intersecting parabolas, as shown in Fig. 7, with the site-to-site transition state located at the intersection point. For a jump from site i to site j, with $i, j = 1, \ldots, M$, the hopping activation energy including guest–guest interactions is given by:

$$E_{\mathrm{a}}(i,j) = E_{\mathrm{a}}^{(0)}(i,j) + \Delta E_{ij}\left(\frac{1}{2} + \frac{\delta E_{ij}^{(0)}}{k_{ij}a_{ij}^2}\right) + \Delta E_{ij}^2\left(\frac{1}{2k_{ij}a_{ij}^2}\right), \tag{16}$$

where $E_{\mathrm{a}}^{(0)}(i,j)$ is the infinite dilution activation energy calculated using the methods of Section 2.3, and a_{ij} is the jump distance. ΔE_{ij} is the shift in the energy difference between sites i and j resulting from guest–guest interactions, and is given by $\Delta E_{ij} = (E_j - E_i) - (\varepsilon_j - \varepsilon_i)$, where $E_k = \varepsilon_k + \sum_{l=1}^{M} J_{kl}n_l$. This method allows the rapid estimation of configuration-dependent barriers during a KMC simulation, knowing only infinite dilution barriers and the nearest-neighbor interactions defined above. The parabolic jump model is most accurate when the spatial paths of jumping molecules are not drastically changed by guest–guest interactions, although the energies can change as shown in Fig. 7. While other lattice models of diffusion in zeolites have been proposed that account for guest–guest attractions [168,226], the

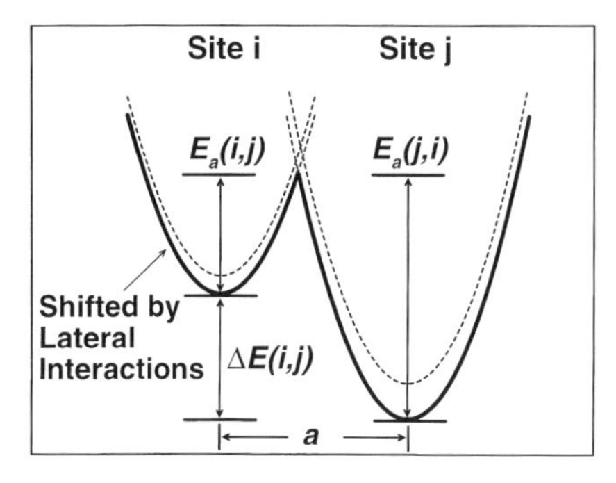

Fig. 7. Site-to-site jump activation energies perturbed by guest–guest interactions, approximated with parabolic jump model.

parabolic jump model has the virtue of being amenable to analytical solution, as discussed in Section 3.3.1.

3.2.4. Recent applications

Infinite dilution. Most KMC simulations of diffusion in zeolites are performed at high guest loadings, to explore the effects on transport of guest–guest interactions. A handful of KMC studies have been reported at infinite dilution, to relate fundamental rate coefficients with observable self-diffusivities for particular lattice topologies. June et al. augmented their TST and RFMD study with KMC calculations of Xe and SF_6 self-diffusivities in silicalite-1 [153]. They obtained excellent agreement among apparent activation energies for Xe diffusion calculated using MD, KMC with TST jump rates, and KMC with RFCT jump rates. The resulting activation energies fall in the range 5–6 kJ mol^{-1}, which is much lower than the experimentally determined values of 15 and 26 kJ mol^{-1} [228,229]. van Tassel et al. reported a similar study in 1994 on methane diffusion in zeolite A, finding excellent agreement between self-diffusivities calculated with KMC and MD [230].

Auerbach et al. reported KMC simulations of benzene diffusion in Na–Y showing that the cation→window jump (see Fig. 4) controls the temperature dependence of diffusion, with a predicted activation energy of 41 kJ mol^{-1} [36]. Because benzene residence times at cation sites are so long, these KMC studies could not be compared directly with MD, but nonetheless yield reasonable agreement with the QENS barrier of 34 kJ mol^{-1} measured by Jobic et al. [45]. Auerbach and Metiu then reported KMC simulations of benzene orientational randomization in various models of Na–Y with different numbers of supercage cations, corresponding to different Si:Al ratios [39]. Full cation occupancy gives randomization rates controlled by intracage motion, whereas half cation occupancy gives rates sensitive to *both* intra- and intercage motion. This finding prompted Chmelka and co-workers to perform exchange-induced sidebands NMR experiments on labeled benzene in the corresponding Ca–Y (Si:Al = 2.0), finding indeed that they were able to measure both the cation→cation and cation→window jump rates within a single experiment [10]. Finally, when Auerbach and Metiu modeled benzene orientational randomization with one-quarter cation occupancy, they found *qualitative sensitivity* to different spatial patterns of cations, suggesting that measuring orientational randomization in zeolites can provide important information regarding cation disorder and possibly Al distributions.

Finite loadings. Theodorou and Wei used KMC to explore a site-blocking model of reaction and diffusion with various amounts of coking [209]. Nelson and co-workers developed similar models, to explore the relationship between the catalytic activity of a zeolite and its lattice percolation threshold [231,232]. In a related study, Keffer, McCormick, and Davis modeled binary mixture transport in zeolites, where one component diffuses rapidly while the other component is trapped at sites, e.g. methane and benzene in Na–Y [202]. They used KMC to calculate percolation thresholds of the rapid penetrant as a function of blocker loading, and found that these thresholds agree well with predictions from simpler percolation theories [233].

Coppens, Bell, and Chakraborty used KMC to calculate the loading dependence of self-diffusion for a variety of lattices, for comparison with MFTs of diffusion (sec Section 3.3.1) [197]. These theories usually predict $D_s(\theta) \cong D_0(1-\theta)$, where θ is the fractional occupancy of the lattice and D_0 is the self-diffusivity at infinite dilution. Coppens et al. found that the error incurred by MFT is greatest for lattices with low-coordination numbers, such as silicalite-1 and other MFI-type zeolites. Coppens et al. then reported KMC simulations showing that by varying the concentrations of weak and strong binding sites [204], their system exhibits most of the loading dependencies of self-diffusion reported by Kärger and Pfeifer [234]. Bhide and Yashonath also used KMC to explore the origins of the observed loading dependencies of self-diffusion, finding that most of these dependencies can be generated by varying the nature and strength of guest–guest interactions [198,199].

Benzene in Na–X. Auerbach and co-workers reported a series of studies modeling the concentration dependence of benzene diffusion in Na–X and Na–Y zeolites [195,203,223,227,235]. These studies were motivated by persistent, qualitative discrepancies between different experimental probes of the coverage dependence of benzene self-diffusion in Na–X [8]. Pulsed field gradient (PFG) NMR diffusivities decrease monotonically with loading for benzene in Na–X [236], while tracer zero-length column (TZLC) data increase monotonically with loading for the same system [237]. We performed KMC simulations using the parabolic jump model to account for guest–guest attractions [203,227]. Our KMC results for benzene in Na–X are in excellent qualitative agreement with the PFG NMR results, and in qualitative disagreement with TZLC. Other experimental methods yield results for benzene in Na–X that also agree broadly with these PFG NMR diffusivities [41,238,239]. Although the evidence appears to be mounting in favor of the PFG NMR loading dependence for benzene in Na–X, it remains unclear just what is being observed by the TZLC measurements. To address this issue, Brandani et al. reported TZLC measurements for benzene in various Na–X samples with different particle sizes. They found tracer exchange rates that exhibit a normal dependence on particle size, suggesting that their diffusivities are free from artifacts associated with unforeseen diffusion resistances at zeolite crystallite surfaces [237].

Noting that molecular transport in TZLC measurements samples longer length scales than that in PFG NMR experiments, Chen et al. have suggested that the TZLC method may be more sensitive than is PFG NMR to electrostatic traps created by random Al and cation distributions [48]. By performing a field theory analysis of an augmented diffusion equation, Chen et al. estimate that such charge disorder can diminish the self-diffusivity by roughly two orders of magnitude from that for the corresponding ordered system. This effect is remarkably close to the discrepancy in absolute magnitudes between PFG NMR and TZLC diffusivities for benzene in Na–X at low loadings [237]. This intriguing prediction by Chen et al. suggests that there should be a striking difference between benzene diffusion in Na–X (Si:Al = 1.2) and in Na–LSX (Si:Al = 1), since the latter is essentially an ordered structure. We are not aware of self-diffusion measurements for benzene in Na–LSX, but we can turn to NMR spin-lattice relaxation data for deuterated benzene in these two zeolites [18,240]. Unfortunately, such data typically reveal only short length scale, intracage dynamics [39], and as a result may not provide such a striking effect. Indeed, the activation energy associated with the NMR correlation time changes only moderately, decreasing from 14.0 ± 0.6 kJ mol^{-1} for Na–X [18] to 10.6 ± 0.9 kJ mol^{-1} for Na–LSX [240], in qualitative agreement with the ideas of Chen et al. [48]. It remains to be seen whether such electrostatic traps can explain the loading dependence observed by TZLC for benzene in Na–X.

By varying fundamental energy scales, the model of Saravanan and Auerbach for benzene in FAU-type zeolites exhibits four of the five loading dependencies of self-diffusion reported by Kärger and Pfeifer [234], in analogy with the studies of Coppens et al. [204] and Bhide and Yashonath [198,199]. However, in contrast to these other KMC studies, we have explored the role of phase transitions [206,207] in determining the loading dependencies of self-diffusion [203]. In particular, we find that Kärger and Pfeifer's type III diffusion isotherm, which involves a nearly constant self-diffusivity at high loadings, may be characteristic of a cluster-forming, subcritical adsorbed phase where the cluster of guest molecules can extend over macroscopic length scales. Such cluster formation suggests a diffusion mechanism involving 'evaporation' of particles from clusters. Although increasing the loading in subcritical systems increases cluster sizes, we surmise that evaporation dynamics remain essentially unchanged by increasing loading. As such, we expect the subcritical diffusivity to obtain its high-loading value at low loadings, and to remain roughly constant up to full loading. In addition, we find that Kärger and Pfeifer's types I, II, and IV are characteristic of supercritical diffusion, and can be distinguished based on the loading that gives the maximum diffusivity, θ_{max}. For example, the PFG NMR results discussed above for benzene in Na–X are consistent with $\theta_{max} \lesssim 0.3$, while the TZLC data give $\theta_{max} \gtrsim 0.5$. Our simulations predict that θ_{max} will decrease with increasing temperature, increasing strength of guest–guest attractions, decreasing free-energy difference between site types, and in general anything that makes sites more equally populated [203].

Reactive systems. Trout, Chakraborty, and Bell applied electronic structure methods to calculate thermodynamic parameters for possible elementary reactions in the decomposition of NO_x over Cu-ZSM-5 [241]. Based on these insights, they developed a KMC model of reaction and diffusion in this system, seeking the optimal distribution of isolated reactive Cu centers [205]. This hierarchical approach to realistic modeling of complex systems presents an attractive avenue for future research.

Open systems. Gladden et al. developed a versatile open-system KMC program that allows them to study adsorption, diffusion, and reaction in zeolites simultaneously [216]. They have applied their algorithm to model ethane and ethene binary adsorption in silicalite-1 [216], finding excellent agreement with the experimental binary isotherm.

Nelson and Auerbach reported open-system KMC simulations of anisotropic diffusion [200] and single-file diffusion [16] (infinitely anisotropic) through zeolite membranes. They defined an anisotropy parameter, η, according to $\eta = k_y/k_x$ where k_x and k_y are the elementary jump rates in the transmembrane and in-plane directions, respectively, as shown in Fig. 8. For example, the $\eta < 1$ case models *p*-xylene permeation through a silicalite-1 membrane (see Fig. 6) oriented along the straight channels (*b*-axis), while $\eta > 1$ corresponds to the same system except oriented along the zig-zag (*a*-axis) or 'corkscrew' channels (*c*-axis) [109]. The limiting case $\eta = 0$ corresponds to single-file diffusion.

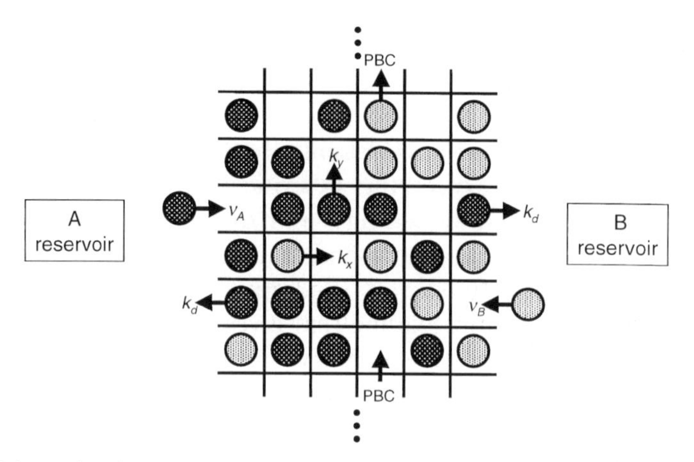

Fig. 8. Schematic of a tracer counter-permeation simulation, with identical but differently labeled particles. Diffusion anisotropy is controlled by the parameter $\eta = k_y/k_x$, with the limiting case $\eta = 0$ corresponding to single-file diffusion.

Nelson and Auerbach have studied how the self-diffusivity depends upon membrane thickness L and anisotropy η. However, the long-time limit of the MSD may not be accessible in a membrane of finite thickness. Furthermore, the natural observable in a permeation measurement is steady-state flux rather than the MSD. To address these issues, they simulated two-component, equimolar counter-permeation of identical, labeled species — i.e. tracer counter-permeation — which has been shown to yield transport identical to self-diffusion [242]. Such a situation is closely related to the tracer zero-length column experiment developed by Ruthven and co-workers [237]. When normal diffusion holds the self-diffusivity is independent of membrane thickness, while anomalous diffusion is characterized by an L-dependent self-diffusivity. For $\eta \gg 1$, Nelson and Auerbach found that diffusion is normal and that MFT becomes exact in this limit [200], i.e. $D_s(\theta) = D_0(1-\theta)$. This is because sorbate motion in the plane of the membrane is very rapid, thereby washing out any correlations in the transmembrane direction. As η is reduced, correlations between the motion of nearby molecules decrease the diffusivity. For small values of η, a relatively large lattice is required to reach the thick membrane limit, such that particle exchange becomes probable during the intracrystalline lifetime. The extreme case of this occurs when $\eta = 0$, for which diffusion is strictly anomalous for all membrane thicknesses.

Nelson et al. applied their open-system KMC algorithm to study the nature of anomalous diffusion through single-file zeolites of finite extent [16]. For times shorter than the vacancy diffusion time, $t_c = L^2/\pi D_0$, particle transport proceeds via the non-Fickian, single-file diffusion mode, with mean-square displacements increasing with the square root of time. For times longer than t_c, however, we find that self-diffusion in single-file systems is completely described by Fick's laws, except that the 'Fickian' self-diffusion constant depends on file length, scaling inversely with L for long files. This gives an intracrystalline residence time, $T_{\text{intra}} = L^2/D_s$, that scales with L^3 for long files, in complete agreement with the mean-field analysis reported by Rodenbeck and Kärger [243]. Nelson and Auerbach found that the fraction of time in the single-file diffusion mode scales inversely with file length for long files, suggesting that Fickian self-diffusion *dominates transport* in longer single-file zeolites. They predicted that the crossover time between (medium time) single-file diffusion and (long time) Fickian diffusion is just above the experimental window for PFG NMR experiments, suggesting that longer-time PFG NMR would observe this transition.

3.3. Mean-field and continuum theories

Mean-field theory (MFT) and continuum theories provide illustrative and efficient means for estimating the results of KMC simulations. MFT reduces the complexity of many-body structure and dynamics to the simplicity of effective one-body properties [155], by averaging over local fluctuations in the instantaneous energy of each adsorption site. Although MFT can give numerical error for lattices with low coordination [197], the theory remains qualitatively reliable except near-critical points, where cooperative fluctuations extend over large distances. MFT can thus serve as a useful launching point for an analytical theory of many-body diffusion in zeolites.

3.3.1. Finite-loading effects

For diffusion at finite loadings on a hypercubic lattice within the site-blocking model, MFT predicts that $D_s(\theta) \cong D_0(1-\theta)$, where θ is the fractional occupancy of the lattice and D_0 is the self-diffusivity at infinite dilution. The factor $(1-\theta)$ is the fraction of jumps that are successful at finite loadings, because they are directed toward vacancies. Spatial correlations between successive jumps, which are accounted for by KMC but ignored by MFT, tend to make $D_s(\theta)$ decrease more rapidly than $(1-\theta)$ [197].

It remains challenging in the general case to apply MFT to diffusion in zeolites, especially when considering a heterogeneous lattice with several distinct site types, such as the lattice of cation and window sites for benzene in FAU-type zeolites. To address this issue for diffusion through cage-type zeolites, Saravanan and Auerbach have shown that a mean-field analysis of cage-to-cage motion yields $D_s(\theta) \cong (1/6)k_\theta a_\theta^2$, where a_θ is the mean intercage jump length, and $1/k_\theta$ is the mean cage residence time [195]. Since a_θ has a very weak temperature and loading dependence [235], e.g. remaining in the range 11–13 Å for FAU-type zeolites, the cage-to-cage rate coefficient carries the interesting T and θ dependencies. Such a formulation is expected to hold for many guests in cage-type zeolites, but not for long-chain alkanes ($>C_8$) in FAU-type zeolites, which are dominated by window-to-window jumps rather than by cage-to-cage jumps [64].

Saravanan and Auerbach have also shown that k_θ is given by $k_\theta = \kappa k_1 P_1$ where P_1 is the probability of occupying a window site between adjacent cages, k_1 is the total rate of leaving a window site, and κ is the transmission coefficient for cage-to-cage motion [195]. This theory provides a picture of cage-to-cage motion involving transition-state theory ($k_1 P_1$) with dynamical corrections (κ). Saravanan and Auerbach have found that P_1 increases with loading when cage sites are more stable than window sites, that k_1 decreases with loading in all cases, and that the balance between k_1 and P_1 controls the loading dependence of self-diffusion. Below we discuss applications of this theory to benzene in FAU-type zeolites [203,227].

3.3.2. Fickian versus Maxwell–Stefan theory

Two theoretical formulations exist for modeling nonequilibrium diffusion, hereafter denoted 'transport diffusion', which ultimately arises from a chemical potential gradient or similar driving force [8,9]. The formulation developed by Fick involves linear response theory relating macroscopic particle flows to concentration gradients, according to $J = -D\nabla\theta$, where J is the net particle flux through a surface $\mathcal{S}$, D is the transport diffusivity, and $\nabla\theta$ is the local concentration gradient perpendicular to the surface $\mathcal{S}$ [155]. While this perspective is conceptually simple, it breaks down qualitatively in remarkably simple cases, such as a closed system consisting of a liquid in contact with its equilibrium vapor. In this case, Fick's law would predict a nonzero macroscopic flux; none exists because the chemical potential gradient vanishes at equilibrium. Fick's law can be generalized to treat very simple multicomponent systems [16,200,242,244–246], such as codiffusion and counter-diffusion of identical, tagged particles.

Despite these shortcomings, Fick's law remains the most natural formulation for transport diffusion through Langmuirian lattice models of zeolite–guest systems. These involve regular lattices of identical sorption sites where guest–guest interactions are ignored, except for exclusion of multiple-site occupancy. Such model systems exhibit Langmuir adsorption isotherms, and give single-component transport diffusivities that are independent of loading [247]. Moreover, for such systems the equation $J = -D\nabla\theta$ is exact for all concentration gradients, i.e. all higher-order terms beyond linear response theory cancel. Nelson and Auerbach exploit this fact in their lattice model studies of counter-permeation through anisotropic [200] and single-file nanoporous membranes [16], described above in Section 3.2.4.

Another formulation of transport diffusion was developed by Onsager, and begins with the equation $J = -L\nabla\mu$, where L is the so-called Onsager coefficient and $\nabla\mu$ is a local chemical potential gradient at the surface $\mathcal{S}$ [8,111]. To make contact with other diffusion theories, the Onsager coefficient is written in terms of the so-called corrected diffusivity, D_c, according to $L = \theta D_c/k_B T$, where θ is the local intracrystalline loading at the surface $\mathcal{S}$. Clearly this formulation does not suffer from the qualitative shortcomings of Fick's law, and can be properly generalized for complex multicomponent systems [248]. The corrected diffusivity depends strongly upon loading for Langmuirian systems, where jump diffusion holds, but depends very weakly on loading for more fluid-like diffusion systems [111], making the Onsager formulation more natural for weakly binding zeolite–guest systems. The relationship between the Fickian and corrected diffusivities is often called the Darken equation, given by [8]:

$$D = D_c \left(\frac{\partial \ln f}{\partial \ln \theta} \right)_T, \tag{17}$$

where f is the fugacity of the external fluid phase. Other versions of the Darken equation often appear, e.g. where D_c is replaced with D_s, the self-diffusivity.

3.3.3. Recent applications

Finite loadings. MFT has been used to explore how site connectivity influences spatial correlations [197], how site energetics control loading dependencies [204], and how system size controls tracer-exchange residence times [243], as discussed above in the context of comparable KMC simulations. Saravanan et al. applied MFT in conjunction with the parabolic jump model to obtain analytical expressions for the cage-to-cage rate constant k_θ, as a function of chemical potential and temperature for the specific example of benzene in FAU-type zeolites [203,227]. Saravanan et al. considered two levels of guest–guest interaction: (1) site blocking alone and (ii) site blocking with nearest-neighbor guest–guest attractions. In what follows, the window and cation sites for benzene in FAU-type zeolites are denoted sites 1 and 2, respectively.

In this site-blocking model, there are only four fundamental rate constants in the problem, $\{k_{i\to j}\}$, where $i, j = 1, 2$. For example, the rate constant $k_{2\to 1}$ is the fundamental rate constant for jumping from a cation site to a window site (see Fig. 4). In the limit where cation sites are very stable compared to window sites, which models benzene in Na–Y, the MFT equations reduce to [195]:

$$\begin{aligned} k_\theta &\cong \frac{3}{2}\left(\frac{2}{2-3\theta}\frac{k_{1\to 1}}{k_{1\to 2}}+1\right)k_{2\to 1} \quad \text{for } \theta < \frac{2}{3} \\ &\cong 3(1-\theta)\left(\frac{3\theta-2}{\theta}\right)k_{1\to 1} \quad \text{for } \theta > \frac{2}{3}. \end{aligned} \tag{18}$$

These MFT formulas agree well with the results of KMC simulations using input rates calculated for benzene in Na–Y [223]. For $T \lesssim 650$ K, Eq. (18) gives diffusion isotherms consistent with Kärger and Pfeifer's type IV diffusion isotherm [234], because jumping out of Na–Y window sites is much faster than jumping off cations, i.e. $k_{1\to 1} \gg k_{2\to 1}$. The type IV isotherm involves a broadly increasing diffusivity at low loadings (as cation sites become occupied, followed by a sharply decreasing diffusivity at high loadings (as all sites become occupied). At present there is a paucity of reliable self-diffusion data for benzene in Na–Y due to the fact that Na–Y crystallites are typically rather small, making intracrystalline diffusion measurements rather challenging. However, QENS data collected at 2 and 4.5 benzenes per cage are consistent with a type I isotherm for benzene in Na–Y [45], which is monotonically decreasing.

Saravanan and Auerbach explored the performance of this MFT in the more general case, for arbitrary guest–guest attractions and for arbitrary cation and window stabilities [203]. It should not be surprising that this lattice model of benzene in FAU-type zeolites with guest–guest attractions supports phase transitions from low to high sorbate density, analogous to vapor–liquid equilibrium of bulk benzene [206,207]. For benzene in Na–X, which involves window sites that are more stable than those in Na–Y, MFT predicts a very broad coexistence region in θ, much broader than that predicted by grand canonical MC simulations [206,207]. This renders MFT pretty useless for benzene in Na–X, because MFT cannot be used to evaluate diffusivities for the wide range of fractional loadings in the MFT coexistence region. On the other hand, MFT predicts a much narrower coexistence curve for benzene in Na–Y, which increases the range of fractional loadings for which MFT can be evaluated. For these values of θ, MFT gives excellent agreement with KMC simulations of benzene in Na–Y. Even with this more sophisticated treatment of benzene in Na–Y, including guest–guest attractions, we still predict [203] a Kärger and Pfeifer type IV diffusion isotherm [234]. Resolving this discrepancy between theory and QENS may require collecting QENS data at more loadings, and may also require more sophisticated simulation approaches.

Lattice topology. The diffusion theory discussed above relies on the tetrahedral topology of FAU-type zeolites. Developing such a theory for general frameworks remains challenging. Braun and Sholl developed a Laplace–Fourier transformation method for calculating exact self-diffusion tensors in generalized lattice gas models [193]. These methods generally involve quite heavy matrix algebra, which can sometimes hide the underlying physical meaning of the parameters. Jousse, Auerbach, and Vercauteren developed an alternative method for deriving analytical self-diffusion coefficients at infinite dilution for general lattices, by partitioning the trajectory of a tracer into uncorrelated sequences of jumps [125]. This approach can be used to analyze both geometric correlations due to the nonsymmetric nature of adsorption sites in zeolite pores, and kinetic correlations arising from insufficient thermalization of a molecule in its final site. This method was applied to benzene diffusion in Na–Y (geometric correlations) and to ethane diffusion in silicalite-1 (geometric and kinetic correlations), yielding quantitative agreement with KMC simulations [125]. The new method was also extended to finite loadings using MFT, yielding a completely analytical approach for modeling diffusion in any guest–zeolite system.

Maxwell–Stefan and Fick. Krishna and van den Broeke modeled the transient permeation fluxes of methane and *n*-butane through a silicalite-1 membrane using both the Fick and Maxwell–Stefan formulations [249]. Transient experiments showed that initially the permeation flux of methane is higher than that of *n*-butane, but that this methane flux eventually reduces to a lower steady-state value. The Maxwell–Stefan formulation succeeded in reproducing this nonmonotonic evolution to steady state for methane; the Fick formulation failed qualitatively in this regard. This is attributed to the fact that multicomponent systems pose a challenge to the Fick formulation of diffusion. van de Graaf, Kapteijn, and Moulijn used the Maxwell–Stefan formulation to interpret permselectivity data for the separations of ethane/methane and propane/methane mixtures with a silicalite-1 membrane [250]. Based only on separately determined single-component adsorption and diffusion parameters, the Maxwell–Stefan model gave permselectivities in excellent agreement with their experimental data.

As discussed in Section 3.2.4, Chen et al. augmented the standard diffusion equation (Fick's second law) [8] with terms representing the effects of static-charge disorder [48]. They analyzed the resulting equation in the hydrodynamic limit using renormalization group theory [155], finding that such disorder can diminish self-diffusivities in zeolites by one to two orders of magnitude. Nelson, Tsapatsis, and Auerbach computed steady-state solutions of the diffusion equation to evaluate the influence of defects, voids, and diffusion anisotropy on permeation fluxes through model zeolite membranes [251]. Nelson et al. augmented the lattice configuration shown in Fig. 8 with various kinds of defect structures, and used a time-dependent, numerical finite-difference approach for computing steady-state fluxes in a variety of situations. They found that with a reasonable anisotropy and with a moderate density of voids in the membrane, permeation fluxes can be controlled by jumps perpendicular to the transmembrane direction. This suggests that oriented zeolite membranes may not behave with the intended orientation if there is a sufficient density of defects in the membrane.

4. Concluding remarks

In this review we have explored recent efforts to model the dynamics of sorbed molecules in zeolites with either atomistic methods or lattice models. We assessed recent approaches for constructing guest–zeolite force fields, as well as atomistic models of aluminosilicate frameworks with charge-compensating cations. We then detailed the techniques and applications of equilibrium MD, transition-state theory, and reactive flux MD to sorbed molecules in zeolites. Changing focus from atomistic methods to lattice models, we discussed the assumptions underlying such lattice models, and analyzed their validity for molecules in zeolites. We then described the techniques and applications of kinetic Monte Carlo, MFT, and other continuum theories to modeling transport in complex guest–zeolite systems.

Over the last several years, a wealth of insight has been gained from studies modeling dynamics in zeolites. Here we summarize some (but not all) of the key ideas that have recently emerged. A useful picture has developed that relates guest size and shape, and zeolite pore size and topology, to the resulting transport properties of hydrocarbons in all-silica zeolites. We have also gained a better understanding of the role of framework flexibility, especially for molecules in tight-fitting guest–zeolite systems. In general, the analysis of both simulation and experimental data in terms of jump diffusion models has proven very useful for developing simplified pictures of dynamics in zeolites. Regarding dynamics in acidic zeolites, we have learned that the inherent disorder in framework charge and proton distributions can produce cage-to-cage jump rates with significant non-Arrhenius behavior. Furthermore, recent studies have suggested that tunneling dominates proton transfer in some acidic zeolites at ambient temperatures, and that true proton-transfer barriers are being underestimated by neglecting tunneling in the interpretation of experimental mobility data.

Recent simulations on lattice models have provided qualitative insights regarding the relationship between fundamental site-to-site jump rates and the resulting loading dependencies of diffusion. Various loading dependencies have also been connected with subcritical and supercritical states of the confined fluid. For the specific example of benzene in Na–X, recent results point toward the validity of the decreasing loading dependence observed by PFG NMR, while the TZLC data remain mysteriously reproducible. At the same time, a possible discrepancy between simulation and experiment may be emerging for the loading dependence of benzene diffusion in Na–Y. Both atomistic methods and lattice models have contributed to our basic understanding of single-file diffusion in zeolites. In particular, we understand more clearly when to expect anomalous mean-square displacements in both simulations and experiments. Finally, while most studies model diffusion in zeolites (translational dynamics), a new emphasis on modeling orientational dynamics has emerged, to reveal more subtle aspects of zeolite structure including possibly Al distributions.

Despite this impressive progress many challenges lay ahead; below we list some desiderata for future modeling of dynamics in zeolites.

To begin, we need better representations of charge disorder, as well as other defect structures in zeolites, to determine their impact on diffusion. In parallel, we require better understanding of the external surfaces of zeolite crystallites, to facilitate realistic grand canonical MD simulations of permeation through zeolites. To facilitate modeling in general, more portable force fields and more tractable ab initio MD would allow simultaneous modeling of diffusion and reaction. We seek more realistic lattice models to bridge the gap between atomistic methods and lattice dynamics [165], to test the presently oversimplified lattice treatments of the loading dependence of activated diffusion. Finally, we hope to understand the loading dependencies of multicomponent diffusion in zeolites, as well as the practical impact of single-file diffusion in zeolite applications [252].

We hope that these computational studies can assist in the design of new materials with advanced performance by elucidating the microscopic factors that control dynamics in zeolites. While this dream is not yet an everyday reality, examples exist today that have the flavor of rational design [253]. We believe that such design will become much more commonplace within the next 10 years, with the advent of better algorithms and faster computers. Perhaps even more significant is the need for enhanced cooperation between experiment and simulation, to inspire the next generation of dynamics models for molecules in zeolites.

Acknowledgements

S.M.A. gratefully acknowledges his research group and collaborators for their invaluable contributions and for many stimulating discussions. This work was supported by the University of Massachusetts at Amherst Faculty Research Grant Program, the Petroleum Research Fund (ACS-PRF 30853-G5), the National Science Foundation (CHE-9403159, CHE-9625735, CHE-9616019 and CTS-9734153), a Sloan Foundation Research Fellowship (BR-3844), a Camille Dreyfus Teacher-Scholar Award (TC-99-041), the National Environmental Technology Institute, and Molecular mulations, Inc.

F.J. wishes to thank particularly Professor D.P. Vercauteren, Director of the Laboratoire de Physico-Chimie Informatique at the University of Namur, for his continuous support, and Prof. A. Lucas, Director of the PAI 4-10, for the attribution of a post-doctoral fellowship. F.J. also acknowledges the FUNDP for the use of the Namur Scientific Computing Facility Center (SCF), and financial support from the FNRS-FRFC, the 'Loterie Nationale' for the convention no. 9.4595.96, and IBM Belgium for the Academic Joint Study on "Cooperative Processing for Theoretical Physics and Chemistry".

References

1. Breck, D.W., *Zeolite Molecular Sieves: Structure, Chemistry and Use*. Wiley, New York, 1974.
2. Barrer, R.M., *Zeolites and Clay Minerals as Sorbents and Molecular Sieves*. Academic Press, London, 1978.
3. van Bekkum, H., Flanigen, E.M. and Jansen, J.C. (Eds.), *Introduction to Zeolite Science and Practice*. Elsevier, Amsterdam, 1991.
4. Meier, W.M. and Olson, D.H., *Atlas of Zeolite Structure Types*, Third Revised Edition. Butterworth-Heinemann, London, 1992.
5. *InsightII 4.0.0 User Guide*. MSI, San Diego, 1996.
6. Herreros, B. (1996), http://suzy.unl.edu/Bruno/zeodat/zeodat.html.
7. Weitkamp, J., In: Olhmann, G., Vedrine, J.C. and Jacobs, P.A. (Eds.), *Catalysis and Adsorption by Zeolites*. Elsevier, Amsterdam, 1991, p. 21.
8. Kärger, J. and Ruthven, D.M., *Diffusion in Zeolites and Other Microporous Solids*. John Wiley & Sons, New York, 1992.
9. Chen, N.Y., Degnan Jr., T.F. and Smith, C.M., *Molecular Transport and Reaction in Zeolites*. VCH Publishers, New York, 1994.
10. Favre, D.E., Schaefer, D.J., Auerbach, S.M. and Chmelka, B.F., *Phys. Rev. Lett.*, **81**, 5852 (1998).
11. Caro, J., Bülow, M., Schirmer, W., Kärger, J., Heink, W., Pfeifer, H. and Ždanov, S.P., *J. Chem. Soc., Faraday Trans. I*, **81**, 2541 (1985).
12. Catlow, C.R.A. (Ed.), *Modelling of Structure and Reactivity in Zeolites*. Academic Press, London, 1992.
13. Demontis, P. and Suffritti, G.B., *Chem. Rev.*, **97**, 2845 (1997) and references therein.
14. Deem, M.W., *AIChE J.*, **44**, 2569 (1998).
15. Auerbach, S.M., *Int. Rev. Phys. Chem.*, **19**, 155 (2000).
16. Nelson, P.H. and Auerbach, S.M., *J. Chem. Phys.*, **110**, 9235 (1999).
17. Webb III, E.B. and Grest, G.S., *Catal. Lett.*, **56**, 95 (1998).
18. Auerbach, S.M., Bull, L.M., Henson, N.J., Metiu, H.I. and Cheetham, A.K., *J. Phys. Chem.*, **100**, 5923 (1996).
19. Hansenne, C., Jousse, F., Leherte, L. and Vercauteren, D.P., *J. Mol. Catal. A: Chemical*, **166**, 147 (2001).
20. Heink, W., Kärger, J., Pfeifer, H., Datema, K.P. and Nowak, A.K., *J. Chem. Soc., Faraday Trans.*, **88**, 3505 (1992).
21. Jousse, F. and Cohen De Lara, E., *J. Phys. Chem.*, **100**, 233 (1996).
22. Akporiaye, D.E., Dahl, I.M., Mostad, H.B. and Wendelbo, R., *J. Phys. Chem.*, **100**, 4148 (1996).
23. Ding, D., Li, B., Sun, P., Jin, Q. and Wang, J., *Zeolites*, **15**, 569 (1995).
24. Vega, A.J., *J. Phys. Chem.*, **100**, 833 (1996).
25. Herrero, C.P. and Ramírez, R., *J. Phys. Chem.*, **96**, 2246 (1992).

26. Herrero, C.P., *J. Phys. Chem.*, **95**, 3282 (1991).
27. Klinowski, J., Ramdas, S., Thomas, J.M., Fyfe, C.A. and Hartman, J.S., *J. Chem. Soc., Faraday Trans. 2*, **78**, 1025 (1982).
28. Feijen, E.J.P., Lievens, J.L., Martens, J.A., Grobet, P.J. and Jacobs, P.A., *J. Phys. Chem.*, **100**, 4970 (1996).
29. Newsam, J.M., *J. Phys. Chem.*, **93**, 7689 (1989).
30. Mellot, C.F., Davidson, A.M., Eckert, J. and Cheetham, A.K., *J. Phys. Chem. B*, **102**, 2530 (1998).
31. Mellot, C.F., Cheetham, A.K., Harms, S., Savitz, S., Gorte, R.J. and Myers, A.L., *J. Am. Chem. Soc.*, **120**, 5788 (1998).
32. Vitale, G., Mellot, C.F., Bull, L.M. and Cheetham, A.K., *J. Phys. Chem. B*, **101**, 4559 (1997).
33. Faux, D.A., *J. Phys. Chem. B*, **102**, 10658 (1998).
34. Jousse, F., Auerbach, S.M. and Vercauteren, D.P., *J. Phys. Chem. B*, **104**, 2360 (2000).
35. Santikary, P. and Yashonath, S., *J. Phys. Chem.*, **98**, 9252 (1994).
36. Auerbach, S.M., Henson, N.J., Cheetham, A.K. and Metiu, H.I., *J. Phys. Chem.*, **99**, 10600 (1995).
37. Jousse, F. and Auerbach, S.M., *J. Chem. Phys.*, **107**, 9629 (1997).
38. Henson, N.J., Cheetham, A.K., Redondo, A., Levine, S.M. and Newsam, J.M., *Computer Simulations of Benzene in Faujasite-type Zeolites*, In: Weitkamp, J., Karge, H.G., Pfeifer, H. and Hölderich, W. (Eds.), *Zeolites and Related Microporous Materials: State of the Art 1994*. Elsevier, Amsterdam, 1994, pp. 2059–2066.
39. Auerbach, S.M. and Metiu, H.I., *J. Chem. Phys.*, **106**, 2893 (1997).
40. Forni, L. and Viscardi, C.F., *J. Catal.*, **97**, 480 (1986).
41. Bülow, M., Mietk, W., Struve, P. and Lorenz, P., *J. Chem. Soc., Faraday Trans. I*, **79**, 2457 (1983).
42. Lorenz, P., Bülow, M. and Kärger, J., *Izv. Akad. Nauk. SSSR, Ser. Khim.*, 1741 (1980).
43. Sousa-Gonçalves, J.A., Portsmouth, R.L., Alexander, P. and Gladden, L.F., *J. Phys. Chem.*, **99**, 3317 (1995).
44. Isfort, O., Boddenberg, B., Fujara, F. and Grosse, R., *Chem. Phys. Lett.*, **288**, 71 (1998).
45. Jobic, H., Fitch, A.N. and Combet, J., *J. Phys. Chem. B*, **104**, 8491 (2000).
46. Burmeister, R., Schwarz, H. and Boddenberg, B., *Ber. Bunsenges. Phys. Chem.*, **93**, 1309 (1989).
47. Bull, L.M., Henson, N.J., Cheetham, A.K., Newsam, J.M. and Heyes, S.J., *J. Phys. Chem.*, **97**, 11776 (1993).
48. Chen, L., Falcioni, M. and Deem, M.W., *J. Phys. Chem. B*, **104**, 6033 (2000).
49. Jaramillo, E. and Auerbach, S.M., *J. Phys. Chem. B*, **103**, 9589 (1999).
50. Newsam, J.M., Freeman, C.M., Gorman, A.M. and Vessal, B., *Chem. Commun.*, **16**, 1945 (1996).
51. Demontis, P., Fois, E.S., Suffritti, G.B. and Quartieri, S., *J. Phys. Chem.*, **94**, 4329 (1990).
52. Catlow, C.R.A., Freeman, C.M., Vessal, B., Tomlinson, S.M. and Leslie, M., *J. Chem. Soc., Faraday Trans.*, **87**, 1947 (1991).
53. Demontis, P., Suffritti, G.B., Fois, E.S. and Quartieri, S., *J. Phys. Chem.*, **96**, 1482 (1992).

54. Smirnov, K.S., *Chem. Phys. Lett.*, **229**, 250 (1994).
55. Bouyermaouen, A. and Bellemans, A., *J. Chem. Phys.*, **108**, 2170 (1998).
56. Fritzsche, S., Wolfsberg, M., Haberlandt, R., Demontis, P., Suffritti, G.P. and Tilocca, A., *Chem. Phys. Lett.*, **296**, 253 (1998).
57. Schrimpf, G., Schlenkrich, M., Brickmann, J. and Bopp, P., *J. Phys. Chem.*, **96**, 7404 (1992).
58. Sastre, G., Catlow, C.R.A. and Corma, A., *J. Phys. Chem. B*, **103**, 5187 (1999).
59. Mosell, T., Schrimpf, G. and Brickmann, J., *J. Phys. Chem. B*, **101**, 9476 (1997).
60. Mosell, T., Schrimpf, G. and Brickmann, J., *J. Phys. Chem. B*, **101**, 9485 (1997).
61. Jousse, F., Vercauteren, D.P. and Auerbach, S.M., *J. Phys. Chem. B*, **104**, 8768 (2000).
62. Thomson, K.T., McCormick, A.V. and Davis, H.T., *J. Chem. Phys.*, **112**, 3345 (2000).
63. Gaub, M., Fritzsche, S., Haberlandt, R. and Theodorou, D.N., *J. Phys. Chem. B*, **103**, 4721 (1999).
64. Clark, L.A., Ye, G.T., Gupta, A., Hall, L.L. and Snurr, R.Q., *J. Chem. Phys.*, **111**, 1209 (1999).
65. Sholl, D.S. and Lee, C.K., *J. Chem. Phys.*, **112**, 817 (2000).
66. Sholl, D.S., *Chem. Phys. Lett.*, **305**, 269 (1999).
67. Schuring, D., Jansen, A.P.J. and van Santen, R.A., *J. Phys. Chem. B*, **104**, 941 (2000).
68. Webb III, E.B., Grest, G.S. and Mondello, M., *J. Phys. Chem. B*, **103**, 4949 (1999).
69. Tepper, H.L., Hoogenboom, J.P., ven der Vegt, N.F.A. and Briels, W.J., *J. Chem. Phys.*, **110**, 11511 (1999).
70. Gergidis, L.N. and Theodorou, D.N., *J. Phys. Chem. B*, **103**, 3380 (1999).
71. Jost, S., Bär, N.-K., Fritzsche, S., Haberlandt, R. and Kärger, J., *J. Phys. Chem. B*, **102**, 6375 (1998).
72. Snurr, R.Q., Bell, A.T. and Theodorou, D.N., *J. Phys. Chem.*, **98**, 11948 (1994).
73. Forester, T.R. and Smith, W., *J. Chem. Soc., Faraday Trans.*, **93**, 3249 (1997).
74. Pellenq, R.J.-M. and Nicholson, D., *J. Phys. Chem.*, **98**, 13339 (1994).
75. Nicholson, D., Boutin, A. and Pellenq, R.J.-M., *Mol. Sim.*, **17**, 217 (1996).
76. Larin, A.V. and Cohen de Lara, E., *Mol. Phys.*, **88**, 1399 (1996).
77. Jousse, F., Larin, A.V. and Cohen De Lara, E., *J. Phys. Chem.*, **100**, 238 (1996).
78. Macedonia, M.D. and Maginn, E.J., *Mol. Phys.*, **96**, 1375 (1999).
79. Jousse, F., Auerbach, S.M., Jobic, H. and Vercauteren, D.P., *J. Phys. IV: France*, **10**, 147 (2000).
80. Gergidis, L.N., Theodorou, D.N. and Jobic, H., *J. Phys. Chem. B*, **104**, 5541 (2000).
81. Lachet, V., Boutin, A., Pellenq, R.J.-M., Nicholson, D. and Fuchs, A.H., *J. Phys. Chem.*, **100**, 9006 (1996).
82. Raj, N., Sastre, G. and Catlow, C.R.A., *J. Phys. Chem. B*, **103**, 11007 (1999).
83. Sastre, G., Raj, N., Catlow, C.R.A., Roque-Malherbe, R. and Corma, A., *J. Phys. Chem. B*, **102**, 3198 (1998).
84. Ewald, P.P., *Ann. Physik*, **64**, 253 (1921).
85. Allen, M.F. and Tildesley, D.J., *Computer Simulation of Liquids*. Clarendon Press, Oxford, 1987.
86. Greengard, L. and Rokhlin, V., *J. Comp. Phys.*, **73**, 325 (1987).
87. Petersen, H.G., Soelvason, D., Perram, J.W. and Smith, E.R., *J. Chem. Phys.*, **101**, 8870 (1994).

88. Remler, D.K. and Madden, P.A., *Mol. Phys.*, **70**, 921 (1990).
89. Galli, G. and Pasquarello, A., *First-Principles Molecular Dynamics*. Kluwer Academic, The Netherlands, 1993, pp. 261–313.
90. Bolton, B., Hase, W. and Peshlerbe, S., Direct dynamics simulations of reactive systems, In: Thompson, D.L. (Ed.), *Modern Methods for Multidimensional Dynamics for Computations in Chemistry*. World Scientific, New York, 1998.
91. Car, R. and Parrinello, M., *Phys. Rev. Lett.*, **55**, 2471 (1985).
92. Schwarz, K., Nusterer, E. and Blöchl, P.E., *Catal. Today*, **50**, 501 (1999).
93. Fois, E. and Gamba, A., *J. Phys. Chem. B*, **103**, 1794 (1999).
94. Štich, I., Gale, J.D., Terakura, K. and Payne, M.C., *J. Am. Chem. Soc.*, **121**, 3292 (1999).
95. Nusterer, E., Blöchl, P.E. and Schwarz, K., *Angewandte Chemie*, **35**, 175 (1996).
96. Nusterer, E., Blöchl, P.E. and Schwarz, K., *Chem. Phys. Lett.*, **253**, 448 (1996).
97. Jeanvoine, Y., Ángyán, J.G., Kresse, G. and Hafner, J., *J. Phys. Chem. B*, **102**, 7307 (1998).
98. Shah, R., Payne, M.C., Lee, M.-H. and Gale, J.D., *Science*, **271**, 1395 (1996).
99. Shah, R., Gale, J.D. and Payne, M.C., *J. Phys. Chem.*, **100**, 11688 (1996).
100. Filippone, F. and Gianturco, F.A., *J. Chem. Phys.*, **111**, 2761 (1999).
101. Alfe, D. and Gillan, M.J., *Phys. Rev. Lett.*, **81**, 5161 (1998).
102. de Wijs, G.A., Kresse, G., Voçadlo, L., Dobson, D., Alfè, D., Gillan, M.J. and Price, G.D., *Nature*, **392**, 805 (1998).
103. Demontis, P., Suffritti, G.B., Alberti, A., Quartieri, S., Fois, E.S. and Gamba, A., *Gazz. Chim. Ital.*, **116**, 459 (1986).
104. Frenkel, D. and Smit, B., *Understanding Molecular Simulations*. Academic Press, San Diego, 1996.
105. Mundy, C.J., Balasubramanian, S., Bogchi, K., Tuckerman, M.E. and Klein, M.L., *Rev. Comp. Chem.*, **14**, 291 (2000).
106. Tuckerman, M.E. and Martyna, G.J., *J. Phys. Chem. B*, **104**, 159 (2000).
107. Hoover, W.G., *Physica A*, **194**, 450 (1993).
108. Hoover, W.G. and Kum, O., *Mol. Phys.*, **86**, 685 (1995).
109. Lovallo, M.C. and Tsapatsis, M., *AIChE J.*, **42**, 3020 (1996).
110. Lin, X., Falconer, J.L. and Noble, R.D., *Chem. Mater.*, **10**, 3716 (1998).
111. Maginn, E.J., Bell, A.T. and Theodorou, D.N., *J. Phys. Chem.*, **97**, 4173 (1993).
112. Heffelfinger, G.S. and van Swol, F., *J. Chem. Phys.*, **100**, 7548 (1994).
113. Nicholson, D., *Supramol. Sci.*, **5**, 275 (1998).
114. Travis, K.P. and Gubbins, K.E., *Langmuir*, **15**, 6050 (1999).
115. Xu, L., Tsotsis, T.T. and Sahimi, M., *J. Chem. Phys.*, **111**, 3252 (1999).
116. Wold, I. and Hafskjold, B., *Int. J. Thermophys.*, **20**, 847 (1999).
117. Travis, K.P. and Gubbins, K.E., *J. Chem. Phys.*, **112**, 1984 (2000).
118. Xu, L., Sedigh, M.G., Tsotsis, T.T. and Sahimi, M., *J. Chem. Phys.*, **112**, 910 (2000).
119. Hahn, K., Kärger, J. and Kukla, V., *Phys. Rev. Lett.*, **76**, 2762 (1996).
120. Kukla, V., Kornatowski, J., Demuth, D., Girnus, I., Pfeifer, H., Rees, L.V.C., Schunk, S., Unger, K.K. and Kärger, J., *Science*, **272**, 702 (1996).
121. van Beijeren, H., Kehr, K.W. and Kutner, R., *Phys. Rev. B*, **28**, 5711 (1983).
122. June, R.L., Bell, A.T. and Theodorou, D.N., *J. Phys. Chem.*, **96**, 1051 (1992).
123. Jousse, F., Leherte, L. and Vercauteren, D.P., *J. Phys. Chem. B*, **101**, 4717 (1997).

124. Kärger, J., Demontis, P., Suffritti, G.B. and Tilocca, A., *J. Chem. Phys.*, **110**, 1163 (1999).
125. Jousse, F., Auerbach, S.M. and Vercauteren, D.P., *J. Chem. Phys.*, **112**, 1531 (2000).
126. Lovesey, S.W., *Theory of Neutron Scattering from Condensed Matter, Vol. 1: Nuclear Scattering*. Clarendon Press, Oxford, 1984.
127. Jobic, H., Bèe, M. and Kearley, G.J., *Zeolites*, **12**, 146 (1992).
128. Bougeard, D., Brémard, C., Dumont, D., Le Maire, M., Manoli, J.-M. and Potvin, C., *J. Phys. Chem. B*, **102**, 10805 (1998).
129. June, R.L., Bell, A.T. and Theodorou, D.N., *J. Phys. Chem.*, **94**, 8232 (1990).
130. Nowak, A.K., den Ouden, C.J.J., Pickett, S.D., Smit, B., Cheetham, A.K., Post, M.F.M. and Thomas, J.M., *J. Phys. Chem.*, **95**, 848 (1991).
131. Goodbody, S.J., Watanabe, K., MacGowan, D., Walton, J.P.R.B. and Quirke, N., *J. Chem. Soc., Faraday Trans.*, **87**, 1951 (1991).
132. Jousse, F., Leherte, L. and Vercauteren, D.P., *J. Mol. Catal. A: Chemical*, **119**, 165 (1997).
133. Runnebaum, R.C. and Maginn, E.J., *J. Phys. Chem. B*, **101**, 6394 (1997).
134. Jobic, H., Bèe, M. and Caro, J., Translational mobility of *n*-butane and *n*-hexane in ZSM-5 measured by quasi-elastic neutron scattering, In: vonr Ballmoos, R., et al. (Ed.), *Proceedings of the 9th International Zeolite Conference*. Butterworth-Heinemann, 1993, pp. 121–128.
135. Nicholas, J.B., Trouw, F.R., Mertz, J.E., Iton, L.E. and Hopfinger, A.F., *J. Phys. Chem.*, **97**, 4149 (1993).
136. Ermoshin, V.A. and Engel, V., *J. Phys. Chem. A*, **103**, 5116 (1999).
137. Hernández, E. and Catlow, C.R.A., *Proc. R. Soc. Lond. A*, **448**, 143 (1995).
138. Maginn, E.J., Bell, A.T. and Theodorou, D.N., *J. Phys. Chem.*, **100**, 7155 (1996).
139. Lachet, V., Boutin, A., Tavitian, B. and Fuchs, A.H., *Faraday Discuss.*, **106**, 307 (1997).
140. Krishna, R., Smit, B. and Vlugt, T.J.H., *J. Phys. Chem. A*, **102**, 7727 (1998).
141. Lachet, V., Boutin, A., Tavitian, B. and Fuchs, A.H., *J. Phys. Chem. B*, **103**, 9224 (1999).
142. Macedonia, M.D. and Maginn, E.J., Grand canonical Monte Carlo simulation of single component and binary mixture adsorption in zeolites, In: *Proceedings of the 12th International Zeolite Conference*. Materials Research Society, 1999, pp. 363–370.
143. Vlugt, T.J.H., Krishna, R. and Smit, B., *J. Phys. Chem. B*, **103**, 1102 (1999).
144. Sholl, D.S. and Fichthorn, K.A., *J. Chem. Phys.*, **107**, 4384 (1997).
145. Fedders, P.A., *Phys. Rev. B*, **17**, 40 (1978).
146. Kärger, J. and Pfeifer, H., Diffusion anisotropy and single-file diffusion in zeolites, In: von Ballmoos, R., et al. (Ed.), *Proc. 9th Int. Zeolite Conf., Montreal 1992*. Butterworth-Heinemann, 1993, pp. 129–136.
147. Kukla, V., Hahn, K., Kärger, J., Kornatowski, J. and Pfeifer, H., Anomalous diffusion in $AlPO_4$-5, In: Rozwadowski, M. (Ed.), *Proceedings of the 2nd Polish-German Zeolite Colloquium*. Nicholas Copernicus University Press, Torun, Poland, 1995, pp. 110–119.
148. Rödenbeck, C., Kärger, J. and Hahn, K., *J. Catal.*, **157**, 656 (1995).
149. Lei, G.D., Carvill, B.T. and Sachtler, W.M.H., *Appl. Catal. A: General*, **142**, 347 (1996).

150. Keffer, D., McCormick, A.V. and Davis, H.T., *Mol. Phys.*, **87**, 367 (1996).
151. Hahn, K. and Kärger, J., *J. Phys. Chem.*, **100**, 316 (1996).
152. Sholl, D.S. and Fichthorn, K.A., *Phys. Rev. Lett.*, **79**, 3569 (1997).
153. June, R.L., Bell, A.T. and Theodorou, D.N., *J. Phys. Chem.*, **95**, 8866 (1991).
154. Chandler, D., *J. Chem. Phys.*, **68**, 2959 (1978).
155. Chandler, D., *Introduction to Modern Statistical Mechanics*. Oxford University Press, New York, 1987.
156. Pechukas, P., Transition state theory, In: Miller, W. (Ed.), *Dynamics of Molecular Collisions*. Plenum, New York, 1976, p. 269.
157. Miller, W.H., *J. Chem. Phys.*, **61**, 1823 (1974).
158. Voter, A.F. and Doll, J.D., *J. Chem. Phys.*, **82**, 80 (1985).
159. Voth, G.A., Chandler, D. and Miller, W.H., *J. Chem. Phys.*, **91**, 7749 (1989).
160. Voter, A., *J. Chem. Phys.*, **82**, 1890 (1985).
161. Vineyard, G.H., *J. Phys. Chem. Solids*, **3**, 121 (1957).
162. Straatsma, T.P., *Rev. Comp. Chem.*, **9** (1996).
163. Sevick, E.M., Bell, A.T. and Theodorou, D.N., *J. Chem. Phys.*, **98**, 3196 (1993).
164. Jonsson, H., Mills, G. and Jacobsen, K.W., Nudged elastic band method for finding minimum energy paths of transitions, In: Berne, B.J., Ciccotti, G. and Coker, D.F. (Eds.), *Classical and Quantum Dynamics in Condensed Phase Simulations*. World Scientific, 1998, p. 385.
165. Voter, A., *J. Chem. Phys.*, **106**, 4665 (1997).
166. Dellago, C., Bolhuis, P.G., Csajka, F.S. and Chandler, D., *J. Chem. Phys.*, **108**, 1964 (1998).
167. Kärger, J., *J. Phys. Chem.*, **95**, 5558 (1991).
168. Jousse, F., Auerbach, S.M. and Vercauteren, D.P., *J. Phys. Chem. B*, **102**, 6507 (1998).
169. Mosell, T., Schrimpf, G., Hahn, C. and Brickmann, J., *J. Phys. Chem.*, **100**, 4571 (1996).
170. Mosell, T., Schrimpf, G. and Brickmann, J., *J. Phys. Chem.*, **100**, 4582 (1996).
171. Tunca, C. and Ford, D.M., *J. Chem. Phys.*, **111**, 2751 (1999).
172. Bigot, B. and Peuch, V.H., *J. Phys. Chem. B*, **102**, 8696 (1998).
173. Sholl, D.S., *Chem. Eng. J.*, **74**, 25 (1999).
174. Murphy, M.J., Voth, G.A. and Bug, A.L.R., *J. Phys. Chem. B*, **101**, 491 (1997).
175. Truong, T.N., *J. Phys. Chem. B*, **101**, 2750 (1997).
176. Wang, Q., Challa, S.R., Sholl, D.S. and Johnson, J.K., *Phys. Rev. Lett.*, **82**, 956 (1999).
177. Fermann, J.T. and Auerbach, S.M., *J. Chem. Phys.*, **112**, 6787 (2000).
178. Fermann, J.T., Blanco, C. and Auerbach, S.M., *J. Chem. Phys.*, **112**, 6779 (2000).
179. Sauer, J., Sierka, M. and Haase, F., In: Truhlar, D.G. and Morokuma, K. (Eds.), *Transition State Modeling for Catalysis*. Number 721 in ACS Symposium Series, Chapter 28, ACS, Washington, 1999, pp. 358–367.
180. Sierka, M. and Sauer, J., *J. Chem. Phys.*, **112**, 6983 (2000).
181. Hernandez, R. and Miller, W.H., *Chem. Phys. Lett.*, **214**, 129 (1993).
182. Ising, E., *Z. Phys.*, **31**, 253 (1925).
183. Gomer, R., *Rep. Prog. Phys.*, **53**, 917 (1990).
184. Panagiotopoulos, A.Z., *J. Chem. Phys.*, **112**, 7132 (2000).
185. van Kampen, N.G., *Stochastic Processes in Physics and Chemistry*. North Holland Publishing Company, New York, 1981.

186. Blanco, C., Saravanan, C., Allen, M. and Auerbach, S.M., *J. Chem. Phys.*, **113**, 9778 (2000).
187. Jacobsen, J., Jacobsen, K.W. and Sethna, J.P., *Phys. Rev. Lett.*, **79**, 2843 (1997).
188. Dobbs, K.D. and Doren, D.J., *J. Chem. Phys.*, **97**, 3722 (1992).
189. Cowell Senft, D. and Ehrlich, G., *Phys. Rev. Lett.*, **74**, 294 (1995).
190. Linderoth, T.R., Horch, S., Lægsgaard, E., Stensgaard, I. and Besenbacher, F., *Phys. Rev. Lett.*, **78**, 4978 (1997).
191. Hershkovitz, E., Talkner, P., Pollak, E. and Geogievskii, Y., *Surf. Sci.*, **421**, 73 (1999).
192. Zhang, Z.Y., Haug, K. and Metiu, H.I., *J. Chem. Phys.*, **93**, 3614 (1990).
193. Braun, O.M. and Sholl, C.A., *Phys. Rev. B*, **58**, 14870 (1998).
194. Deem, M.W., Newsam, J.M. and Creighton, J.A., *J. Am. Chem. Soc.*, **114**, 7198 (1992).
195. Saravanan, C. and Auerbach, S.M., *J. Chem. Phys.*, **107**, 8120 (1997).
196. Mitchell, M.C., McCormick, A.V. and Davis, H.T., *Z. Phys. B*, **97**, 353 (1995).
197. Coppens, M.O., Bell, A.T. and Chakraborty, A.K., *Chem. Eng. Sci.*, **53**, 2053 (1998).
198. Bhide, S.Y. and Yashonath, S., *J. Chem. Phys.*, **111**, 1658 (1999).
199. Bhide, S.Y. and Yashonath, S., *J. Phys. Chem. B*, **104**, 2607 (2000).
200. Nelson, P.H. and Auerbach, S.M., *Chem. Eng. J.*, **74**, 43 (1999).
201. Nelson, P.H., Kaiser, A.B. and Bibby, D.M., *J. Catal.*, **127**, 101 (1991).
202. Keffer, D., McCormick, A.V. and Davis, H.T., *J. Phys. Chem.*, **100**, 967 (1996).
203. Saravanan, C. and Auerbach, S.M., *J. Chem. Phys.*, **110**, 11000 (1999).
204. Coppens, M.O., Bell, A.T. and Chakraborty, A.K., *Chem. Eng. Sci.*, **54**, 3455 (1999).
205. Trout, B.L., Chakraborty, A.K. and Bell, A.T., *Chem. Eng. Sci.*, **52**, 2265 (1997).
206. Saravanan, C. and Auerbach, S.M., *J. Chem. Phys.*, **109**, 8755 (1998).
207. Dukovski, I., Saravanan, C., Machta, J. and Auerbach, S.M., *J. Chem. Phys.*, **113**, 3697 (2000).
208. Karger, J., Single-file diffusion in zeolites, In: Karge, H.G. and Weitkamp, J. (Eds.), *Molecular Sieves — Science and Technology, Vol. 7: Sorption and Diffusion*. Springer-Verlag, Berlin, New York, 1999 and references therein.
209. Theodorou, D.N. and Wei, J., *J. Catal.*, **83**, 205 (1983).
210. Gladden, L.F., Sousa-Gonçalves, J.A. and Alexander, P., *J. Phys. Chem. B*, **101**, 10121 (1997).
211. Fichthorn, K.A. and Weinberg, W.H., *J. Chem. Phys.*, **95**, 1090 (1991).
212. Metiu, H.I., Lu, Y.T. and Zhang, Z.Y., *Science*, **255**, 1088 (1992).
213. Saravanan, C., PhD Thesis, University of Massachusetts at Amherst, 1999.
214. Maksym, P., *Semicond. Sci. Technol.*, **3**, 594 (1988).
215. Krishna, R., Vlugt, T.J.H. and Smit, B., *Chem. Eng. Sci.*, **54**, 1751 (1999).
216. Gladden, L.F., Hargreaves, M. and Alexander, P., *Chem. Eng. J.*, **74**, 57 (1999).
217. Barrer, R.M., *J. Chem. Soc., Faraday Trans.*, **86**, 1123 (1990).
218. Ford, D.M. and Glandt, E.D., *J. Phys. Chem.*, **99**, 11543 (1995).
219. Ford, D.M. and Glandt, E.D., *J. Membrane Sci.*, **107**, 47 (1995).
220. Schmidt-Rohr, K. and Spiess, H.W., *Multidimensional Solid-State NMR and Polymers*. Academic Press, London, 1994.
221. Boddenberg, B. and Beerwerth, B., *J. Phys. Chem.*, **93**, 1440 (1989).

222. Schaefer, D.J., Favre, D.E., Wilhelm, M., Weigel, S.J. and Chmelka, B.F., *J. Am. Chem. Soc.*, **119**, 9252 (1997).
223. Saravanan, C. and Auerbach, S.M., *J. Chem. Phys.*, **107**, 8132 (1997).
224. Snurr, R.Q., *Chem. Eng. J.*, **74**, 1 (1999).
225. Kuscer, I. and Beenakker, J.J.M., *J. Stat. Phys.*, **87**, 1083 (1997).
226. Rodenbeck, C., Kärger, J. and Hahn, K., *Phys. Rev. E*, **55**, 5697 (1997).
227. Saravanan, C., Jousse, F. and Auerbach, S.M., *Phys. Rev. Lett.*, **80**, 5754 (1998).
228. Bülow, M., Härtel, U., Müller, U. and Unger, K.K., *Ber. Bunsen-Ges. Phys. Chem.*, **94**, 74 (1990).
229. Shen, D. and Rees, L.V.C., *J. Chem. Soc., Faraday Trans.*, **86**, 3687 (1990).
230. van Tassel, P.R., Somers, S.A., Davis, H.T. and McCormick, A.V., *Chem. Eng. Sci.*, **49**, 2979 (1994).
231. Nelson, P.H., Kaiser, A.B. and Bibby, D.M., *Zeolites*, **11**, 337 (1991).
232. Nelson, P.H. and Bibby, D.M., *Stud. Surf. Sci. Catal.*, **68**, 407 (1991).
233. Stauffer, D. and Aharony, A., *Introduction to Percolation Theory*. Taylor & Francis, Inc., Bristol, PA, 1991.
234. Kärger, J. and Pfeifer, H., *Zeolites*, **7**, 90 (1987).
235. Saravanan, C., Jousse, F. and Auerbach, S.M., *J. Chem. Phys.*, **108**, 2162 (1998).
236. Germanus, A., Kärger, J., Pfeifer, H., Samulevic, N.N. and Zdanov, S.P., *Zeolites*, **5**, 91 (1985).
237. Brandani, S., Xu, Z. and Ruthven, D., *Microporous Mater.*, **7**, 323 (1996).
238. Jobic, H., Bée, M., Kärger, J., Pfeifer, H. and Caro, J., *J. Chem. Soc., Chem. Commun.*, 341 (1990).
239. Shen, D.M. and Rees, L.V.C., *Zeolites*, **11**, 666 (1991).
240. Vitale, G., Bull, L.M., Morris, R.E., Cheetham, A.K., Toby, B.H., Coe, C.G. and MacDougall, J.E., *J. Phys. Chem.*, **99**, 16087 (1995).
241. Trout, B.L., Chakraborty, A.K. and Bell, A.T., *J. Phys. Chem.*, **100**, 17582 (1996).
242. Nelson, P.H. and Wei, J., *J. Catal.*, **136**, 263 (1992).
243. Rodenbeck, C. and Kärger, J., *J. Chem. Phys.*, **110**, 3970 (1999).
244. Kehr, K.W., Binder, K. and Reulein, S.M., *Phys. Rev. B*, **39**, 4891 (1989).
245. Qureshi, W.R. and Wei, J., *J. Catal.*, **126**, 126 (1990).
246. Binder, K., *Rep. Prog. Phys.*, **60**, 487 (1997).
247. Kutner, R., *Phys. Lett. A*, **81**, 239 (1981).
248. Krishna, R., *Chem. Eng. Sci.*, **48**, 845 (1993).
249. Krishna, R. and van den Broeke, L.J.P., *Chem. Eng. J. Biochem. Eng. J.*, **57**, 155 (1995).
250. van de Graaf, J.M., Kapteijn, F. and Moulijn, J.A., *AIChE J.*, **45**, 497 (1999).
251. Nelson, P.H., Tsapatsis, M. and Auerbach, S.M., *J. Membrane Sci.*, **184**, 245 (2001).
252. Rödenbeck, C., Kärger, J. and Hahn, K., *J. Catal.*, **176**, 513 (1998).
253. Freyhardt, C.C., Tsapatsis, M., Lobo, R.F., Balkus, K.J., Davis, M.E., et al., *Nature*, **381**, 295 (1996).

Computer Modelling of Microporous Materials
C.R.A. Catlow, R.A. van Santen and B. Smit (editors)

Chapter 4

Dynamic Monte Carlo simulations of diffusion and reactions in zeolites

Frerich J. Keil*

Hamburg University of Technology, Chair of Chemical Reaction Engineering, Eissendorfer Str. 38, D-21073 Hamburg, Germany

Marc-Olivier Coppens**

Department of Chemical Technology (DelftChemTech), Delft University of Technology, Julianalaan 136, NL-2628 Delft, The Netherlands

1. Introduction

This chapter continues the theme of simulation of dynamic processes in zeolites, but concentrating on dynamical Monte Carlo (MC) methods, which are widely used in physics and engineering [1–6] and which were introduced in the previous chapter. They are useful to probe the kinetics of heterogeneously catalyzed reactions, e.g. which often exhibit very complex nonlinear dynamic behavior including multiple steady states, oscillations, chemical waves, bifurcation, and chaos. Mean field rate equations of chemical kinetics, on the other hand, typically neglect self-established correlations between the reactant species. A broad variety of surface reactions have been studied by MC methods. Reviews on this subject were published, e.g. by Albano [7], Kang and Weinberg [8], Gelten et al. [9], and Gelten [10]. A problem associated with MC simulations is the dynamic interpretation of the MC results. MC simulations use 'jumps' instead of the real time. Fichthorn and Weinberg [11] have demonstrated that within the theory of Poisson processes, both static and dynamic properties of Hamiltonian systems may be consistently

*E-mail: keil@tuhh.de
**E-mail: m.o.coppens@tnw.tudelft.nl

simulated with the benefit that an exact correspondence between MC time and real time can be established in terms of the dynamics of individual species comprising the ensemble. Molecules should not all be moved at the same time during the simulations, one after the other, or according to some other fixed scheme, as this is not only unrealistic but may also lead to persistent errors in the asymptotically obtained results. Instead, one should think in terms of events that are Poisson-distributed in time. Reaction probabilities are replaced by rate constants. With an event-oriented method and knowledge of the times needed for each microscopic event, problems can be avoided, because the clock is advanced every time an event occurs, and the event-list is updated. In cases where the microscopic events are unknown, some reference time needs to be defined. There are different approaches to implement a dynamic MC simulation. Fichthorn and Weinberg [11] introduced the variable step-size method, and Gillespie [12] and Jansen [13] the discrete event simulation.

Recently, we reviewed the subject of simulation of diffusion in zeolites [14]. In the present chapter, we therefore focus exclusively on situations involving chemical reactions, a field that is in full development, thanks to increasing hardware and software capabilities, and simultaneous theoretical advances in molecular dynamics, transition-state theory and quantum chemistry, the results of which can be used as input in dynamic MC simulations. Results for such problems as diffusion-limited isomerization and cracking reactions, using realistic zeolite lattice representations, are finally within reach.

Reactions typically are 'rare events' compared to diffusion processes, so that the exclusive use of microscopic molecular dynamics simulations is unfeasible or at least uneconomical. The computing times needed to sample sufficient configurations in order to obtain accurate ensemble averages of effective reaction rates, yields, or selectivities, are prohibitively long. In some cases macroscopic differential equations can be employed, but this assumes that long-range correlations in the system are absent, and that diffusion and reactions are decoupled. Microscopic methods need to be employed on scales smaller than the correlation length. However, Vlachos and Katsoulakis [15] have recently presented a mesoscopic theory that includes intermolecular interactions; results were in good agreement with gradient MC simulations.

2. Simulation results for reactions in zeolites

About 20 years ago, Theodorou and Wei [16] were probably the first to apply MC calculations to diffusion and reaction problems in zeolites.

As in similar surface reaction problems, the zeolite crystal was modeled as a finite, two-dimensional rectangular grid of intersecting channels. Pore segments between successive intersections, or 'sites', were considered to have the same length. This grid was used as a simplified model of the ZSM-5 channel network, which retained the feature of regularity as well as the feature of interconnection of pores (coordination of each site = 4). The grid had a size between 11×11 and 21×21. The following assumptions were made:

- Diffusion of a molecule was considered to be a succession of discrete jumps from intersection to intersection. Only single jumps of length l were considered.
- The time interval between successive jumps of a molecule j was set constant and equal to τ_j. It was assumed that no blocking or occupancy effects occurred in this case.
- A molecule moved to one of the four adjacent sites with equal probability (no blocking or occupancy assumed).
- Surface diffusion was included by lateral movements of molecules from site to site along the grid border.
- The diffusion rates in the different channels of ZSM-5 were set to be equal.
- Pore intersections were assumed to be sorption sites of equal strength.

The authors examined two modes of pore blocking: random blocking of pores in the interior of the catalyst crystal ('bulk blocking') and random blocking of pore entrances along the border. At the beginning of each random walk computer experiment, a random-blocking configuration of blocked pores, or pore entrances, was generated. At time $t = 0$ the particle was placed at the center of the grid. The particle was subsequently allowed to diffuse through the grid. In case of a totally unblocked grid, this random walk experiment is nothing more than a finite-difference approximation to the two-dimensional continuum diffusion problem, which can be described by a differential equation. In mathematical terms, 'blocking' means a reduced transition probability for movement through a pore. By introducing an effective diffusivity, the solution of the above-mentioned differential equation could be fitted to random-walk simulations of a single particle in a blocked grid, with bulk pores blocked at random. At high border blocking, fitting becomes worse, which means that the continuum diffusion equation is no longer adequate for describing intracrystalline transport. As long as there exist enough open sites along the border, a product molecule can migrate by surface diffusion to a nonadsorbing blocked site, and from there on to the surroundings. The simulations allowed to study these effects in detail. The effects of bulk pore blocking are presented in Figs. 1 and 2. As can be seen from these figures, bulk pore blocking led to a much sharper, convex decrease than border blocking, for which the diffusivity curve is concave. Higher occupancies led to more retardation, and a comparison with the continuum model (differential equation) revealed that the decrease with occupancy of the effectiveness factor in the isomerization of xylene can only be described by such a model for occupancies up to approximately 35% (see Fig. 3). At higher

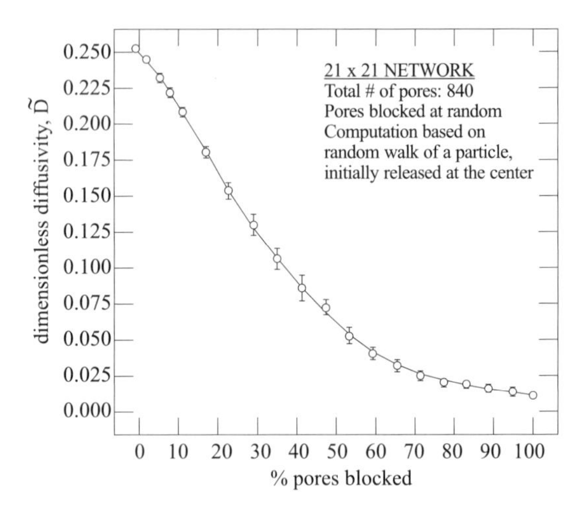

Fig. 1. Effect of bulk blocking of pores on diffusivity /16/.

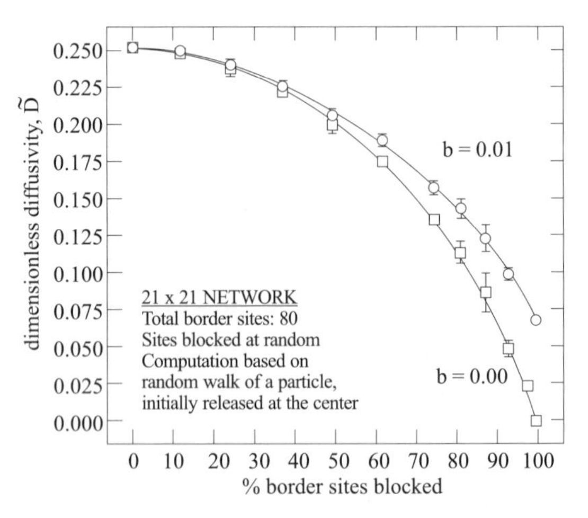

Fig. 2. Effects of border blocking on diffusivity /16/.

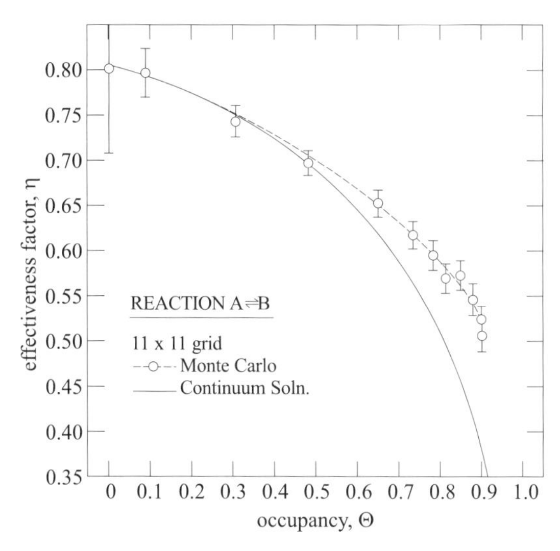

Fig. 3. Variation of effectiveness factor with global occupancy (comparison between Monte Carlo and continuum approaches) /16/.

occupancies the continuum, mean field approximation underestimates the effectiveness factor obtained from MC simulations, showing the importance of the latter.

Border blocking causes several effects on diffusion and reaction that were classified by Theodorou and Wei as follows:

- Decrease in the rate of entrance from the gaseous phase into the crystal, due to unavailability of pore entrances.
- Increase in the 'tortuosity' of diffusion paths due to surface blocking.
- Decrease in the intrinsic catalytic activity of the crystal because blocked sites are catalytically inactive.
- Decrease in occupancy with increasing degree of blocking.

Sundaresam and Hall [17] developed a mathematical model for diffusion and reaction in blocked zeolites that takes into account nonidealities arising from interaction between sorbed molecules as well as the effect of pore and surface blocking. Their model combined a microscopic approach, in which expressions for chemical potential and diffusive fluxes are calculated within a lattice-gas framework, with a continuum approach, which takes into account the effect of surface blocking. The effects of pore blocking on the diffusive fluxes were modeled by an effective medium approximation. The simulations showed that the effect of pore blocking on the selectivity is more pronounced in the presence of repulsive lateral interaction than in its absence. In the reaction scheme a set of reactants *R* formed two isomers

A and B. Furthermore, the authors detected that molecules may be more likely to jump to a neighboring vacant site if some of its other neighbors are occupied. The activation energy for migration of a molecule is proportionally reduced by the interaction with molecules occupying an adjacent site.

As a three-dimensional extension to Theodorou and Wei's model, Tsikoyiannis and Wei [18] represented the zeolite crystal pore space as a finite cubic grid of intersecting channels exhibiting spatial periodicity. The crystal was assumed exposed to a gaseous environment of chemical species. Gaseous particles bombarded and attached themselves onto the surface of the crystal and subsequently diffused to the interior and reacted according to specific rules that were given as inputs to the model. One type of site was employed. The molecular motion within the crystal consisted of a succession of random discrete jumps from one lattice site to the other. The lattice sites were assumed to coincide with the channel intersections. Chemical reaction was treated as an instantaneous molecular transformation, taking place at a particular lattice site. All elementary events were independent Poisson processes. Each event was associated with a parameter ρ, which was called the rate of occurrence, so that the time interval to the event was an independently distributed random variable with an exponential probability density $F(t)$:

$$F(t)\,\mathrm{d}t \equiv P(t < \tau < t + \mathrm{d}t) = \rho \mathrm{e}^{-\rho t}\,\mathrm{d}t, \tag{1}$$

which is normalized to one. The average time interval is given by:

$$\langle \tau \rangle = \int_0^{\infty} tF(t)\,\mathrm{d}t = 1/\rho \tag{2}$$

A particle located at a site of the lattice stood in place for some time interval τ_{d}, which is a random variable exponentially distributed with a certain jump-rate parameter, and then attempted to jump to another site. The jump-rate parameter was characteristic of the particle's diffusive behavior since it assigns the timescale of its motion. For reacting systems, a particle located at a site could undergo a monomolecular reaction to a different species. This event occurred after a random time interval τ_{r} exponentially distributed with a reaction-rate parameter characteristic of the reactants and products. The assumed Poisson character of the elementary events was justified on physical grounds: for activated processes, such as a molecular jump, or a chemical transformation, it is consistent with the theory of absolute reaction rates, and has been confirmed by molecular dynamics simulations. The authors developed a Markovian formalism, which allowed to derive approximate solutions to the master equation describing diffusion and reaction in the zeolite. The master equation was solved by an MC approach. In the beginning, the grid was completely empty, and then a long sequence of elementary events was started, which described the time evolution of the system. For each system configuration, a particular event to be executed next and the time of its occurrence were chosen according to the probability structure built into the model. The system configuration was then updated. After an initial time, in which the grid was allowed to relax to a steady state, statistical averages such as site occupancies and the fluxes of the components into the crystal were computed. First, the simplest possible mechanism for molecular motion and reaction was treated, namely that there is no interaction between adsorbed molecules, except the hard force which excludes two molecules from occupying the same site. For single-component diffusion the authors found the following results. The transient uptake diffusivity was independent of occupancy; the Wicke–Kallenbach diffusivity, when the gradient of occupancy is the driving force, was independent of occupancy; when the gradient of pressure was the driving force, it was proportional to $(1-\theta)^2$, whereby θ represents the occupancy. The tracer diffusion in 1-D channels did not follow Einstein's equation in two- and three-dimensional channels, but was anomalous [19,20]. For multi-component diffusion the transient-uptake diffusion matrix depended on occupancy. During co-diffusion of two species across a barrier, there could be entrainment effects. In case of soft interaction between adjacent adsorbed molecules that interacted by weak attractive or repulsive forces, the transient-uptake diffusivity of single- and multi-component systems was dependent on occupancy.

Frank et al. [21,22] and Wang et al. [23,24] studied xylene isomerization in ZSM-5. A two-dimensional lattice was used in all cases. Frank et al. [21,22] executed three different types of simulations: first, simulations of transport diffusion of a single component as well as a binary mixture; second, simulations of self-diffusion under comparable assumptions as above; and third, a simulation of diffusion combined with reaction. Diffusion was simulated as a hopping mechanism under the following assumptions: a molecule in a cage could migrate to one of four neighboring cages with equal probability. No memory effects or interactions between two molecules or molecules and zeolites were included. All adsorption sites were equal and isothermal. Each zeolite cage could contain at least one molecule. A maximum number of molecules per cage were fixed. To convert MC time steps into real time, the average adsorption time on a site was approximated by a harmonic oscillator of which the frequency was chosen in such a way that the oscillator's energy equals the activation energy. To perform a single jump event the following procedure was chosen: a cage was selected at random, and, if the cage contained at least one molecule, a direction of diffusion was assigned to the molecule with equal probability of 0.25 in each of the four possible directions. A successful jump occurred if the neighboring cage in the direction of diffusion contained at least one empty adsorption site and the window between the two cages was not blocked by a molecule diffusing from the neighboring cage into the selected cage in the same MC step. When both prerequisites were fulfilled the molecule left the cage, it was placed in the neighboring cage and the window between the two cages marked as blocked for the MC step. If the neighboring cage was completely filled or another molecule diffused through the same window, four different possibilities to determine how the diffusion process should go on were distinguished. In case of the simulation of transport diffusion two ranges of different constant concentrations were attached to the left and to the right side of the simulation grid. Thus, a concentration gradient existed over the zeolite grid from one side to the other. The diffusion coefficient was calculated according to Fick's first law. The self-diffusion coefficient was evaluated via the Einstein's relation. Frank et al. tuned the hopping ratios of *ortho*-, *meta*-, and *para*-xylene to measure their effect on the reaction path. This enabled them to get an idea of the ratios of the hopping rates needed to match the experimental observations. They found that a ratio *ortho*:*meta*:*para* of 1:1:10 of diffusion coefficients led to a better correspondence with experiments than the often assumed ratio 1:1:1000 (see Figs. 4 and 5). Since there is a debate on this ratio, simulations may yield better insights.

A similar study was carried out by Wang et al. [23] on the isomerization of xylene. The zeolite crystal was modeled as a finite, two-dimensional rectangular grid of intersecting channels. The adsorption and desorption of molecules took place at border sites only, and the diffusion of sorbed molecules in the channels was described as a random-walk process. The reaction occurred in the sorbed phase. A 21×21 network was employed. The selectivity towards *meta*-xylene decreased strongly as a function of the Thiele modulus for *ortho*- and *para*-xylene isomerization (see Fig. 6). On the other hand, the selectivity toward *para*-xylene increased sharply with increasing Thiele modulus

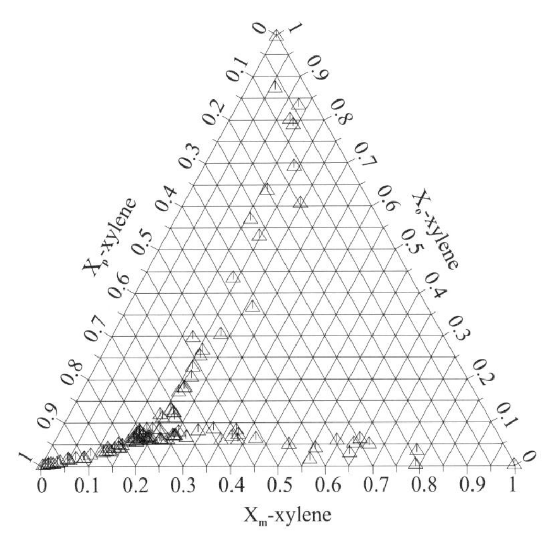

Fig. 4. Reaction path of isomerization of xylenes: $P_{\text{Do-xylene}} : P_{\text{Dm-xylene}} : P_{\text{Dp-xylene}} =$ 1:1:10 /21/.

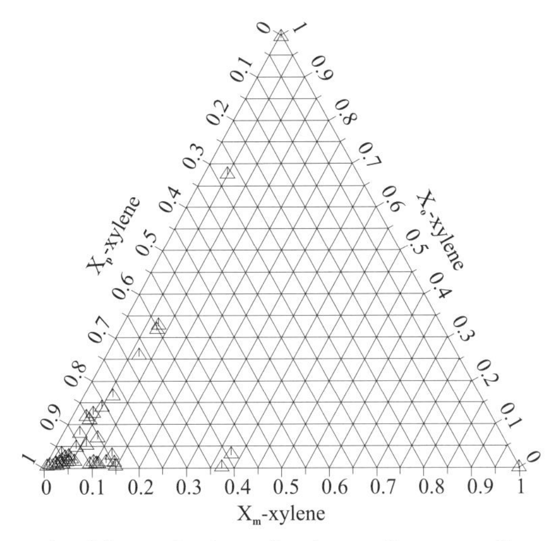

Fig. 5. Reaction path of isomerization of xylenes: $P_{\text{Do-xylene}} : P_{\text{Dm-xylene}} : P_{\text{Dp-xylene}} =$ 1:1:100 /21/.

for *meta*-xylene isomerization. The results were satisfactorily consistent with those of experiments. Wang et al. [24] also investigated the alkylation of toluene with methanol and ethanol over a modified ZSM-5 zeolite. The MC results were consistent with a large amount of experimental data for the reactions. It could be seen that *para*-selectivity

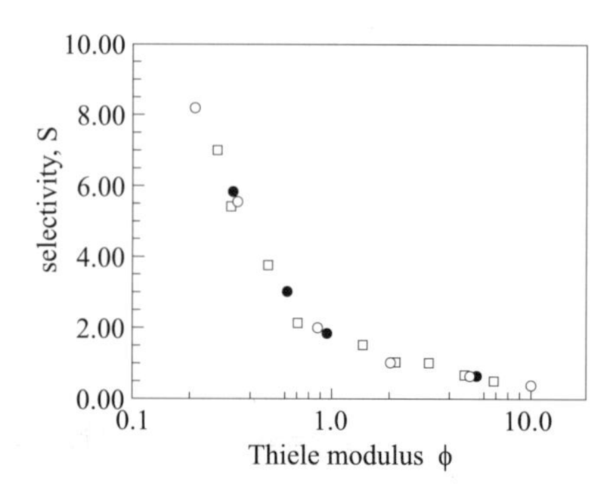

Fig. 6. Dependence of the selectivity of *meta*-xylene upon the Thiele modulus for *ortho*-xylene isomerization. (○) continuum model; (●) experimental; (□) /23/.

in the alkylation of toluene with methanol and ethanol could be attributed to the diffusivity, the equilibrium adsorption constant, and the intrinsic rate constant of xylenes. Both the equilibrium constant and the intrinsic rate constant are related to the surface acidity of zeolites, and the intracrystalline structure is pore structure dependent; therefore, the shape selectivity of the reactions is the result of the channel structure and the surface acidity of zeolite catalysts.

Ethylation and disproportionation of ethylbenzene on ZSM-5 was studied by Klemm et al. [25] to explain experiments and to provide suggestions to enhance *para*-selectivity. *Para*-Selectivity could be mainly attributed to product-phase selectivity rather than transition-state phase selectivity, while the opposite holds for the disproportionation reaction. Products with a bulky structure cannot diffuse out quickly enough. These findings were deduced from the simulations and a comparison with experimental results. The MC and a continuum approach gave similar results in this case (see Fig. 7).

While providing interesting qualitative information, caution needs to be exercised with the quantitative interpretation of the former simulation results, since both diffusion experiments and simulations revealed that diffusivities depend strongly on the pore network topology, so that results obtained with a two-dimensional grid may not be generally valid [25,26].

Trout et al. [27] performed dynamic MC simulations to investigate diffusion and reaction in ZSM-5 in the absence of concentration gradients. Their model utilized a physically realistic, coarse-grained three-dimensional representation of the distribution of adsorption and reaction sites in ZSM-5 in order to capture the inherent anisotropy of diffusion in this zeolite, and incorporated both weakly adsorbing,

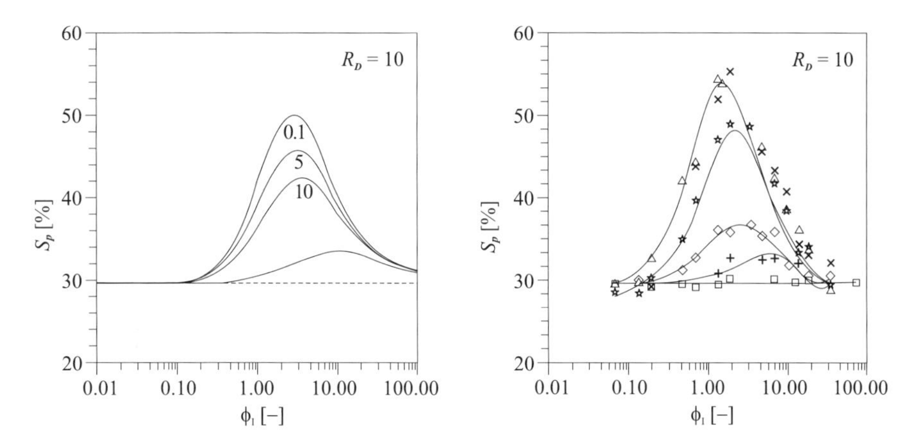

Fig. 7. Dependence of *para*-selectivity S_p on Thiele modulus ϕ_I of isomerization at different values of ϕ_F /25/.

catalytically inactive sites and strongly adsorbing, catalytically active sites. The effects of adsorbate concentration and of the fraction of blocked sites on the self-diffusion coefficient were investigated. Also the effects of the average number of catalytically active sites per unit cell and of the concentration of reactant within the zeolite on the dynamics of simultaneous diffusion and reaction for an isomerization $A \Leftrightarrow B$ were studied. It was assumed that A and B had identical adsorption and diffusion properties. In the absence of active sites, the initial position of particles on the lattice was determined by sampling from a uniform random distribution. During the simulation, particles were allowed to move from site to site via discrete hops. Each attempted move of a particle corresponded to an MC step, and was performed as follows. The net rate constant for any attempt was calculated as $k = hN$, where h is the rate constant for a single hop of a particle and N is the number of particles that can hop at the start of an iteration. The number N was a constant when only diffusion occurred. The time interval to the next step, Δt, was sampled from a Poisson distribution, $\Delta t = -(1/k)\ln(x)$, where x is a random number between 0 and 1. The clock was advanced by that time. The particle that attempted to hop, as well as the direction in which that particle attempted to hop, were randomly selected. Before the adsorbate motion was sampled, the positions of the particles were initialized by allowing them to hop successfully an average of 3000 times per particle, since it was found that after this number of successful hops, the diffusivity did not change as a function of the number of hops. Periodic boundary conditions were used. In the second set of simulations, the reaction $A \Leftrightarrow B$ was assumed to occur on a quenched set of reaction sites distributed throughout the lattice by sampling their location randomly from a uniform distribution. These sites were called 'quenched' sites, because they did not move during the course of a simulation. The reaction sites strongly adsorbed A and B, whereas the other lattice sites weakly adsorbed both components. In order to maximize the efficiency of the simulation, two separate clocks were used, a diffusion and a reaction clock. The iteration time was sampled from a Poisson distribution in both cases, but while the diffusion clock was reset after each MC step, the reaction clock was reset only after the reaction had reached equilibrium. The simulations were performed under steady-state conditions, so that the net adsorption rate of A or B onto reaction sites was equal to the net desorption rate. The authors found a very rapid decline of the reduced diffusivity with increasing fractional occupancies. This effect was less pronounced when the percentage of blocked sites was greater than about 20%. A comparison of the amount of product formation as a function of time obtained by a homogeneous model (without diffusion limitations) with ZSM-5 showed that the formation of B took five times longer for ZSM-5. Figure 8 presents the effects of increasing the fraction of reaction sites for cases in which 33.33% of all sites not containing reaction centers were occupied by weakly adsorbed species (A or B). As the fraction of reaction sites increased, the scaled value of the time required to reach 90% $[B]_{eq}$ decreased and then reached a minimum. This behavior reflects a trade-off between an increase in the rate of reaction

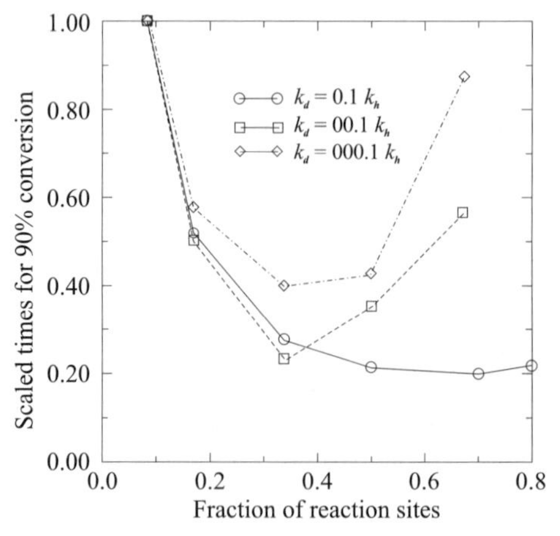

Fig. 8. Effects of the strength and fraction of reaction sites on the dynamics of diffusion and reaction when 33.33% of the weakly adsorbing sites are occupied. The graph shows times for the product to reach 90% of the steady-state concentration; k_d and k_h are rate constants for a single particle to desorb or hop, respectively /27/.

and a decrease in the rate of diffusion as a consequence of the increasingly larger fraction of blocked sites. Decreasing the value of k_d (see Fig. 8) increased the minimum value of time. The reduced diffusivities decreased as the fraction of reaction sites increased and the value of k_d decreased. In summary, one can say that there is an optimal concentration of reaction sites. This optimum is a result of the trade-off between increased ability for the reaction to proceed and decreased effectiveness of reaction sites as a consequence of reduced diffusive transport of species to and from the reaction sites. A lower occupancy of the zeolite enhanced the reaction rate by reducing the fraction of reaction sites occupied by adsorbed species.

Hinderer and Keil [28], and Keil et al. [29] used a three-dimensional network with a fine grid to simulate the conversion of methanol to olefins (MTO) in a ZSM-5 catalyst; their MC approach enables to include complex chemical networks between a large number of components, and does not assume the knowledge of the single-component diffusivities. The latter is important, because even the single-component diffusivities are not available in the temperature range of interest. The network consisted of a fine cubic grid of the pore space (see Fig. 9). Each cube could be one of four lattice elements: rigid zeolite framework, free intracrystalline pore space, active site, or window site. A molecule could occupy any vacant element in the pore space or an active site. Sorbate–sorbate interactions were taken into consideration. During one step of MC simulation each molecule was chosen in a random sequence. If the movement of a molecule was possible in the direction chosen at random, the movement continued until the molecule collided with another molecule or the rigid zeolite framework. When the saturation capacity was reached, a channel or intersection could not be occupied by more molecules. A molecule residing on an

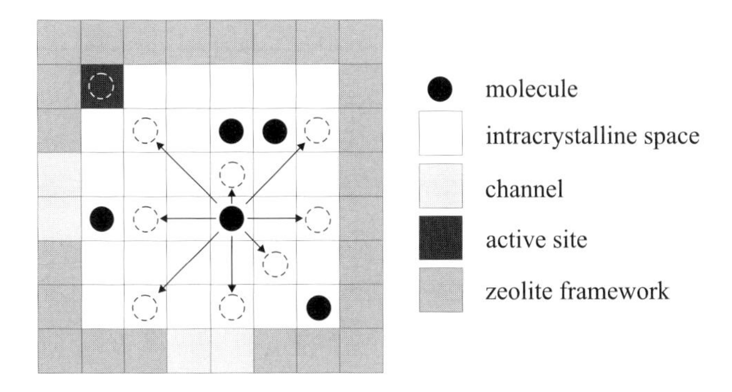

Fig. 9. Movements of molecules during MC calculations /28/.

active site had to surmount a potential barrier. Likewise, a channel intersection could be left only when the molecule had passed a potential barrier. Pairwise additive Lennard-Jones potentials were employed. The MC time step was set equal to the mean residence time on an active site, which is given by the reciprocal of the product of the vibration frequency, ν, and a Boltzmann factor. MC simulations of the MTO synthesis were based on a lattice structure consisting of 4096 channel intersections. Rigid framework elements formed the boundary of the lattice structure except for one side, where a region of constant concentration of methanol molecules was attached. A simplified reaction scheme was used for the description of the reaction kinetics. The fluxes of molecules of each component leaving the lattice structure were balanced in order to simulate the product distribution. Reaction probabilities were tuned according to measurements. Alternatively, they could have been obtained by quantum chemical calculations. In Fig. 10, the composition of the reaction mixture is presented. After one million time steps a good agreement between simulations and experiments could be obtained.

Keil et al. [29] also developed a model for MTO synthesis in composite catalysts (see Fig. 11). Diffusion and reaction in the zeolite particles were modeled as before, with the aid of MC calculations. Multicomponent diffusion in the inert matrix, in which the zeolite crystals were embedded, was calculated according to the dusty-gas model. Volume change during the reaction was taken into account. In Fig. 12 effectiveness factors as a function of particle size and zeolite fraction are presented. The volume change during the reaction influences the effectiveness factor considerably.

In the former examples, MC simulations revealed the significance of short-range correlations that could not be captured efficiently using

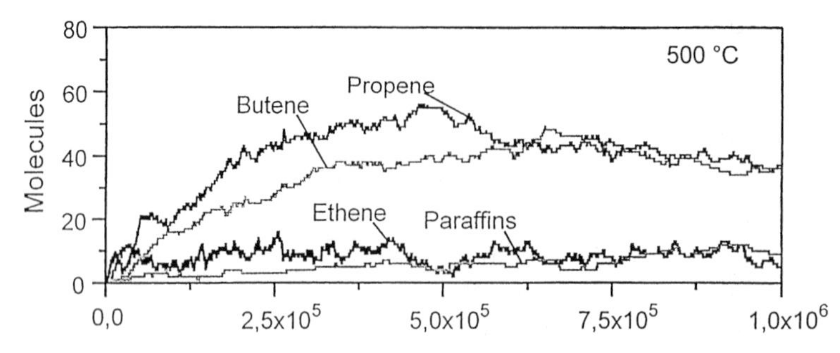

Fig. 10. Number of olefin and paraffin molecules within a lattice of 4096 intersections /28/.

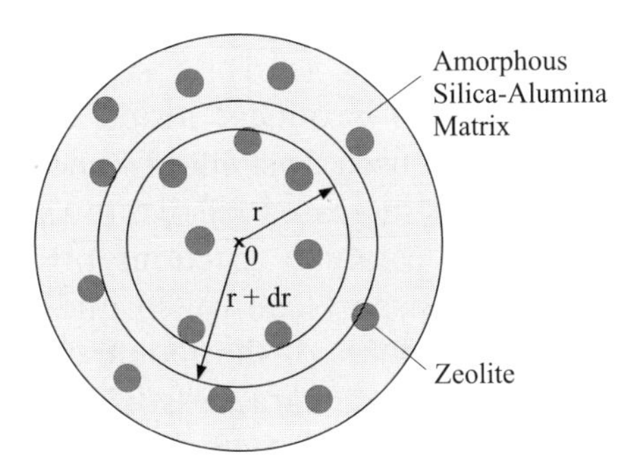

Fig. 11. Section of a spherical composite pellet /29/.

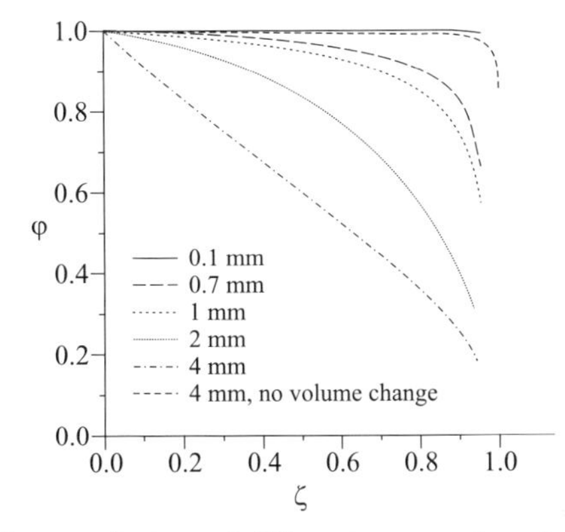

Fig. 12. Effectiveness factors of different composite catalyst pellets /29/.

a classical continuum approach. Even more drastic deviations occur when the correlations are long-range, so that Einstein's law of diffusion fails. This special situation occurs in so-called 'one-dimensional' zeolites, where molecules diffuse mainly along one single direction, because the interconnections between parallel pore channels are too small for molecules to move from one channel to another. If molecules cannot pass each other, the variance of the position of a molecule is no longer proportional to time (Einstein's law: $\langle r^2 \rangle \sim t$), but proportional to the square root of time (anomalous diffusion: $\langle r^2 \rangle \sim t^{1/2}$), as Richards [20] and Fedders [19] have shown. In real zeolite crystals, this 'single-file' diffusion may be difficult to observe over long distances, as a result of defects or occasional passing of molecules. In any case, the study of single-file diffusion with reaction is a topic of great interest, to which dynamic MC simulations contribute.

Kärger et al. [30] found that, unlike adsorption and desorption kinetics, tracer exchange and chemical reaction are dramatically retarded in single-file systems, compared to systems with normal diffusion. Molecular confinement tends to significantly increase the accumulation of the reaction products inside a sample. A Thiele modulus approach could be used, just like for normal diffusion; the values of the Thiele modulus increased by two to three orders of magnitude, suggesting that the catalytic reactions observed with single-file systems are strongly transport controlled and are most likely to occur in the vicinity of the orifices of the zeolite channels. Rödenbeck et al. [31] studied the influence of characteristic parameters such as file length, jump rates, and reaction rate coefficient on the tracer exchange and the effectiveness factor of a first-order reaction in a single-file system, and compared the results with experiments in zeolite L. An analytical treatment of this problem, based on the molecular residence time distribution, which can be inferred from the tracer exchange, was introduced in later work [32], and the work was extended [33] to investigate the influence of exchange limitations at the margins and intermolecular attractions. It was demonstrated that the apparent activation energy of the conversion can exceed the intrinsic activation energy, a phenomenon that cannot occur for systems undergoing normal diffusion but was found experimentally for some single-file zeolites. This is the case when the channels are crowded, the reaction is diffusion controlled, and the activation energies of the elementary diffusion processes fulfil certain relations (e.g. neighboring particles can pass each other, yet with an activation energy higher than the intrinsic activation energy of the conversion).

Okino et al. [34] simulated diffusion and reaction in a single-file system, by considering all possible configurations of M species in a length N one-dimensional pore. A deterministic model consisting of $(M+1)^N$ variables could be constructed for the system. By considering only pairs of adjacent cells, or $(M+1)^2(N-1)$ doublets, a simplified, reduced-order model emerged that, nevertheless, was able to capture the most important correlations between cells when the dominant mode of transport was through single-site hops. This model could be used for reversible reactions including molecules of different mobilities, and could be extended to higher-dimensional pores and more complex molecular interactions.

Very recently, Nedea et al. [35] used MC methods and analytical techniques to extend earlier studies of single-file diffusion with reaction. They also considered cases where only part of the sites is reactive. Comparisons between mean-field predictions and MC simulations for the occupancy profiles and reactivity again reveal substantial differences when diffusion rates are high, since mean-field results only include single-file behavior by changing the diffusion rate constant, but effectively allow passing of particles. The reactivity converges to a limit value when there are more reactive sites, because centrally located sites have little effect on the kinetics. Occupancy profiles showed approximately exponential behavior from the ends to the middle of the system.

Molecular Traffic Control (MTC) is another unusual phenomenon that could be studied using dynamic MC simulations. MTC was postulated by Derouane and Gabelica [36] to describe chemical reactions in zeolites and other nanoporous materials where the reactant and product molecules prefer 'traffic' along different diffusion pathways into and out of the catalyst particles, leading to an enhancement of the effective reaction rate. That this phenomenon is indeed theoretically possible could only recently be confirmed thanks to both molecular dynamics [37,38] and MC simulations [39,40], using a network of single-file channels with selective adsorption affinities to either the reactant or product molecules. The molecular dynamics simulations demonstrated the molecular confinement along different pathways, while the dynamic MC simulations showed an increased reaction rate as a result of the increased diffusion times along single-file channels. Molecular distributions were compared to those of a reference lattice where all channels were equally accessible by reactant and product molecules.

3. Outlook

Dynamic MC simulations of reactions in zeolites are still in their infancy. Although we have limited ourselves to dynamic MC methods in this short review, it is important to realize that these methods cannot really be isolated from other computational techniques: in the future, combination with quantum chemical computations of the elementary reaction steps, as well as with transition-state theory and/or short-time molecular dynamics for diffusion, will increasingly contribute to providing deeper insights into the parameters that determine the selectivity and yield of reactions in zeolites [14,41]. The continued increase in computational power and the availability of faster and more robust algorithms is the basis for advances in the practical use of statistical mechanics to simulate diffusion and reaction in zeolites. We would also like to stress, however, that computer is merely a (increasingly powerful) tool; new analytical–theoretical developments, such as mesoscopic and extended mean-field theories, including correlation effects, and solutions using a lattice Green's function approach [42] are of equal importance in reaching progress in this field. The methodology described here for zeolites can be extended to other nanoporous materials in physics, chemistry, and biology, such as (carbon) nanotubes, structured mesoporous sieves (e.g. MCM-41, MCM-48, SBA-15), superionic conductors, ion channels in biological membranes, and protein crystals.

References

1. Binder, K. (Ed.), *The Monte Carlo Method in Condensed Matter Physics.* Springer-Verlag, Berlin, 1992, pp. 121–204.
2. Binder, K. (Ed.), *Applications of the Monte Carlo Method in Statistical Physics.* Springer-Verlag, Berlin, 2nd ed., 1987, pp. 181–221.
3. Binder, K. and Heermann, D., *Monte Carlo Simulation in Statistical Physics.* Springer-Verlag, Berlin, 4th ed., 2002, pp. 29–33.
4. Allen, M.P. and Tildesley, D.J. (Eds.), *Computer Simulation in Chemical Physics.* Kluwer Academic Publishers, Dordrecht, 1993, pp. 110–139.
5. Gubbins, K.E. and Quirke, N. (Eds.), *Molecular Simulation and Industrial Applications.* Gordon & Breach Science Publ., Amsterdam, 1996, pp. 411–432.
6. Frenkel, D. and Smit, B., *Understanding Molecular Simulation — From Algorithms to Applications.* Academic Press, New York, 2nd ed., 2001.
7. Albano, E.V., *Heterogeneous Chem. Rev.*, **3**, 389 (1996).
8. Kang, H.C. and Weinberg, W.H., *Chem. Rev.*, **95**, 667 (1995).
9. Gelten, R.J., van Santen, R.A. and Jansen, A.P.J., In: Balbuena, P.B. and Seminario, J.M. (Eds.), *Molecular Dynamics: From Classical to Quantum Methods. Theoretical and Computational Chemistry.* Elsevier, Amsterdam, Vol. 7, 1999, pp. 737–787.
10. Gelten, R.J., *Monte Carlo Simulations of Catalytic Surface Reactions.* Ph.D. Thesis, Technical University of Eindhoven, 1999, pp. 17–38.
11. Fichthorn, K. and Weinberg, W.H., *J. Phys. Chem.*, **95**, 1030 (1991).
12. Gillespie, D.T., *J. Comput. Phys.*, **22**, 403 (1975).
13. Jansen, A.P.J., *Comput. Phys. Commun.*, **86**, 1 (1995).
14. Keil, F.J., Krishna, R. and Coppens, M.-O., *Rev. Chem. Eng.*, **16**, 71 (2000).
15. Vlachos, D.G. and Katsoulakis, M.A., *Phys. Rev. Lett.*, **85**, 3898 (2000).
16. Theodorou, D. and Wei, J., *J. Catal.*, **84**, 205 (1983).
17. Sundaresam, S. and Hall, C.K., *Chem. Eng. Sci.*, **41**, 1631 (1986).
18. Tsikoyiannis, J.G. and Wei, J., *Chem. Eng. Sci.*, **46**, 233 (1991).
19. Fedders, P.A., *Phys. Rev. B*, **17**, 40 (1978).
20. Richards, P.M., *Phys. Rev. B*, **16**, 1393 (1977).
21. Frank, B., Dahlke, K., Emig, G., Aust, E., Broucek, R. and Nywit, M., *Microporous Mat.*, **1**, 43 (1993).
22. Frank, B., Dahlke, K. and Emig, G., *Chem.-Ing. Techn.*, **64**, 1104 (1992) (in German).
23. Wang, J.G., Li, Y.W., Chen, S.Y. and Peng, S.Y., *Catal. Lett.*, **26**, 189 (1994).
24. Wang, J.G., Li, Y.W., Chen, S.Y. and Peng, S.Y., *Zeolites*, **15**, 288 (1995).
25. Klemm, E., Wang, J.-G. and Emig, G., *Chem. Eng. Sci.*, **52**, 3173 (1997).

26. Coppens, M.-O., Bell, A.T. and Chakraborty, A.K., *Chem. Eng. Sci.*, **53**, 2053 (1998).
27. Trout, B.L., Chakraborty, A.K. and Bell, A.T., *Chem. Eng. Sci.*, **52**, 2265 (1997).
28. Hinderer, J. and Keil, F.J., *Chem. Eng. Sci.*, **51**, 2667 (1996).
29. Keil, F.J., Hinderer, J. and Garayhi, A.R., *Catal. Today*, **50**, 637 (1999).
30. Kärger, J., Petzold, M., Pfeifer, H., Ernst, S. and Weitkamp, J., *J. Catal.*, **136**, 283 (1992).
31. Rödenbeck, C., Kärger, J. and Hahn, K., *J. Catal.*, **157**, 656 (1995).
32. Rödenbeck, C., Kärger, J. and Hahn, K., *Phys. Rev. E*, **55**, 5697 (1997).
33. Rödenbeck, C., Kärger, J. and Hahn, K., *J. Catal.*, **176**, 513 (1998).
34. Okino, M.S., Snurr, R.Q., Kung, H.H., Ochs, J. and Mavrovouniotis, M.L., *J. Chem. Phys.*, **111**, 2210 (1999).
35. Nedea, S.V., Jansen, A.P.J., Lukien, J.J. and Hilbers, P.A.J., *Phys. Rev. E*, **65**, art. No. 066701 Part 2 (2002).
36. Derouane, E.G. and Gabelica, Z., *J. Catal.*, **65**, 486 (1980).
37. Snurr, R.Q. and Kärger, J., *J. Phys. Chem. B*, **101**, 6469 (1997).
38. Clark, L.A., Ye, G.T. and Snurr, R.Q., *Phys. Rev. Lett.*, **84**, 2893 (2000).
39. Neugebauer, N., Bräuer, P. and Kärger, J., *J. Catal.*, **194**, 1–3 (2000).
40. Bräuer, P., Neugebauer, N. and Kärger, J., *Coll. Surf. A*, **187–188**, 459 (2001).
41. Broadbelt, L.J. and Snurr, R.Q., *Appl. Catal. A*, **200**, 23–46 (2000).
42. Barzykin, A.V. and Hashimoto, S., *J. Chem. Phys.*, **113**, 2841 (2000).

Computer Modelling of Microporous Materials
C.R.A. Catlow, R.A. van Santen and B. Smit (editors)

Chapter 5

Planewave pseudopotential modelling studies of zeolites

Julian D. Gale

Nanochemistry Research Institute, Department of Applied Chemistry, P.O. Box U1987, Perth 6845, Western Australia

1. Introduction

Our attention now turns to the application of quantum mechanical methods which are needed as much of the extensive research that has been performed in the field of zeolite science is to comprehend the catalytic properties of such materials which depend on bond breaking and making. Not only do we wish to deduce how activation barriers are lowered to accelerate reactions, but also there is a desire to be cognizant of all the reasons for specificity and shape selectivity. To achieve this requires a detailed atomic level picture of what occurs as molecules, adsorb and undergo chemical processes within the confines of microporous materials. Here quantum mechanical methods have a pivotal role to play as the only means of directly addressing such questions and providing a framework for the interpretation of experimental spectroscopic data.

If we go back a decade and view where theoretical techniques had reached in the study of zeolite reactivity, one will find there was only a single approach that was dominant. Nearly all studies were based on the use of small cluster models, in which a fragment of the aluminosilicate framework is cleaved from the bulk and dangling bonds appropriately terminated [1,2]. Often these molecular entities were just generic fragments, not specifically designed to represent any particular material, though not in all cases [3,4]. For the determination of reaction pathways, the only approach in common use was to locate the saddle point statically based on the curvature of the internal energy surface and then to apply statistical mechanics in the form of transition state or activated complex theory.

In this chapter we will concentrate on the events of the last 5 years of the twentieth century which have seen a proliferation of new studies with far greater realism in terms of the description of specific zeolites and the influence of dynamical motion. All this has become possible because of the advances in computer hardware, but also very much due to the advent of new algorithms for solid-state quantum mechanical calculations. Here we will highlight the technique that has had a major impact in this revolution — the total energy planewave pseudopotential method [5].

2. Periodic boundary conditions versus cluster methods

We know from the fact that diffraction patterns can be readily measured from most microporous materials that they have a fundamental underlying periodicity, though this does not mean to imply they can be readily solved, which is a different matter altogether. Having said this, we know that they are not truly regular, as this averaging often hides disordering of framework dopants and counter cations. Furthermore, we know that such materials are not infinitely periodic as crystals have a finite size to them — in some cases they may only extend for a few unit cells in each direction.

Regardless of the aforementioned reservations, it is quite a reasonable approximation to treat a microporous material as being infinitely repeated in all three crystallographic directions. This periodicity can be exploited to offer a way of dealing with the quantum mechanics of zeolites which is very attractive. Clearly, there is no problem in representing the difference between particular zeolite topologies as the structure can be taken straight from the crystallographic data.

Given the advantages of such an approach, one might ask why the majority of calculations to date have not followed this route — the answer being practical considerations. Quite naturally, molecular quantum chemistry has evolved much faster than equivalent techniques in the solid state because of its greater simplicity of implementation, for a particular level of theory. Furthermore, the rapid advancement of molecular quantum chemistry was aided by the ready availability of standard packages, especially the Gaussian code. Although periodic calculations on zeolites have been possible since the early nineties [6,7], through the also widely distributed CRYSTAL program [8], the lack of analytical first derivatives and significant computational demands has largely restricted its use to high-symmetry materials. This obviously precludes the study of complex chemical reactivity.

As a consequence of the above considerations, the dominant approach until recently has been the cluster approach in zeolite science because it was possible with the available off the shelf software. However, it is not without its own difficulties. There is always the debate concerning how large a cluster must be in order to mimic accurately the true environment of the zeolite [9]. If we consider that zeolites are at least semi-ionic materials then the interactions will be long ranged due to electrostatic potential. The rate at which the energy of interaction will decay can be approximated by the ratio of the surface area to volume. For a spherical cluster, this will give a decay with the inverse distance to only the first power — this is too slow to converge with presently feasible sizes of cluster. Beyond the issue of the size, there is also the question of how to saturate dangling bonds. In early studies, hydride moieties were used due to computational restrictions, though now hydroxyl groups are the preferred choice. This still leaves the issue of where to place the hydrogen atoms and what degrees of freedom can be allowed to vary during geometry optimisation.

Many of the problems of the finite cluster representation of zeolites can be overcome through the use of embedding schemes [10,11] — a topic that is described in detail in the next chapter. Although we will not discuss the question of embedding schemes here, there will always be an element of uncertainty still due to the influence of the boundary region. Conversely, the strength of embedded cluster methods is that they permit high-level quantum mechanical techniques to be utilised for the study of chemical reactivity. If we desire very accurate activation energies then we will be forced to resort to post-Hartree-Fock methods, such as configurational interaction, that cannot be currently applied to complex materials in the solid state.

When considering the accuracy of quantum mechanical methods, there are two key factors — the accuracy of the particular Hamiltonian being used and the accuracy with which you are representing the chemical problem that you are aiming to study. Both of these considerations can give rise to errors of similar magnitudes if not properly controlled. However, given that the majority of quantum mechanical studies currently being performed on zeolites operate within the framework of density functional theory (DFT) where there are no problems in exploiting periodicity, solid state approaches offer clear benefits in terms of realism of description of the porous environment. The next chapter will return to the consideration of cluster as opposed to periodic methods.

3. The planewave pseudopotential method

In the previous section we highlighted the virtues of periodic boundary conditions as a means for performing electronic structure calculations. However, such techniques will be of no use unless calculations can be performed efficiently on modern computers. Over a decade ago the main obstacles to this were overcome and within a few years the first calculations on zeolites began to appear. In this section we shall give a brief description of the underlying method, with emphasis on why the technique has become feasible. A more detailed description can be found in the review by Payne et al. [5].

Over the last few decades there have been two distinct strands to the development of ab initio quantum mechanics. Within molecular quantum chemistry the main approach followed was to start from the Hartree-Fock method and then to systematically improve the answers through the inclusion of electron correlation effects via techniques such as perturbation theory or configurational interaction. By contrast, the mainstay of solid-state physics was the theorem of Hohenberg and Kohn [12] that stated that the ground-state energy of a system was a function of the electron density, giving rise to DFT. Unfortunately, the exact form of this functional is unknown and so approximations must be made. Most practical DFT calculations operate within the framework of Kohn–Sham theory [13] that introduces an orbital basis, in an analogous manner to Hartree-Fock theory, so that the kinetic energy can be more accurately determined. The remaining unknown term is the so-called exchange-correlation energy. Here the starting point is to take the solution for the uniform electron gas, leading to the local density approximation (LDA).

Density functional theory was largely shunned by the chemistry community, until relatively recently, since LDA calculations tend to systematically over-bind, which is a particular hindrance when contemplating intermolecular interactions. With the advent of non-local DFT, in which the gradient of the density as well as its absolute value determines the exchange-correlation potential, the accuracy for many chemical problems was improved. This has now reached the point where this approach is competitive with the first level of post-Hartree-Fock correlated methods, in particular MP2 theory. Thus, it was the introduction of these so-called generalised gradient approximations (GGAs) [14–16] that led to the widespread use of DFT in zeolite modelling.

Beyond the choice of the particular quantum mechanical technique, there is the crucial selection of an appropriate basis set with which we express the wavefunction of the system. The relative simplicity of DFT, in comparison to Hartree-Fock-based methods, allows much greater flexibility in the implementation, in practice, if not in principle. When dealing with periodic systems a natural basis set to choose is one that already encapsulates this feature and thus leads inevitably to the planewave. Here a planewave is just a sinusoidal wave, which can be represented by $\exp(ik \cdot r)$, that satisfies Blochs theorem by containing an integral number of wavelengths within the unit cell. In essence, the wavefunction is just represented as a Fourier series. However, in principle, there are an infinite number of planewaves that satisfy this constraint and so some restriction must be introduced. We know that the true wavefunction will be a smooth continuous function and therefore low-frequency planewaves will be more important than high-frequency ones. Hence, we can define an upper bound to the frequency, or in the general parlance used a planewave kinetic energy cut-off.

One of the great strengths of the planewave basis set is that the accuracy can be systematically improved by increasing a single number, within the limits of the underlying theory of course. This makes them much simpler to work with than a Gaussian basis where systematic convergence is more complex to achieve. Furthermore, because the basis set is not associated directly with any given atom, there is no basis set superposition error (the artificial lowering of energies through the use of other atoms basis functions) which can cause uncertainties when considering intermolecular forces and is particularly relevant to studies of adsorption in zeolites. On the downside, this disconnection of the atoms from the basis set means that many of the concepts traditionally used by chemists, such as Mulliken charges and bond orders, are not naturally yielded by the calculations. However, it has been demonstrated how these may be obtained through projection onto a localised basis [17].

While the above makes the planewave approach sound ideal, there are further complications that must be highlighted. If one attempted to perform an all-electron quantum mechanical calculation with planewaves on most solids, this could easily require millions of basis functions for something approaching a converged calculation — well beyond the bounds of what would be generally possible. Hence fundamental to the use of planewaves is normally the use of a pseudopotential to represent the effective potential of the nucleus and core electrons experienced by the valence electrons [18]. This allows only the softer valence electron wavefunction to be treated, which is usually the part that controls the chemistry. Through the use of pseudopotentials the planewave kinetic energy cut-off can be reduced to the point where only several thousand basis functions are required for a small unit-cell zeolite, for example.

The details of pseudopotential construction are beyond the scope of this chapter, except to state that the aim is to reproduce the all-electron results. The extent to which this is achieved for all chemical environments is a matter of compromise. Highly transferable pseudopotentials are typically more expensive to use because they require increased characteristic planewave cut-offs. Hence there is a trade-off between accuracy and computational speed, as is often the case. The current state of the art is to use so-called ultrasoft pseudopotentials, developed by Vanderbilt [19], that typically require about half the cut-off energy of the more traditional norm-conserving variants. We should also mention the projector-augmented wave (PAW) method of Blöchl [20] that has similar benefits, while retaining the frozen core electrons. However, it has been less widely implemented and therefore more sparsely used to date.

Although the use of pseudopotentials has been around for many decades now, practical planewave calculations on complex materials were not possible until relatively recently because of the shear number of basis functions. Indeed it might be hard to understand how a calculation involving thousands of planewaves can be competitive with an equivalent localised basis function calculation which might require only a few tens to hundreds of Gaussians. The answer is for much the same reason that Gaussian functions are chosen over Slater ones — the speed of evaluation of matrix elements. In a planewave basis set, certain matrix elements are diagonal in reciprocal space, such as kinetic energy and the Coulomb energy, while others can be rapidly computed following a real-space transformation.

The key bottleneck for planewave calculations in the early days was solving for the wavefunction by matrix diagonalisation. The cost of this process scales as the cube of the number of planewaves and so rapidly becomes prohibitive, not only in CPU time, but even more so in memory usage. The major break through came when Car and Parrinello [21] proposed a combination of several developments that made large-scale planewave calculations practical. Most significantly they replaced matrix diagonalisation by allowing the wavefunction to evolve according to a molecular dynamics scheme with the wavefunctions being assigned a fictious mass, chosen so that the frequency spectrum for the nuclear and electronic parts were sufficiently decoupled. An alternative formulation of this solution, which is at least as widely used today, if not more so, is to just minimise the energy with respect to the planewave coefficients while maintaining orthogonality of the bands using conjugate gradients [22,23]. This second scheme ensures that the electronic wavefunction remains in the ground state at all nuclear configurations.

Beyond the calculation of the self-consistent field, and thus the energy, there is another reason why planewave calculations have made such rapid progress in being applied to complex systems, such as zeolites. It is relatively trivial to compute the forces on the atoms because only a few terms actually depend on the nuclear co-ordinates — a further benefit of non-atom centred basis functions. Given computationally inexpensive forces it becomes possible to optimise geometries in a straightforward manner, using either conjugate gradient or Newton–Raphson techniques (usually based on a guessed initial Hessian), and even to perform ab initio molecular dynamics when one desires to go beyond finding the local energy minimum.

The final factor that has led to the wide application of planewave techniques is the increasing availability of massively parallel computers. Because this particular section of the scientific community was quick to embrace the new technology, due to not being reliant on matrix diagonalisation which is known to parallelise relatively inefficiently, it was able to exploit the growing number of multiprocessor supercomputers more rapidly than some alternative technologies. Consequently many of the calculations performed to date in this area have been possible due to parallel computing.

4. Applications

Planewave-based methods have now been used to explore a variety of different aspects of zeolite chemistry, with the number rapidly expanding. Here we will consider a few of the problems that have been tackled to date in order to highlight what can be achieved. Of course there have been many other studies, using the same approach, of dense silica phases, especially quartz and stishovite [24–26], and also other silicate minerals, including clay materials [27]. However, the focus here will be confined to microporous materials.

4.1. Brønsted acid sites

This topic will be considered in considerable detail in the next chapter focussing more on cluster calculations. To the best of our knowledge, the first planewave pseudopotential study of zeolite to be published was due to Campana et al. [28], though several more papers began to appear in the following year. The question to be tackled in this chapter was along the lines of many previous cluster studies — the detailed nature of the Brønsted acid site in an aluminosilicate material. However, due to the use of periodic boundary conditions there was the new added dimension that the calculation was for a specific framework topology. Although there had been earlier periodic studies using localised basis functions, there were much greater restrictions in this prior work as to the extent to which the system could be relaxed and the ability to consider low defect concentrations.

For their study, Campana et al. chose the zeolite offretite which has a 54 atom hexagonal unit cell. The structure is composed of orthogonal 12-ring and 8-ring channels, with the larger channels running parallel to the *c* axis. As a consequence of symmetry there are two unique tetrahedral sites and four distinct oxygen atoms in the asymmetric unit. Prior to considering the various permutations of defects that were possible, the purely siliceous structure was optimised with the unit-cell dimensions fixed at those determined experimentally. This led to Si–O distances in the range of 1.62–1.65 Å as opposed to the experimental values of 1.59–1.72 Å [29]. Direct comparison between theory and experiment is not meaningful here though, since the experimental structure corresponds to a material where up to a third of the tetrahedral sites contained aluminium. The lower degree of scatter for the theoretical bond lengths is almost certainly realistic for the purely siliceous framework, while the experimental variation reflects the influence of the impurities, as will be further discussed later.

For offretite, Campana et al. estimated that the energy difference between the two non-equivalent tetrahedral sites for the uncompensated substitution of aluminium amounted to 15.9 $\mathrm{kJ\,mol^{-1}}$, with T2 being the more stable. From interatomic potential studies we know that the energy difference between sites is small when extensive framework relaxation is allowed for. Hence in this sense the number may represent an upper bound, given that using a supercell would permit even more lowering of both substitution energies. However, because the aluminium is normally inserted during synthesis the conditions of this process will have more control over where the impurities are sited within the framework than the thermodynamics of the final material. More relevant are the energy differences between different hydrogen-binding sites, since protons are known to be able to undergo exchange from one oxygen to another. For this material the proton affinities were found to span a range of 36.4 $\mathrm{kJ\,mol^{-1}}$, while the highest absolute value of 1253 $\mathrm{kJ\,mol^{-1}}$ lies on the low side of typical cluster calculation estimates. Subsequent studies have generally found much smaller differences in relative proton affinities. This may be because the first work was performed within the Local Density Approximation while most later works have used gradient-corrected functionals.

Since the above first study of offretite, acid sites in several other zeolites have also been studied, including chabazite [30,31], sodalite [32], gmelinite [33] and mordenite [34]. From all of these we can begin to build a coherent picture as to the performance of planewave methods for structural parameters and energetic trends. Often one of the limiting factors in assessing the validity of theoretical results for zeolites is the lack of diffraction data for purely siliceous forms which is the natural starting point for any modelling study. For one of the studied materials, chabazite, there has recently been a successful synthesis of a pure SiO_2 polymorph which makes a full comparison possible.

Detailed results for purely siliceous chabazite have been reported by both Shah et al. [35,36] and Jeanvoine et al. [31]. A comparison of the bond lengths from these two works against experiment [37] is given in Table 1. To complicate matters, different approximations were made for the unit cell of the material in the two theoretical works. In the older work of Shah et al., the cell was constrained to be that obtained from a shell model optimisation of the pure siliceous form since this approach has generally been found to be reliable for silica polymorphs with errors typically of the order of better than 2% [38]. One might wonder why the unit cell was not optimised directly with the quantum mechanical technique. There is a complication with planewaves in the optimisation of unit-cell dimensions because of the fact that the basis set quality is related to the reciprocal lattice vectors. Hence as the cell expands or contracts, so the effective planewave cut-off changes. This can now be overcome to a reasonable extent by estimating a correction based on the variation of the system energy with planewave cut-off to yield a compensating term in the stresses [39].

Table 1

Comparison of the structural properties of purely siliceous chabazite according to experiment and theory

	Experiment	Shah et al. [41]	Jeanvoine et al. [31]
r(Si–O1) (Å)	1.6030	1.603	1.617
r(Si–O2) (Å)	1.5990	1.593	1.600
r(Si–O3) (Å)	1.5987	1.595	1.615
r(Si–O4) (Å)	1.6105	1.606	1.613
Volume ($Å^3$)	779.3	766.6	792.3

In the work of Jeanvoine et al. [31], they chose to use the experimental cell as a constraint based on a highly siliceous form of chabazite, SSZ-13 [40], but not the pure form since this was only synthesised after the former work had been completed. Consequently, the cell volume was slightly too large with respect to the final experimental value due to the presence of a small, but significant, amount of aluminium. Also reported was the theoretical-optimised cell volume, which was estimated to be 807.9 Å^3. Not surprisingly, this is larger than the true value since it is apparent that gradient-corrected functionals generally overestimate unit-cell parameters by a few percent. Given that the unit-cell volumes of the two planewave studies differ from the experimental observable, it is not surprising to discover that the bond lengths are not perfectly reproduced. However, the calculated values do bracket the experimental quantities, which indicates that for matching constrained cell parameters DFT should be capable of reasonable reproduction of the structure, though in the unconstrained case we would expect a small systematic overestimation of bond lengths.

Turning now to consider the influence of the introduction of aluminium into the framework with a proton charge compensating on an adjacent oxygen, there is a considerable amount of theoretical information available for the zeolites previously mentioned. Unfortunately, it is very difficult to obtain experimental corroboration for all of this detail because of the disordered nature of most defects. What is known is that for aluminosilicates there is a wide variation in the T–O bond length distribution, such as that highlighted for offretite earlier, which extends from values below those of typical Si–O bond lengths to those greater than for a typical Al–O bond. Theoretical studies have been able to verify the description of how these extremes arise.

Let us take just one example, that of gmelinite as recently studied by Benco et al. [33]. Here the bond lengths for the purely siliceous material range from 1.624 to 1.638 Å, which is a particularly narrow range due to the fact there is only one symmetry unique T site. On the introduction of aluminium, this yields Al–O distances of 1.684–1.718 Å to unprotonated oxygens. The longest bond lengths are found for the bridging hydroxyl group where the ranges are 1.703–1.715 and 1.877–1.923 Å for Si–OH and Al–OH, respectively. The presence of the proton clearly accounts for the existence of long T–O bonds in the diffraction average, but this leaves the question of how the distances shorter than for the purely siliceous material arise. Inspection of the second co-ordination sphere of tetrahedra reveals that the Si–O bonds are compressed to values between 1.58 and 1.61 Å due to the presence of the larger aluminium. While the effect may be overestimated here due to the particular unit-cell constraints chosen, it is likely that this is the source of the short distances.

A further aspect of Brønsted acid sites that has been addressed through the application of planewave calculations is the relative acidity of different materials, in particular when comparing materials with the same topology, but different chemical composition. Such an example is again provided by the high-silica form of chabazite, H-SSZ-13, and its silicoaluminophosphate analogue, H-SAPO-34. In the latter material the acid site is created by isomorphous replacement of phosphorous by silicon, again leading to the necessity of a charge-compensating proton.

Shah et al. [41] have studied the properties of the four possible proton-binding sites at the oxygens that surround the framework impurity for both H-SSZ-13 and H-SAPO-34. The results of these calculations are summarised in Table 2. The same two systems were later also studied by Jeanvoine et al. [31] who found values broadly consistent with those that had gone before, but with some quantitative differences that at least in part arise from the different choice of unit-cell dimensions.

Comparing the relative energies of the proton binding at the four unique oxygens, we see that the first two sites, in order of stability, are the same for both materials with O1 being the more favourable one. However, in all cases the energy differences are of the order of only a few $kJ\,mol^{-1}$ which is certainly at the limit of the accuracy of the under lying method. Furthermore, these values are of the same order of magnitude as thermal energies at standard conditions and therefore must be considered negligible. Preferences for particular sites could also

Table 2
Calculated properties of the Brønsted acid sites for H-SSZ-13 (Z) and H-SAPO-34 (S) and experimental values [89,40] for the hydroxyl-stretching frequency for comparison

Proton site	Relative energy ($kJ\,mol^{-1}$)	r(O–H) (Å)	Harmonic frequency (cm^{-1})	Anharmonic frequency (cm^{-1})	Observed frequency (cm^{-1})
O1 (Z)	0	0.972	3845	3590	3603
O2 (Z)	7	0.971	3840	3580	3579
O3 (Z)	4	0.973	3825	3565	
O4 (Z)	9	0.972	3830	3570	
O1 (S)	0	0.970	3870	3610	3625
O2 (S)	5	0.970	3860	3600	3601
O3 (S)	2	0.971	3860	3600	
O4 (S)	4	0.970	3865	3605	3630

be easily altered by the interaction between neighbouring acidic defects as they will not be truly periodically repeating. With energy differences of this magnitude it would be necessary to consider the dipole–dipole coupling energy of repeating images to be more precise. In short, it could be considered beyond the accuracy of present day simulations to a priori predict a site preference, when the energy differences are so small.

Recently, a study of the relative binding energies of the four different oxygens in faujasite was performed using DFT using a localised numerical basis set [42]. These results again indicate that only small differences exist, with the values spanning just less than 10 $kJ\,mol^{-1}$ from a GGA calculation using an LDA-optimised geometry. Interestingly, they also tabulate estimates for the same system from other techniques, including an embedded cluster study [43] which predicts differences up to three times as large, though the order is consistent. This emphasises again that the precise answer that is obtained in zeolite problems may be as much dictated by the atomic representation of the problem as by the Hamiltonian.

The theoretical vibrational frequencies determined for the two materials with the chabazite structure were calculated using the numerical finite-difference approach for the hydroxyl group alone. This assumes that the hydroxyl-stretching mode is sufficiently decoupled from the other framework vibrations, which is a reasonable first approximation due to the mass of the proton. However, this procedure does mean that there will be some degree of error in comparison to a full phonon calculation, which would now be possible through the linear response approach [44]. It can be seen from the results that it is imperative to allow for anharmonicity if a meaningful comparison to experimental values is to be made from lattice dynamical calculations on modes such as this. Apart from determining the anharmonic corrections through fitting the energy surface, it is also possible to directly obtain anharmonic frequencies via ab initio molecular dynamics, as will be illustrated later. However, this approach is more time consuming and is complicated by the issue of statistical noise.

While there is some correlation of the calculated frequencies with the tentative experimental assignment of values, this is clearly not perfect. Again, given the uncertainties in the values relative to the degree of separation, it would be hard to make a definitive assignment.

One thing that does emerge from the comparison of the aluminosilicate and silicoaluminophosphate is that there are some small systematic differences. In particular, the hydroxyl bond length is longer in the aluminosilicate and the anharmonic-stretching frequency is lower. Furthermore, the projected charge of the proton is higher for the zeolite (+0.55 vs. +0.53). All this suggests that H-SSZ-13 should be more acidic than H-SAPO-34 as a consequence of its chemical composition. Of course acidity can be calibrated in many ways leading to different scales, especially with regard to whether one considers loss of a proton to effectively infinite separation or just the local transfer to an adsorbed molecule, where the electrostatic potential in the cavity will play a role. Evidence from the adsorption of hydrogen-bonding molecules, which will be discussed later, also supports the greater acidity of zeolites over aluminophosphates for a given framework topology and defect concentration. More importantly, this appears to be borne out by experimental observation.

4.2. Extra-framework cations

A natural extension of the study of zeolite acid sites is to consider the effect of exchanging the proton for a metal cation. There have been a small number of studies to date using planewaves in this area and most have concentrated on sodium or potassium, since these are the predominant species found in many zeolites. There are, of course, several successful studies of cation location in zeolites using interatomic potential-based methods as discussed in Chapter 9.

Filippone et al. [45] performed a combined ab initio molecular dynamics and experimental study of sodium-containing sodalites. Three different materials were studied, one with a composition of $Na_6(Al_6Si_6O_{24})$, one of the naturally occurring form, chloro-sodalite ($Na_8Cl_2(Al_6Si_6O_{24})$), and finally a hypothetical structure in which all the sodium is removed, while retaining the framework aluminium, but without a negative charge. This final material, which is unlike any real zeolite, was found to be metallic and locally resilient with respect to collapse, even at finite temperatures, which the authors claim demonstrates the intrinsic stability of the framework without the steric influence of the cations. The structural trends of framework distortions with composition were reproduced satisfactorily, within the limitations of the local density approximation, as were the general features of the experimental infrared and Raman spectra. Subsequently, the sodalite family of minerals has received further attention for the sodium hydroxosodalite dihydrate form [46,47], though here the emphasis is more on the properties of the hydroxyl groups and the influence of isotopic substitution of the framework.

Cation substitution in offretite has also been studied for both potassium and sodium [48]. The site preferences for both cations were examined over a range of different initial starting positions and for aluminium situated in both of the two symmetry unique tetrahedral sites. Potassium is found to have a definite preference for the relatively inaccessible cancrinite cages, in accord with experiment, with aluminium specifically located at the T1 site. In contrast, sodium has several possible binding sites all within ambient thermal energy of each other, with aluminium mutually exclusively doped on both sites. The overall pattern found is that as the size of the cation increases, so does the degree of specificity for particular sites.

The first paper to appear concerning alkali metal cations in zeolites [49] was actually rather novel and concerned a more unusual material than a routine alkali metal zeolite. It has been observed experimentally that several zeolites undergo a dramatic transformation in properties when alkali metals are adsorbed [50]. For instance, dehydrated sodium zeolite-Y on treatment with sodium vapour turns a deep red colour [51]. Evidence from ESR suggests that this is due to the formation of Na_4^{3+} clusters based around the sodalite cages.

Ursenbach et al. [49] wished to validate this hypothesis through performing ab initio molecular dynamics. However, at the time, it was not feasible to do this for the relatively large unit cell of zeolite-Y. Hence they employed a combination of techniques in which the sodium atoms and their associated electrons were treated using DFT, including an effective pseudopotential due to the framework oxygens, while the rest of the system interacts via interatomic potentials. Their work confirms that the excess electron is indeed delocalised at the centre of the sodalite cage between four sodium cations. This highlights another strength of the planewave approach — when dealing with such delocalised electrons it would have been necessary to include floating functions in a localised basis set description in order to obtain the correct result. However, without the advance knowledge of the solution it may have been difficult to select the location and extent of these functions. Interest in such electride-like systems continues because of their novel properties and band structure calculations have been performed on a sodalite system with Self-Interaction Correction [52].

4.3. Adsorption of polar molecules

Over the years there has been one topic that has been the subject of intense interest within the field of quantum mechanical studies of zeolites and that has been concerning the nature of adsorbed polar molecules at Brønsted acid sites. This began with many cluster studies, performed at different levels of approximation from semi-empirical methods to post-Hartree-Fock ab initio and for a wide variety of model sizes as computer power evolved [53]. Inevitably this has carried on with the arrival of planewave techniques in the field, though now there is a high level of agreement on many aspects that did not exist in the days of purely cluster-based calculation.

The principal topic of debate in this arena has been concerning the bound state of polar molecules, and in particular water, methanol and ammonia, at acid sites. All three entities are capable of acting as both hydrogen-bond acceptors and donors. This gives rise to the possibility that the molecules may physisorb, through the formation of a six-membered ring complex with two hydrogen bonds, or that proton transfer may occur to form a chemisorbed cationic species with a similar structure. Both of these possibilities are schematically illustrated in Fig. 1. The question that has been posed is whether these two states both exist as local minima on the energy surface in equilibrium with each other, or whether one extreme or the other is the only possible form.

Rather than trying to discuss all three cases together, we will first consider the case of methanol, which has been the most widely studied because of its relevance to catalysis, as we shall see in the next section. Furthermore, it turns out that methanol is perhaps the most finely balanced of the trio of species mentioned, since its basicity

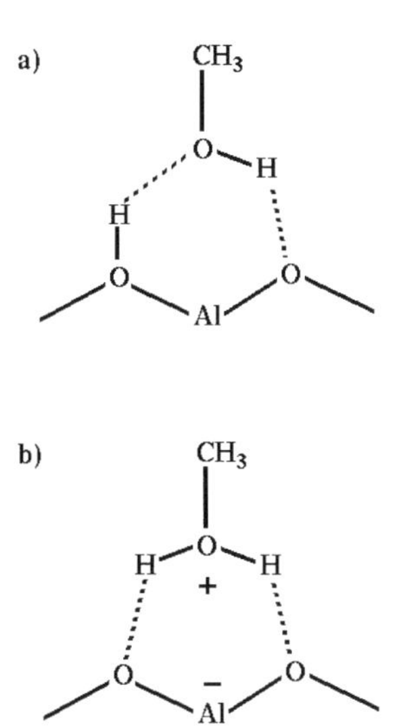

Fig. 1. Schematic illustration of methanol adsorbed (a) in a physisorbed state and (b) in a chemisorbed state as a methoxonium cation.

is intermediate between that of the other two, though only slightly greater than that of water.

Cluster studies have indicated at various times that both methanol and methoxonium were the only minimum. However, it was discovered that the preference for one form or the other could be altered according to the choice of zeolite fragment and the constraints placed upon it. For instance, it appears that constraining a cluster model, consisting of three tetrahedral sites, to have a mirror plane of symmetry in one plane for the binding of methanol and at right angles for methoxonium, is enough to reverse the order of stability compared to an unconstrained calculation [54]. This resulted in making methoxonium the stable species. Subsequently it has become feasible to run cluster calculations without the need for symmetry [55], which was used originally out of computational necessity, and it now appears that physisorbed methanol is always the local minimum. The methoxonium species instead represents the transition state for proton transfer between two oxygens adjacent to aluminium.

Despite the increasing size of cluster calculation that is feasible, even today, it is not possible to be sure that a model is large enough to achieve a converged result. In particular, one can never be sure that the electrostatic potential in the region of the acid site is correctly reproduced and this may be critical to the issue of whether the proton can transfer or not to molecules of intermediate basicity, such as methanol. Hence, planewave calculations have a valuable role to play here as a further check.

The first examination of the issue of the adsorbed state of methanol to appear using planewaves was the work of Nusterer et al. (1996), who performed first principles molecular dynamics within sodalite. This work used the PAW method, rather than pseudopotentials as utilised in most of the other works, since to represent the core electrons and nuclei. Sodalite is a hypothetical system for methanol adsorption since it is actually too large to diffuse into the material in practice. However, it is computationally convenient, having a small cubic unit cell, and obviously contains sodalite cages which occur as a structural motif in more open porous materials that were beyond the scope of what was possible at the time.

These first ab initio dynamical calculations in this area revealed something hither to unseen in cluster studies. Prior to this study it was always assumed that methanol would hydrogen bond across one edge of the tetrahedron-containing aluminium. However, during the dynamics Nusterer et al. [92] found that methanol rotated so as to hydrogen bond across the six-ring as illustrated in Fig. 2. This structure

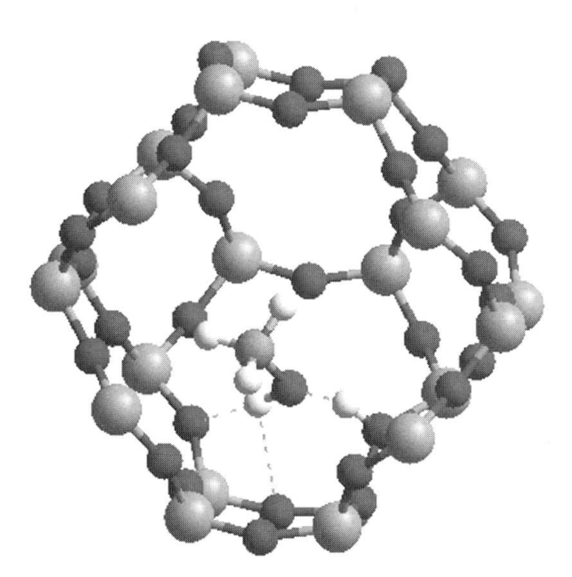

Fig. 2. Methanol adsorbed in sodalite (Nusterer et al., 1996), depicting the more stable configuration in which methanol hydrogen bonds across a six-ring, rather than across the aluminium defect site.

was estimated to be roughly 10 $kJ\,mol^{-1}$ more stable than the original starting configuration. Clearly in cluster models of the day this minimum could not have been located as it is specific to the particular topology of sodalite, though is indicative of a general principle that will be more widespread. Furthermore, it would have required more tetrahedra to be explicitly included than was generally the case in that era.

Concurrently with the above work, Shah et al. [56] were studying methanol adsorption in a different zeolite, that of chabazite. In common with sodalite it possesses a relatively small unit cell of 36 atoms, though in this case it is slightly distorted away from cubic to give rhombohedral symmetry. The framework topology of this material is quite different though, in that it possesses elliptical cavities formed from double 6-ring units and connected via 8-ring windows along all three crystallographic axes. Chabazite is also known to readily adsorb methanol and importantly acts as a catalyst for its conversion to other products [57].

In this initial study of methanol in chabazite only static calculations were initially performed. However, several distinct local minima were discovered. Most striking of these minima, with the exception of the unstable minimum in which the methyl group is closest to the framework, is the one in which methoxonium is found to be stable

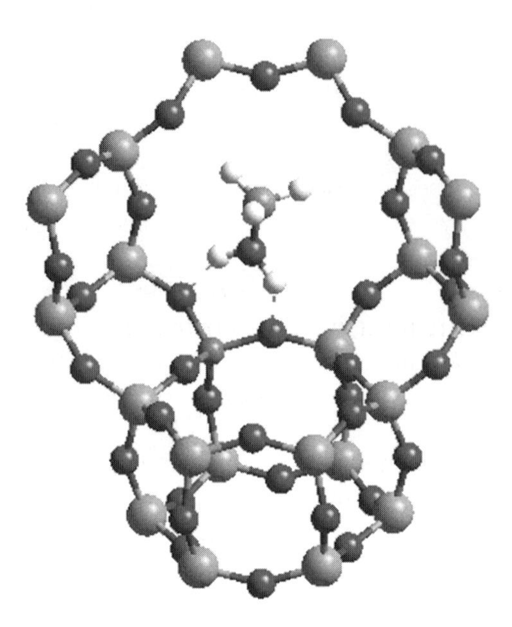

Fig. 3. The metastable structure of the methoxonium cation within the 8-ring window of the chabazite framework [56].

(Fig. 3), rather than methanol. Furthermore, this minimum was found to be the lowest in energy of the four located. The existence of this charge-separated local minimum can be ascribed to the fact that the configuration lies within the plane of the confined 8-ring window where the electrostatic field will be more extreme than in the open-cage environment of sodalite. Initially this seemed to suggest that the balance between the two modes of adsorption could be so finely balanced that it depended on the particular microporous material in question.

The case of methanol in chabazite was subsequently revisited by Haase et al. [58] using ab initio molecular dynamics, rather than energy minimisation. While they confirmed that the methoxonium cation was indeed a local minimum on the energy surface, it was not actually the global one. A configuration lying about 18 $kJ\,mol^{-1}$ lower in energy was identified that had been missed in the original static study. Here the methanol remained unprotonated, though with one very short hydrogen-bonded distance of 1.283 Å to the framework proton, whose hydroxyl bond length was concomitantly stretched to 1.177 Å. Although the methanol was still predominantly within the 8-ring window, this lowest minimum located to date differed from the methoxonium configuration by forming an 8-ring hydrogen-bonded geometry, rather than a 6-ring one. This extends further the trend seen for sodalite, suggesting that such polar molecules have a preference for not bridging directly across aluminium.

This case study of methanol adsorption in zeolites emphasises the impact that ab initio molecular dynamics has had in this field. When performing energy minimisation one is to some extent limited by the preconceptions of the moment, particularly in such complex situations. The arrival of feasible molecular dynamics has allowed the barriers to be overcome, in more sense than one, and can reveal unexpected findings. However, it must be remembered that this is just a partial solution, since only real times of the order of a few picoseconds are currently accessible, which is insufficient for molecules to explore the full conformational space.

The range of zeolites in which methanol adsorption has been studied has continued to expand and now also includes ferrierite, silicalite (ZSM-5) and theta-1, all of which have been studied using ab initio molecular dynamics. Based on the work of Stich et al. [59], we can examine the distribution of O–H distances averaged over the length of the dynamics for four simulations in three different zeolites (Fig. 4). Here the two sets of data for ferrierite correspond to runs in which methanol was started either in the 8-ring side channel or in the main 10-ring channel. Examination of these plots indicates that the issue of the state of methanol is even more complex than first imagined and indeed that trying to say what state the molecule is in may not always be a meaningful question. In the case of chabazite there are two broad peaks that are characteristic of both extremes, while ZSM-5, which has a larger channel aperture, shows a peak predominantly associated with the presence of methanol. The particularly interesting case is the 8-ring channel of ferrierite which shows mainly one peak characteristic of methoxonium. If the trajectory of the simulation is followed, the initial configuration in which methanol is hydrogen bonded to the framework across aluminium first acquires the proton and then undergoes a 90° rotation. This creates a methoxonium species that hydrogen bonds across the channel that persists for the remainder of the simulation as shown in Fig. 5. In effect this takes the previously observed behaviour for sodalite and the global minimum of chabazite to a new extreme, where the hydrogen bond is not just across a single ring, but the flexibility of the framework allows it to be accommodated across a narrow channel.

Recently, Haase and Sauer [60] have also examined ferrierite with methanol adsorbed and found no stable methoxonium species. They postulate that this may be due to the use of a supercell in one direction where there is a particularly short lattice parameter. However, this had

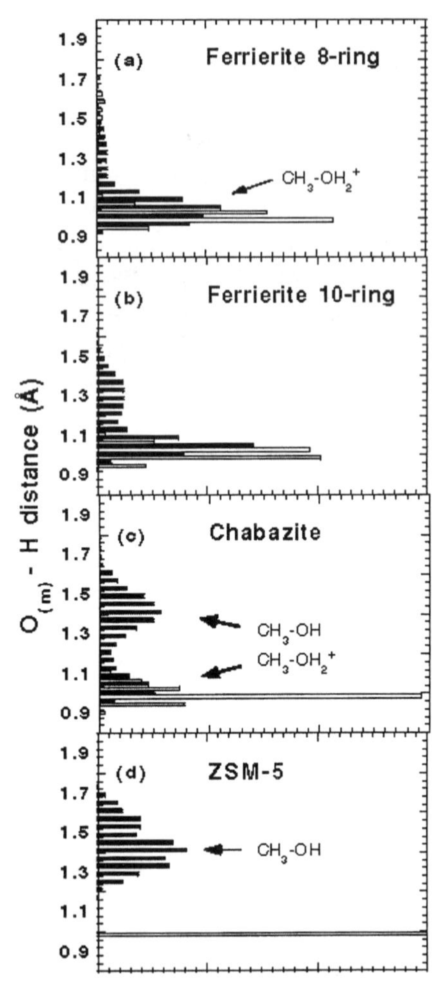

Fig. 4. Bond length distributions of the O–H distances for a single methanol adsorbed in (a) the 8-ring channel of ferrierite, (b) the 10-ring channel of ferrierite, (c) chabazite and (d) H-ZSM-5 [59]. The distributions in each panel correspond to the two protons.

already been discounted in the previous paper of Stich et al. [59] where a supercell calculation showed no change in the behaviour of the system with respect to a single cell. The most likely cause is the use of different initial configurations, which highlights the limitations of the timescales accessible to ab initio molecular dynamics. Hence, the true nature of methanol in ferrierite may remain uncertain for some time to come. In reality the debate is probably academic, since under typical conditions of commercial significance the loading of methanol is much higher which leads to different properties again, as we shall see later.

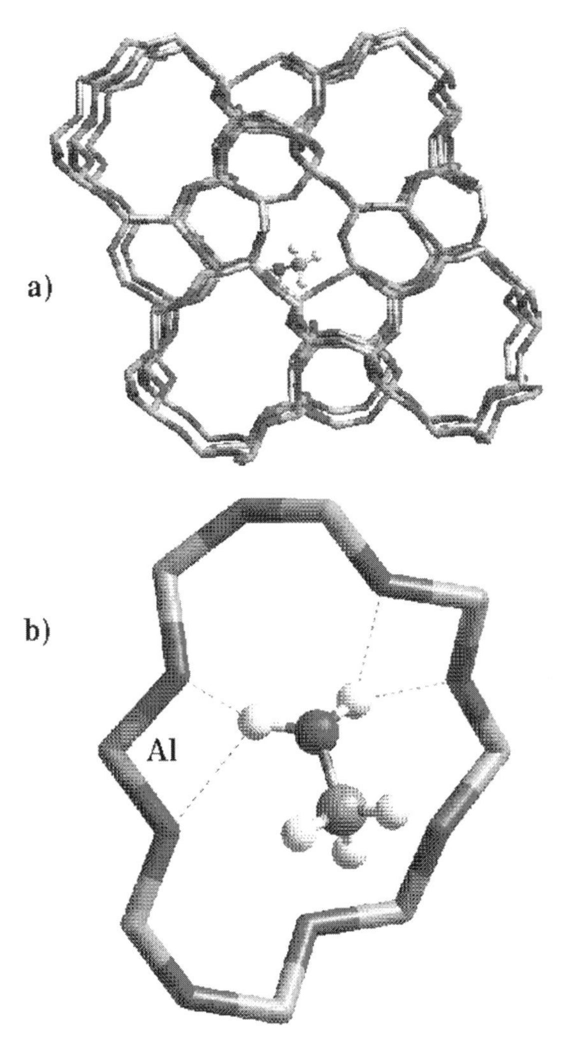

Fig. 5. Example of a typical configuration from the ab initio molecular dynamics of methanol in ferrierite showing the methoxonium cation bridging the 8-ring channel. Two different views are given showing (a) the position of methoxonium when looking along the channel system, and (b) the local environment within the channel.

One thing that is clear from all the planewave dynamical simulations to date is that protons appear to be very mobile, even on a picosecond timescale, when polar molecules are adsorbed. It is likely that the timescale for such processes is a little too rapid in comparison to reality, since density functional theory tends to underestimate activation barriers. Recent NMR experiments on chabazite [61] suggest that the frequency of proton exchange for water is only of the order of

$10–10^3\ s^{-1}$, which translates to a significantly greater barrier for hydrogen migration. Hence there are still some issues to be addressed here.

One quantity that can be compared against experiment more readily than structural detail is the heat of adsorption for methanol at low coverage. A collection of reported theoretical values are given in Table 3 for a range of materials. For consistency we have taken values that are uncorrected for zero-point energy, as this represents the majority of published values. Also they are strictly speaking for absolute zero, which is obviously not the case for experimental measurements. Hence these factors add a degree of uncertainty to the theoretical numbers, though the combined magnitude of the errors will be small compared with the scatter of values and will systematically decrease the binding energies.

Experimentally, the best estimate of the heat of adsorption of m thanol is for ZSM-5 with a fractional coverage of 0.8 of all acid sites [62]. Here the value obtained is $-115 \pm 5\ \mathrm{kJ\,mol^{-1}}$, which is in the region of the energies of adsorption for several zeolites as obtained from planewave calculations. If we specifically consider the prediction for ZSM-5 then the theoretical adsorption is too strong, though the zero-point and thermal corrections will improve matters. In addition, the silicon to aluminium ratio for the experimental measurements was almost certainly much lower which will influence the comparison.

If we compare different theoretical approaches it can be readily seen that periodic planewave calculations consistently give much more exothermic binding energies than those obtained from cluster

Table 3

Predicted energies of adsorption for methanol within zeolites according to planewave and cluster calculations

Zeolite	Method	Energy of adsorption ($\mathrm{kJ\,mol^{-1}}$)	Reference
Chabazite	PW91/Planewave	−85	[63]
Chabazite	PW91/Planewave	−129	[60]
Ferrierite	PW91/Planewave	−121	[60]
Theta-1	PW91/Planewave	−131	[60]
ZSM-5	PW91/Planewave	−139	[60]
3T cluster	BP/Gaussian	−70	[90]
3T cluster	BLYP/Gaussian	−64	[91]
3T cluster	HF/Gaussian	−49	[55]
3T cluster	MP2/Gaussian	−79	[55]
Embedded cluster	BLYP/Gaussian	−115	[64]

calculations, in some cases by a factor of almost two. The lower energy for binding in the case of chabazite according to Shah et al. [63] was in part due to the fact that this was not the lowest energy minimum for this system, though methanol was still less strongly bound than in the work of Haase et al. [58] for the equivalent minimum. Adsorption energies from clusters involving three tetrahedral sites tend to be closer to the strength of binding seen for methanol in silicalite (i.e. with no acid sites) where the heat of adsorption is -65 ± 10 kJ mol^{-1} [62]. Embedding of the cluster in order to account for the long-range electrostatic potential greatly improves the performance of cluster techniques, bringing the energies more into line with those from periodic methods as shown by the work of Greatbanks et al. [64]. Finally, for this topic, we should note that uncertainties due to the particular choice of quantum mechanical technique or density functional can be significant, as illustrated from cluster methods.

Perhaps the most direct comparison that can be made between theory and experiment comes from vibrational spectroscopy. Although it is difficult to accurately reproduce the intensities for infrared and particularly Raman spectra, the determination of peak positions is more routine. There are two approaches that are widely used for calculation of vibrational spectra. The first is to either analytically or numerically determine the dynamical matrix for the system and thereby to obtain the frequencies of oscillation about the equilibrium geometry. This is appropriate for well-localised minima, though corrections for anharmonicity will still often be necessary, especially for hydrogen-stretching modes. The second approach is to Fourier transform the velocity auto-correlation function from an ab initio molecular dynamics run, which directly accounts for the anharmonicity, and is suitable for situations where the energy surface contains multiple minima that are thermally accessible.

The infrared spectrum for methanol adsorbed on H-ZSM-5 can be regarded as containing four main features. The O–H stretch of isolated methanol that occurs at 3680 cm^{-1} vanishes and is replaced by bands at 3545, 2993, 2440 and 1687 cm^{-1}, that are believed to be associated with the hydroxyl moiety [65], plus the expected modes from the methyl group, which are slightly shifted and broadened by a weak interaction with the framework. In other materials the spectrum can differ. For example, in H-mordenite bands are observed at 2700–3100, 2400–2600 and 1700–1750 cm^{-1} — i.e. all the hydroxyl modes are more strongly shifted [66].

For the local minimum in which methoxonium exists within chabazite, the vibrational frequencies have been obtained statically [56].

Here the hydroxyl-stretching modes were found to be at 2009 and 2757 cm^{-1}, with the H–O–H bend at 1583 cm^{-1}. While this fails to explain the modes observed in quantitative fashion, it demonstrates that very large shifts relative to gas-phase methanol should be expected. It also indicates that the mode at 3500 cm^{-1} seen for H-ZSM-5 is unlikely to be associated with methoxonium. From subsequent work, we now know that there are several minima that are thermally accessible to this system and that methanol represents the dominant adsorbed state. Hence, the method of choice for vibrational calculations on such systems is actually via ab initio molecular dynamics.

In the case of methanol adsorbed in sodalite, Nusterer et al. (1996) extracted frequencies from their dynamical simulations by using the elapsed time between points of zero velocity for the hydrogen atoms alone. They obtained two main bands for the hydroxyl modes, one very broad, one from 1600 to 2600 cm^{-1} and one at 3100–3600 cm^{-1} that has a distinct maximum in the region of 3500 cm^{-1}. This later band was assigned to the O–H of methanol, while the much broader peak at lower frequency was found to project on to the zeolite proton stretch and the $H \cdots O(CH_3)$–H intermolecular bending mode. In general, the spectrum obtained is much more consistent with that seen experimentally for H-ZSM-5, than the modes from methoxonium. Hence it can be proposed that the vibrational evidence is indicative of physisorbed methanol, though the variations in spectra between materials means that this may not be true of all environments.

Stich et al. [59] examined the power spectrum resulting from ab initio molecular dynamics for a number of different zeolites, including chabazite, ferrierite and H-ZSM-5. The result for two molecules in ferrierite, in which one is present as methanol and one as methoxonium, is shown in Fig. 6 as an illustration of what is typically obtained. Because of the short lengths of time for which it is possible to run such simulations and the high degree of proton anharmonicity, the power spectra obtained are invariably extremely broad. Indeed methanol which is adsorbed in active regions of the zeolite, where it is readily protonated, are likely to be lost in the background, such is the spread of frequencies. Contrastingly, methanol that is localised, and likely to be less reactive, will yield sharper, better-defined vibrational peaks. As the timescales that are feasible increase, so will the resolution of predicted vibrational data. For now we can simply state that the large shifts of hydroxyl modes have been reproduced by planewave calculations and that the zeolite proton is the atom that undergoes the greatest degree of perturbation.

To complete our discussion of methanol adsorption as an illustration of how polar molecules may be studied in zeolites, we should consider

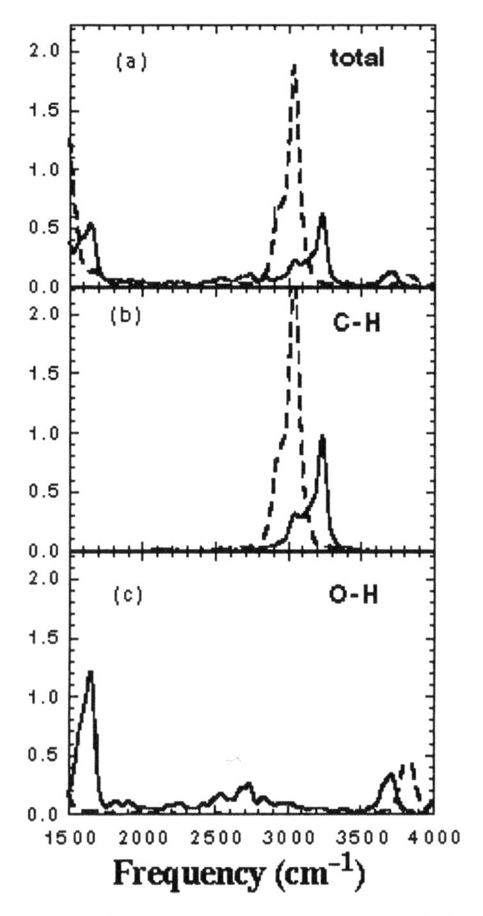

Fig. 6. The power spectrum for a methanol molecule (dashed line) and methoxonium cation (solid line) within ferrierite indicating the vibrational frequencies determined according to the velocity autocorrelation function.

the influence of loading on the behaviour of the system. While there is still some debate about whether methoxonium species can be formed at loadings of up to one molecule per acid site, all planewave calculations appear to concur that proton transfer happens without a barrier once there is a second molecule. Here the methoxonium cation is found to possess one hydrogen bond to the framework and one to the second methanol as shown in Fig. 7. The second methanol also forms a hydrogen bond back to the framework, though dynamical simulations suggest that this is relatively weak as it is seen to jump between different oxygens rather rapidly. This configuration is reminiscent of the chains of hydrogen-bonded methanols, proposed by Mirth et al. [65], to explain

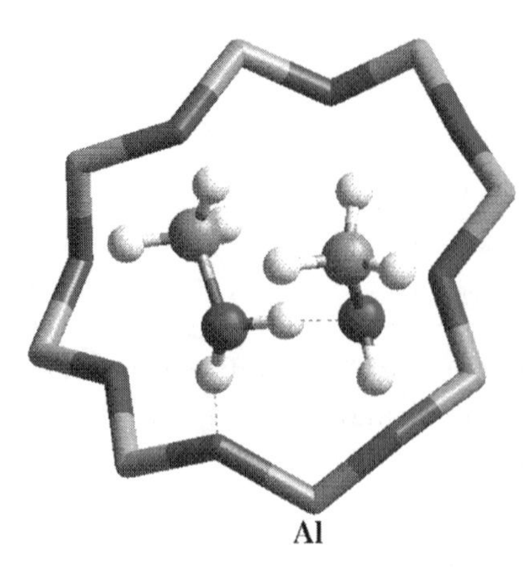

Fig. 7. A typical configuration for a loading of two methanols per acid site in chabazite, indicating the formation of methoxonium that is solvated by both the framework and the second methanol.

their NMR data. On the timescale accessible to current planewave calculations, no exchange of the proton between the framework and methoxonium is seen once it is solvated by a further methanol molecule.

If the simulation of loadings of methanol is taken to even higher levels then further interesting observations can be made in such dynamical calculations. The highest loading studied to date is that of four methanol molecules within ferrierite at a temperature of 700 K, which is in the region of typical temperatures used for the reactions of methanol in zeolites [67]. Under these conditions it is found that the carbon–oxygen bond of methanol shows very anharmonic behaviour with the mean bond length over the course of the simulation being 1.6 Å. This compares to a gas-phase equilibrium value of 1.45 Å. Even more striking is the fact that for short periods of time it is excited sufficiently to vibrate to a length of 1.8 Å. All of this suggests that the combination of protonation to form methoxonium, plus the solvation by further methanol molecules within the zeolitic environment is enough to render the C–O bond more susceptible to nucleophilic attack. If confirmed, this would offer a partial explanation for the catalytic influence of zeolites. This is a point we will consider further in the next section.

Having considered methanol in some detail to illustrate what progress has been made through the use of planewave techniques, brief mention should be made that water has been studied in a parallel fashion, though less extensively. Many of the issues that have been debated are identical, for instance whether water can exist as hydroxonium within a zeolite. Given the foregoing discussion for methanol and the relative magnitude of their basicities, it will be no surprise to learn that water is found to exist purely as a physisorbed hydrogen-bonded complex at coverages of up to one water per Brønsted acid site [93].

Water adsorbed within the confines of H-SAPO-34 turned out to be a particularly significant case in the history of planewave studies of zeolites. After a range of cluster calculations had all demonstrated that water was unprotonated for zeolite models, only for a neutron diffraction study [68] to have located both water and hydroxonium within the aforementioned material, a comment was famously published under the banner "Quantum mechanics proved wrong" [69]. This article drew the sweeping conclusion that the failure to predict the formation of hydroxoxium meant that such studies were fundamentally flawed. However, the planewave calculations of Termath et al. [70] (and also those of Jeanvoine et al. [71] who published concurrently) indicate that the demise of quantum mechanics may have been somewhat exaggerated. They demonstrated that with a loading of at least three water molecules per unit cell it is possible to form $H_3O^+(H_2O)_n$ clusters within H-SAPO-34, and therefore it is quite feasible for water and hydroxonium to co-exist in the cages of this material. Consequently theory is quite consistent with the picture that comes from the neutron diffraction average.

In general, it appears that water, just like methanol, will protonate when the loading exceeds two molecules per acid site in a zeolite on the basis of results for the sodalite system [93]. However, as we have already seen, silicoaluminophosphates are less acidic than zeolites of the same topology which is borne out by the fact that an extra molecule of solvating water is needed to stabilise proton transfer. Other forms of hydrated zeolites have also been studied using the planewave pseudopotential technique, including those with hydroxyl groups present, such as sodium hydroxo sodalite dihydrate [46,47].

From the foregoing examples, it is apparent that density functional theory, allied to planewave techniques, is a powerful tool for the study of the adsorption of polar molecules in zeolites. If we had to select the key feature that has had the greatest single impact, it would have to be the ease with which it has become feasible to perform ab initio molecular dynamics. Through offering a means to look beyond the expected result by sampling a wider section of the energy surface it has revealed a number of new aspects of molecule–zeolite binding. The remaining challenge is to be able to study molecules where dispersion is the principal cohesive term. Fortunately, the adsorption of many such molecules, including hydrocarbons, can be treated surprisingly well with interatomic potentials. Thus, it only becomes an issue for studies of hydrocarbon reactivity.

4.4. Chemical reactivity

The ultimate aim of many theoretical zeolite studies is to understand the mechanisms for catalysis. Although the majority of work to date has been concerned with the adsorbed state of reactive molecules, as a necessary prerequisite, progress is now being made in determining reaction pathways. Here the planewave pseudopotential technique can be just as effective as it has been for the study of adsorption of polar molecules. Indeed, because of the greater spatial extent of most transition states it becomes even harder to define a cluster that is adequate to describe the encapsulating environment.

One process has again dominated the theoretical literature to date in this area, which is the methanol to gasoline (MTG) reaction [72]. It was found that if methanol is passed through an acidic zeolite under the correct conditions then a series of chemical transformations occur. The first product to appear is always dimethyl ether, as a result of the condensation of two methanol molecules. However, after an induction period, C_2 products begin to appear and ultimately hydrocarbons in the gasoline fraction. This has become the basis of a commercial process, though admittedly one which is not currently widely used, where the chosen catalyst is H-ZSM-5.

Despite extensive experimental work, the sequence of reactions that form the MTG process are still not fully understood because of the difficulties of probing transient intermediates within the zeolitic environment while being certain that the observed species is actually on the mechanistic path. Quantum mechanical methods can, in principle, map out all the intermediates and the activation energies that link them, thus giving direct information about how the reaction proceeds. In practice, this is currently a challenging task and so there is some way to go to achieve this aim.

Much of the theoretical work to date has been centred on the initial transformation of methanol to dimethyl ether, since this is likely to have the lowest activation barrier being the first formed product. On the basis of experiment, there are two proposed routes by which this reaction might occur. In the first, methanol (or methoxonium) is nucleophilically attacked by an oxygen of the zeolite framework adjacent to aluminium, displacing water to yield a framework methoxy-group [73,74]. This species, which has been believed to have been observed spectroscopically [75], then reacts with a second methanol to form dimethyl ether. In the second pathway, the zeolite acts largely as a spectator to two methanols directly undergoing a nucleophilic displacement of water, though it does provide a proton and solvates the ejected molecule.

The above two pathways were first studied in detail by Blaszkowski and van Santen [76,77] using the cluster approach. They found that the route involving formation of the framework methoxy-groups was disfavoured and that direct reaction represented the pathway with the lowest overall activation energy. The intermediates for both mechanisms were also characterised using the planewave approach for chabazite, which is also an active catalyst for the MTG reaction, by Shah et al. [63]. The sequence of intermediates for both routes is illustrated schematically in Fig. 8, along with the energy differences between species. Although the activation energies were not determined in this work, the order of stability of the intermediates again favours the direct condensation of methanol. While the main conclusion was consistent with the cluster work, there are differences in detail, such as to the precise structure of some intermediate configurations, and the adsorption energies are systematically lower for the periodic study. In their work, Shah et al. also examined the stability of other reactive

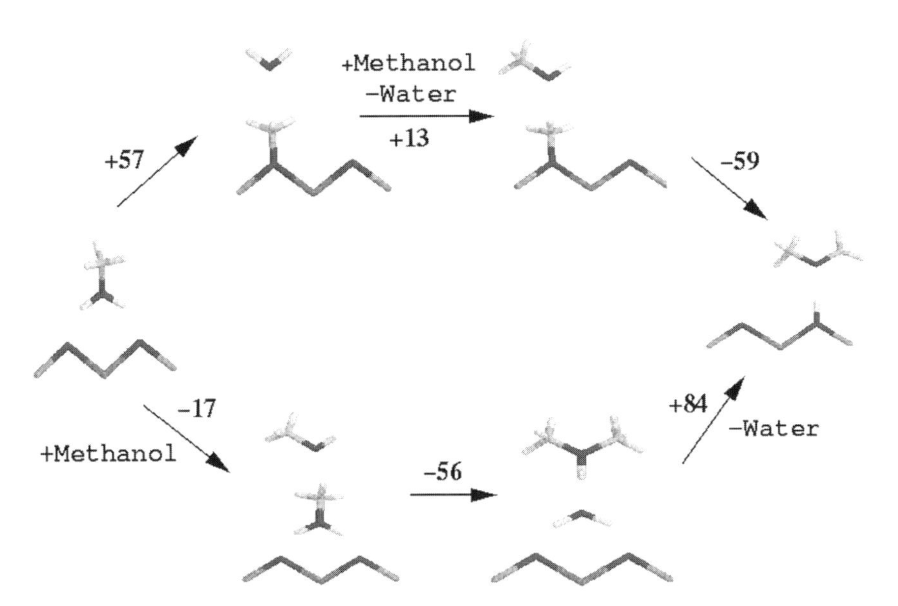

Fig. 8. Schematic illustration of the intermediates along two possible pathways for dimethyl ether formation [63]. Indicated are the energies of reaction (in $kJ\,mol^{-1}$), uncorrected for zero-point energy effects, in the direction given by the corresponding arrow.

intermediates that have been proposed to exist within a microporous material on route to gasoline. In particular, the formation of a carbene from the framework methoxy-group and of an ylide from the trimethyl oxonium cation were examined. Despite exploring several configurations under favourable conditions for trapping local minima of the desired form, neither the carbene nor the ylide were found to be stable. The carbene always abstracted the framework hydrogen to form a methyl group again and the ylide likewise protonated itself to return to the trimethyl oxonium cation. Hence while it is impossible to rule out these species completely at this stage, it seems unlikely that they have even a transient existence as discrete entities within the acidic environment of a zeolite.

The greatest challenge for planewave calculations is the location of reaction pathways. Unlike cluster methods, where the ease of using analytical second derivatives with eigenvector following algorithms makes transition-state location reasonably routine, in the condensed phase the majority of transition-state searches have been performed by a sequence of constrained optimisations. Typically, from the knowledge of the reactant and product configurations a reaction co-ordinate can be selected, such as a bond that forms or breaks, and the energy is profiled along this pathway, with all other degrees of freedom being optimised. Obviously, this process can be performed statically, in which case the activation internal energy is obtained, or alternatively it can be carried out with constrained ab initio molecular dynamics, wherein the free energy of activation is determined.

The first determination of a transition state for a reaction within a zeolite using the periodic planewave approach, to the best of our knowledge, was for the direct condensation of two methanols to form dimethyl ether in chabazite [78], as suggested by the earlier studies. Starting from a configuration for two methanols that lies 61 $kJ\,mol^{-1}$ higher in energy than the most favourable one, to avoid the computational expense of studying the transition to this further intermediate, the C–O distance for the bond that was forming was constrained to obtain the energy profile. The final transition-state configuration is shown in Fig. 9. The geometry of this structure has the appearance of a classic S_N2 barrier point, with the two oxygens being almost equidistant and the CH_3 group being planar. Interestingly, the activated complex is orientated across the elliptical cage of chabazite and there is no significant directional interaction with the framework, except for the water molecule that is being formed.

The activation energy calculated for the formation of dimethyl ether in this static study was found to be 71 $kJ\,mol^{-1}$, which when allowing

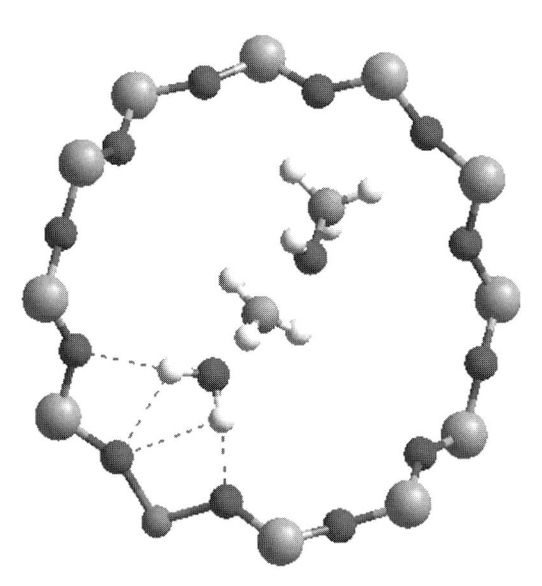

Fig. 9. The S_N2 transition state for condensation of two methanol molecules to form dimethyl ether within the elliptical cage of chabazite.

for the rearrangement energy to reach the starting configuration yields a true barrier height of 132 kJ mol^{-1}. This compares to a cluster estimate of 145 kJ mol^{-1} [76]. Experimentally, the activation energy for dehydration of methanol in H-mordenite has been estimated at 80 kJ mol^{-1} [79], which is surprisingly considerably lower than the theoretical estimates, given the tendency of density functional theory to underestimate barrier heights. The same work also suggests that two distinct activated complexes are involved. If the equivalent reaction is performed in the gas phase, where the mechanism is actually different, then the barrier height obtained is in excess of 100 kJ mol^{-1} higher, so clearly that the zeolite has a genuinely catalytic effect.

One of the factors that might contribute to the discrepancy between theory and experiment for the activation energy for formation of dimethyl ether is the static calculation of the transition state at 0 K, as well as the low effective loading of methanol. Very recently, Hytha et al. [80] have performed constrained ab initio molecular dynamics to determine the free energy of activation based on four methanols per acid site within ferrierite at 700 K, based on the previous simulations of Stich et al. [67]. This represents a situation much closer to the true experimental one, and a more facile reaction might be expected because of the dynamical weakening of the C–O bond of methoxonium observed previously. Indeed the free energy of activation for condensation is found to be approximately 100 $kJ\,mol^{-1}$ and represents a significant reduction over the static value for chabazite, though still higher than the experimental estimate. One thing is clear, which is that the role of the entropy is non-negligible as it leads to a slightly earlier transition state with a lower barrier than that based on the internal energy profile. The inclusion of entropy also leads to an increase in stability of the products over the reactants. Only as further simulations of this type are performed will we begin to be able to understand the influence of the change of framework topology and Si/Al ratio on the results, and whether these factors can account for the difference between theory and experiment in this respect.

From the mechanistic work performed to date, a clear picture is emerging that the zeolitic framework fails to become involved in reactions as a nucleophilic entity. Instead it acts as a strong acid capable of protonating clusters of reactant molecules and stabilising intermediates via dielectric screening. We can also speculate that perhaps one of the main factors that contributes to the catalytic effect of zeolites is the fact that there is no energy penalty associated with rearrangement of the solvent cage at the transition state.

5. Future directions

While planewave methods will remain an effective tool for the study of periodic materials for some time to come, if we look to the future there are other developments that may challenge them. Currently there is considerable interest in methods whose computational expense only increases linearly with the number of atoms in the system [81]. This has clear advantages for extending the scope of quantum mechanical methods to larger and more realistic models. However, in order to do this real-space localisation of the basis functions is required.

Interest in linear scaling, as well as other aspects, has led to new types of basis sets for this purpose including grids [82], finite elements [83] and radially confined numerical pseudo-atomic orbitals [84]. Of course similar technologies can be developed based on the more traditional Gaussian basis sets as well [85]. All of this will lead to increasing diversity within the techniques employed as it is possible to trade basis set quality, in particular systematic convergence, against computational efficiency. When one considers that a fairly high proportion of a microporous material's unit-cell volume is almost empty space (or low electron density to be more precise) then it might be anticipated that atom-centred functions will prove more efficacious in the long term.

Even within the framework of the planewave method there are likely to be advances that improve the utility of the technique for zeolite studies. One direction forward is the avoidance of pseudopotentials through mixed basis set calculations, which makes all electron studies feasible, while retaining the benefits of planewaves in the valence region [86]. Apart from avoiding a degree of uncertainty due to transferability of the pseudopotentials, it will allow properties that are strongly influenced by, or actually involve, core electrons to be determined for comparison against experiment, such as chemical shifts in NMR.

There will also be more widespread application of methods that are computationally demanding, but already being utilised in planewave framework, as computer power increases. An example of this includes the inclusion of quantum behaviour of hydrogen in ab initio molecular dynamics via the Path Integral technique. In one case where it has been applied to zeolites so far, this was found not to alter the qualitative conclusions, but definitely has an influence if quantitative information is desired [59]. It is also likely that the method of linear response, currently largely used for calculation of phonon spectra and mechanical properties, will be used to accelerate the static location of transition states in chemical reactivity studies. The next chapter will return to several of the issues discussed here and in particular, will extend the discussion of reactivity and the identification of intermediates.

Acknowledgements

I would like to thank the many people that I have had the pleasure of collaborating with in this area of work and who have contributed so much, particularly Rajiv Shah, Mike Payne, Ivan Stich, Marek Hytha, Eric Sandré and Kiyoyuki Terakura.

References

1. Sauer, J., *Chem. Rev.*, **89**, 199–255 (1989).
2. Sauer, J., Kölmel, C., Hill, J.-R. and Ahlrichs, R., *Chem. Phys. Lett.*, **164**, 193–198 (1989).
3. Kassab, E., Seiti, K. and Allavena, M., *J. Phys. Chem.*, **92**, 6705–6709 (1988).
4. Ahlrichs, R., Bär, M., Häser, M., Kölmel, C. and Sauer, J., *Chem. Phys. Lett.*, **164**, 199–204 (1989).
5. Payne, M.C., Teter, M.P., Allan, D.C., Arias, T.A. and Joannopoulos, J.D., *Rev. Mod. Phys.*, **64**, 1045–1097 (1992).
6. Aprà, E., Dovesi, R., Freyria-Fava, C., Pisani, C., Roetti, C. and Saunders, V.R., *Modelling Simul. Mater. Sci.*, **1**, 297–306 (1993).

7. White, J.C. and Hess, A.C., *J. Phys. Chem.*, **97**, 6398–6404 and 8703–8706 (1993).
8. Dovesi, R., Pisani, C., Roetti, C., Causa, M. and Saunders, V.R., *Quantum Chemistry Program Exchange*. Publication 577, University of Indiana, Bloomington, IN, 1988.
9. Brand, H.V., Curtiss, L.A. and Iton, L.E., *J. Phys. Chem.*, **96**, 7725–7732 (1992).
10. Teunissen, E.H., Jansen, A.P.J. and van Santen, R.A., *J. Phys. Chem.*, **99**, 1873–1879 (1995).
11. Eichler, U., Kölmel, C.M. and Sauer, J., *J. Comput. Chem.*, **18**, 463–477 (1996).
12. Hohenberg, P. and Kohn, W., *Phys. Rev.*, **136**, B864–B871 (1964).
13. Kohn, W. and Sham, L.J., *Phys. Rev.*, **140**, A1133–A1138 (1965).
14. Lee, C., Yang, W. and Parr, R.G., *Phys. Rev. B*, **37**, 785–789 (1988).
15. Becke, A.D., *J. Chem. Phys.*, **96**, 2155–2160 (1992).
16. Perdew, J.P. and Wang, Y., *Phys. Rev. B*, **45**, 13244–13249 (1992).
17. Sánchez-Portal, D., Artacho, E. and Soler, J.M., *J. Phys.: Condens. Matter*, **8**, 3859–3880 (1996).
18. Kleinman, L. and Bylander, D.M., *Phys. Rev. Lett.*, **48**, 1425–1428 (1982).
19. Vanderbilt, D., *Phys. Rev. B*, **41**, 7892–7895 (1990).
20. Blöchl, P.E., *Phys. Rev. B*, **50**, 17953–17979 (1994).
21. Car, R. and Parrinello, M., *Phys. Rev. Lett.*, **55**, 2471–2474 (1985).
22. Teter, M.P., Payne, M.C. and Allan, D.C., *Phys. Rev. B*, **40**, 12255–12263 (1989).
23. Gillan, M.J., *J. Phys.: Condens. Matter*, **1**, 689–711 (1989).
24. Lin, J.S., Payne, M.C., Heine, V. and McConnell, J.D.C., *Phys. Chem. Miner.*, **21**, 150–155 (1994).
25. Lee, C. and Gonze, X., *Phys. Rev. B*, **51**, 8610–8613 (1995).
26. Holm, B. and Ahuja, R., *J. Chem. Phys.*, **111**, 2071–2074 (1999).
27. Bridgeman, C.H., Buckingham, A.D., Skipper, N.T. and Payne, M.C., *Mol. Phys.*, **89**, 879–888 (1996).
28. Campana, L., Selloni, A., Weber, J., Pasquarello, A., Papai, I. and Goursot, A., *Chem. Phys. Lett.*, **226**, 245–250 (1994).
29. Gard, J.A. and Tait, J.M., *Acta Crystallogr. B*, **28**, 825–834 (1972).
30. Shah, R., Payne, M.C., Lee, M.-H. and Gale, M.C., *Science*, **271**, 1395–1397 (1996a).
31. Jeanvoine, Y., Ángyán, J.G., Kresse, G. and Hafner, J., *J. Phys. Chem. B*, **102**, 5573–5580 (1998a).
32. Fois, E., Gamba, A. and Tabacchi, G., *J. Phys. Chem. B*, **102**, 3974–3979 (1998).
33. Benco, L., Demuth, T., Hafner, J. and Hutschka, F., *J. Chem. Phys.*, **111**, 7537–7545 (1999).
34. Demuth, T., Hafner, J., Benco, L. and Toulhoat, H., *J. Phys. Chem. B*, **104**, 4593–4607 (2000).
35. Shah, R., Gale, J.D. and Payne, M.C., *Int. J. Quant. Chem.*, **61**, 393–398 (1997a).
36. Shah, R., Gale, J.D. and Payne, M.C., *J. Chem. Soc. Chem. Commun.*, 131–132 (1997b).
37. Díaz-Cabañas, M.-J., Barrett, P.A. and Camblor, M.A., *Chem. Commun.*, 1881–1882 (1998).
38. Henson, N.J., Cheetham, A.K. and Gale, J.D., *Chem. Mater.*, **6**, 1647–1650 (1994).
39. Francis, G.P. and Payne, M.C., *J. Phys.: Condens. Matter*, **2**, 4395–4404 (1990).
40. Smith, L.J., Davidson, A. and Cheetham, A.K., *Catal. Lett.*, **49**, 143–146 (1997).
41. Shah, R., Gale, J.D. and Payne, M.C., *Phase Transitions B*, **61**, 67–81 (1997d).

42. Hill, J.-R., Freeman, C.M. and Delley, B., *J. Phys. Chem. A*, **103**, 3772–3777 (1999).
43. Eichler, U., Brändle, M. and Sauer, J., *J. Phys. Chem. B*, **101**, 10035–10050 (1997).
44. Giannozzi, P., de Gironcoli, S., Pavone, P. and Baroni, S., *Phys. Rev. B*, **43**, 7231–7242 (1991).
45. Filippone, F., Buda, F., Iarlori, S., Moretti, G. and Porta, P., *J. Phys. Chem.*, **99**, 12883–12891 (1995).
46. Fois, E. and Gamba, A., *J. Phys. Chem. B*, **101**, 4487–4489 (1997).
47. Fois, E. and Gamba, A., *J. Phys. Chem. B*, **103**, 1794–1799 (1999).
48. Campana, L., Selloni, A., Weber, J. and Goursot, A., *J. Phys. Chem.*, **99**, 16351–16356 (1995).
49. Ursenbach, C.P., Madden, P.A., Stich, I. and Payne, M.C., *J. Phys. Chem.*, **99**, 6697–6714 (1995).
50. Edwards, P.P., Anderson, P.A. and Thomas, J.M., *Acc. Chem. Res.*, **29**, 23–29 (1996).
51. Armstrong, A.R., Anderson, P.A., Woodall, L.J. and Edwards, P.P., *J. Am. Chem. Soc.*, **117**, 9087–9088 (1995).
52. Blake, N.P. and Metiu, H., *J. Chem. Phys.*, **109**, 9977–9986 (1998).
53. Sauer, J., Ugliengo, P., Garrone, E. and Saunders, V.R., *Chem. Rev.*, **94**, 2095–2160 (1994).
54. Sauer, J., Kölmel, C., Haase, F. and Ahlrichs, R., In: von Ballamoos, R., Higgins, J.B. and Treacey, M.M.J. (Eds.), *Proc. 7th Int. Conf. Zeolites, Montreal.* Butterworths, London, 1993, pp. 679–686.
55. Haase, F. and Sauer, J., *J. Am. Chem. Soc.*, **117**, 3780–3789 (1995).
56. Shah, R., Gale, J.D. and Payne, M.C., *J. Phys. Chem.*, **100**, 11688–11697 (1996b).
57. Tintskaladze, G.P., Nefedova, A.R., Gryaznova, Z.V., Tsitsishvili, G.V. and Charkviani, M.K., *Zh. Fiz. Khim.*, **58**, 718–721 (1984).
58. Haase, F., Sauer, J. and Hutter, J., *Chem. Phys. Lett.*, **266**, 397–402 (1997).
59. Stich, I., Gale, J.D., Terakura, K. and Payne, M.C., *J. Am. Chem. Soc.*, **121**, 3292–3302 (1999).
60. Haase, F. and Sauer, J., *Microporous Mesoporous Mat.*, **35–36**, 379–385 (2000).
61. Afanassyev, I.S., Moroz, N.K. and Belitsky, I.A., *J. Phys. Chem. B*, **104**, 6804–6808 (2000).
62. Lee, C.-C., Gorte, R.J. and Farneth, W.E., *J. Phys. Chem. B*, **101**, 3811–3817 (1997).
63. Shah, R., Gale, J.D. and Payne, M.C., *J. Phys. Chem. B*, **101**, 4787–4797 (1997c).
64. Greatbanks, S.P., Hillier, I.H., Burton, N.A. and Sherwood, P., *J. Chem. Phys.*, **105**, 3770–3776 (1996).
65. Mirth, G., Lercher, J.A., Anderson, M.W. and Klinowski, J., *J. Chem. Soc. Faraday Trans.*, **86**, 3039–3044 (1990).
66. Izmailova, S.G., Karetina, I.V., Khvoshchev, S. and Shubaeva, M.A., *J. Colloid Interface Sci.*, **165**, 318–324 (1994).
67. Stich, I., Gale, J.D., Terakura, K. and Payne, M.C., *Chem. Phys. Lett.*, **283**, 402–408 (1998).
68. Smith, L.J., Cheetham, A.K., Morris, R.E., Marchese, L., Thomas, J.M., Wright, P.A. and Chen, J., *Science*, **271**, 799–802 (1996b).
69. Chemistry and Industry, 117 (1996).

70. Termath, V., Haase, F., Sauer, J., Hutter, J. and Parrinello, M., *J. Am. Chem. Soc.*, **120**, 8512–8516 (1998).
71. Jeanvoine, Y., Ángyán, J.G., Kresse, G. and Hafner, J., *J. Phys. Chem. B*, **102**, 7307–7310 (1998b).
72. Stöcker, M., *Micro. Meso. Mater.*, **29**, 3–48 (1999).
73. Ono, Y. and Mori, T., *J. Chem. Soc. Faraday Trans. 1*, **77**, 2209–2221 (1981).
74. Kubelková, L., Nováková, J. and Nedomová, K., *J. Catal.*, **124**, 441–450 (1990).
75. Forester, T.R. and Howe, R.F., *J. Am. Chem. Soc.*, **109**, 5076–5082 (1987).
76. Blaszkowski, S.R. and van Santen, R.A., *J. Am. Chem. Soc.*, **118**, 5152–5153 (1996).
77. Blaszkowski, S.R. and van Santen, R.A., *J. Phys. Chem. B*, **101**, 2292–2305 (1997).
78. Sandré, E., Payne, M.C. and Gale, J.D., *Chem. Commun.*, 2445–2446 (1998).
79. Bandiera, J. and Naccache, C., *Appl. Catal.*, **69**, 139–148 (1991).
80. Hytha, M., Stich, I., Gale, J.D., Terakura, K. and Payne, M.C., *Chem.-Eur. J.*, **7**, 2521–2527 (2001).
81. Ordejón, P., Artacho, E. and Soler, J.M., *Phys. Rev. B*, **53**, R10441–R10444 (1996).
82. Chelikowsky, J.R., Troullier, N., Wu, K. and Saad, Y., *Phys. Rev. B*, **50**, 11355–11364 (1994).
83. Tsuchida, E. and Tsukada, M., *Phys. Rev. B*, **52**, 5573–5578 (1995).
84. Artacho, E., Sánchez-Portal, D., Ordejón, P., García, A. and Soler, J.M., *Phys. Stat. Sol. (b)*, **215**, 809–817 (1999).
85. Scuseria, G.E., *J. Phys. Chem. A*, **103**, 4782–4790 (1999).
86. Krack, M. and Parrinello, M., *Phys. Chem. Chem. Phys.*, **2**, 2105–2112 (2000).
87. Gale, J.D., Catlow, C.R.A. and Cheetham, A.K., *J. Chem. Soc. Chem. Commun.*, 178–179 (1991).
88. Gale, J.D., Shah, R., Payne, M.C., Stich, I. and Terakura, K., *Catal. Today*, **50**, 525–532 (1999).
89. Smith, L.J., Cheetham, A.K., Marchese, L., Thomas, J.M., Wright, P.A., Chen, J. and Gianotti, E., *Catal. Lett.*, **41**, 13–16 (1996a).
90. Gale, J.D., *Topics Catal.*, **3**, 169–194 (1996).
91. Gale, J.D., Catlow, C.R.A. and Carruthers, J.R., *Chem. Phys. Lett.*, **216**, 155–161 (1993).
92. Nusterer, E., Blöchl, P.E. and Schwarz, K., *Angew Chem. Int'l Ed. Engl.*, **35**, 175 (1996).
93. Nusterer, E., Blöchl, P.E. and Schwarz, K., *Chem. Phys. Lett.*, **253**, 448 (1996).

Computer Modelling of Microporous Materials
C.R.A. Catlow, R.A. van Santen and B. Smit (editors)

Chapter 6

Reaction mechanisms in protonic zeolites

Xavier Rozanska and Rutger A. van Santen

Laboratory of Inorganic Chemistry and Catalysis,
Schuit Institute of Catalysis, Technical University of Eindhoven,
P.O. Box 513, NL-5600 MB Eindhoven, The Netherlands

1. Introduction

This chapter continues the theme of the application of quantum mechanical methods to the study of microporous materials with a strong focus on mechanistic concepts related to hydrocarbon activation catalyzed by acidic zeolites. A strong emphasis is also placed on the relationship between transition-state (TS) structures and their energies as a function of the zeolite micropore structure. We show that reactivity and selectivity of zeolite-catalyzed reactions is characterized by limited solvation of the reactant, and that the match between transition-state shape and size with the zeolite micropore dimensions is critical. Reaction mechanisms strongly resemble those of classical electrophylic substitution reactions. However, we find that a unique feature of zeolite-catalyzed reactions is that protonated intermediates considered in homogeneous catalysis become converted to transition states and activated complexes.

The relevant background to the chapter is given in earlier chapters and in references [1]. We note that as zeolites are silicate crystals, they show good thermal and mechanical properties, which is of major relevance to their use as catalysts in heterogeneous catalysis. And we recall that their well-defined microporous structure is essential to the molecular sieve capability of zeolites [4–6], and that when used as a catalyst or catalyst support, it will induce product selectivity when reactions are diffusion limited [7,8].

It is also of relevance to our discussion that zeolite crystals are insulators. They are ionic crystals, and the radii of the zeolite oxygen and silicon atoms is close to that of the ones observed for O^{2-} and Si^{4+} ions in other crystals [9]. However, bonding in zeolites has a strong component of covalence [1,10–12]. Zeolite crystals have a small dielectric constant: ε_1 is generally between 2 and 7 for full silicon zeolites [13,14]. Dispersive van der Waals contributions also play a major role [15,16], stemming mainly from interaction with the highly polarizable oxygen anions.

We also recall that substitution of framework Si atoms can lead to the incorporation of extra framework cations or of protons [1–21] and that additionally, silicon substitution introduces sites that can show catalytic activity [1–18]. The incorporated protons show acidic properties [1–19]. Owing to the classic Brønsted acid site, discussed previously as shown in Fig. 1 in which the proton is covalently bonded to the oxygen [19–21].

Proton-exchanged zeolites have been extensively investigated [1–27] and indeed induce reactions similar to those known from superacids [28,29], although being only moderately acidic [1–6]. Because of their very limited solvation power, the enthalpies of reaction catalyzed by zeolites compare more closely to gas-phase reactions than to reactions catalyzed by homogeneous high-dipolar acids [12]. Proton-exchanged zeolites can, for instance, induce alkylation, transalkylation, isomerization, and cracking reactions [30,31]. Moreover, they are also applied to achieve organic fine chemical reactions [30].

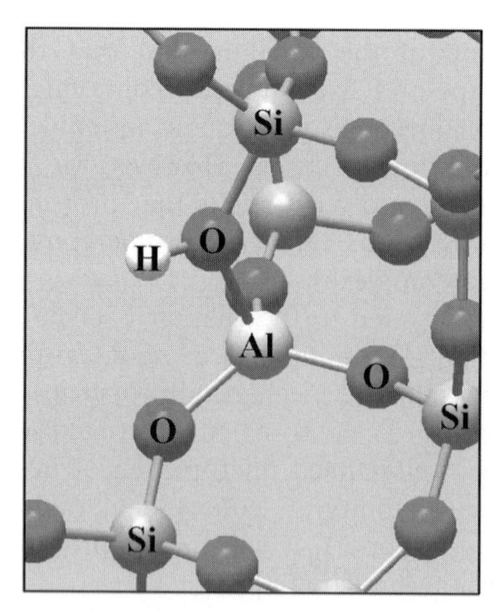

Fig. 1. Proton bonded to bridging O-atom in the neighborhood of a silicon atom substitution with an aluminum atom in H-mordenite.

This chapter will initially discuss and compare some of our earlier computational results [32–39]. More recent calculations then provide further support to the general discussion. Many of the methods used have been discussed in the previous chapter and are further described in Refs. [32–40]. We will discuss general concepts that characterize the zeolite-catalyzed reactions. These concepts are of key importance to a proper rationalization of the mechanism of such reactions [41].

2. Reaction mechanisms of catalytic reactions activated by proton-exchanged zeolites

In this section, we recall some of the most important mechanistic aspects of reactions induced by proton-exchanged zeolites, which will allow us to sketch the general scene, before focusing on more specific topics. We will use a smaller number of reactions to illustrate these points. General aspects relating to reaction mechanisms in zeolite catalyst are also discussed in Ref. [41].

2.1. The carbocationic nature of the transition states

Clearly, if one has to define transition-state structures as they are formed in hydrocarbon reactions catalyzed by proton-exchanged zeolites, the best way to visualize them is as carbocation-like transition-state structures [8,41]. An example that illustrates characteristic elementary reaction steps that occur in proton-exchanged zeolite is the chemisorption of olefin (see Fig. 2). This reaction has been the subject of many experimental [23,42] and theoretical studies [25,26]. The reaction, which starts from an olefin physisorbed to the acidic proton, leads to the formation of a more stable chemisorbed olefin or alkoxy species. We show the geometries of olefins (viz. propylene and isobutene) in the transition states in Fig. 3. In the transition-state structure, the olefin becomes protonated; meanwhile the alkoxy bond is associatively being formed. We note how close the hydrocarbon molecule geometries are to those of the carbocations. In this figure, the zeolite atoms are not shown but full periodic zeolite crystals were considered in the calculations on which they are based [35,43].

Experimental studies show that hydrocarbon activation energies follow the same energy ordering as carbocation structure energies [31], which allows for a convenient shortcut in the description of the reaction pathway as the transition-state structure could be substituted with their close-corresponding carbocation [30,31].

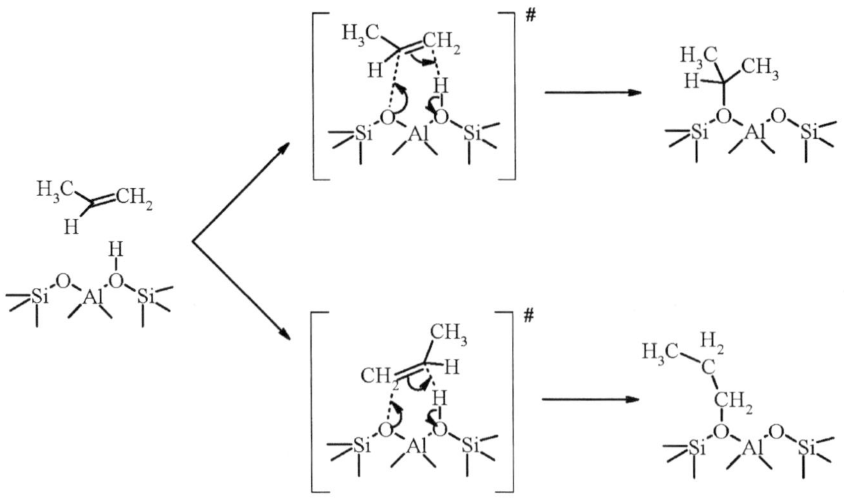

Fig. 2. Reaction mechanisms of olefin chemisorption catalyzed by proton-exchanged zeolite.

The activation energies of the protonation reaction follow the energy ordering of the carbocation structures, namely, primary < secondary < tertiary. Therefore, a reaction in which a tertiary carbocation-like transition-state structure is involved is more likely to occur than a reaction that proceeds via a secondary or primary carbocation-like transition-state structure. The activation energies to protonate olefins are 100, 60, and 20 kJ/mol for primary, secondary, and tertiary carbocations, respectively, according to periodic DFT calculation [35,43] and refer to situations with no steric constraints, to which we return in later sections. Interestingly, the activation energies for chemisorption of isobutene and propylene through a primary carbocation-like transition state are similar (i.e. 120 and 128 kJ/mol for isobutene and propylene, respectively) [35,43].

The simple relationship between activation energy and the type of the transition-state carbocation is also basic to the concept of transition-state selectivity [31]. Here, we do not refer to transition-state shape selectivity, but to the selectivity that arises from the intrinsic energetic differences, which we illustrate for the alkylation of benzene with propylene. According to the experimental literature and in agreement with theory, an olefin is readily protonated in a protonic zeolite, which is not the case for benzene that is more difficult to protonate [42]. Thermodynamic considerations explain this result [44], and the aromaticity of benzene is of course the main reason. Protonation

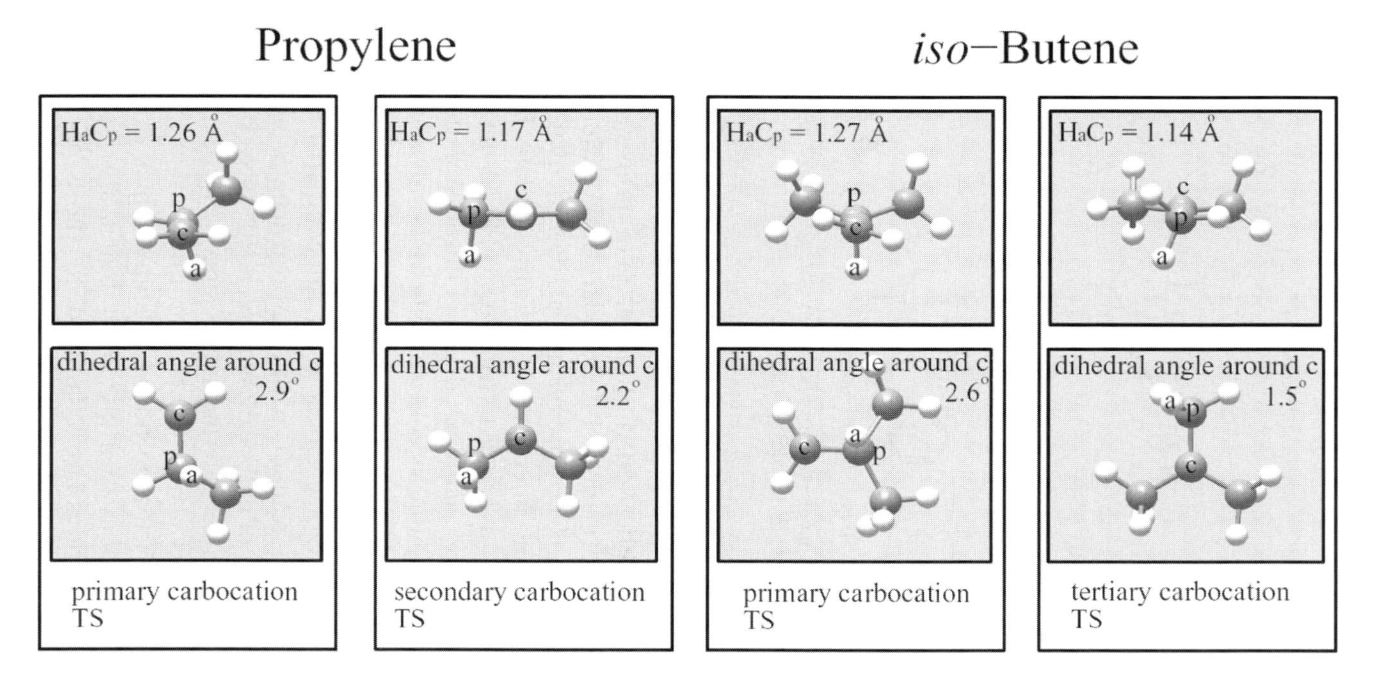

Fig. 3. Geometric details of propylene and isobutene chemisorption transition states in acidic chabazite. C_p is the carbon atom that is being protonated. C_c is the carbon atom that will connect to the zeolitic oxygen atom. H_a is the zeolitic acidic proton. These geometries were obtained using periodic DFT calculations.

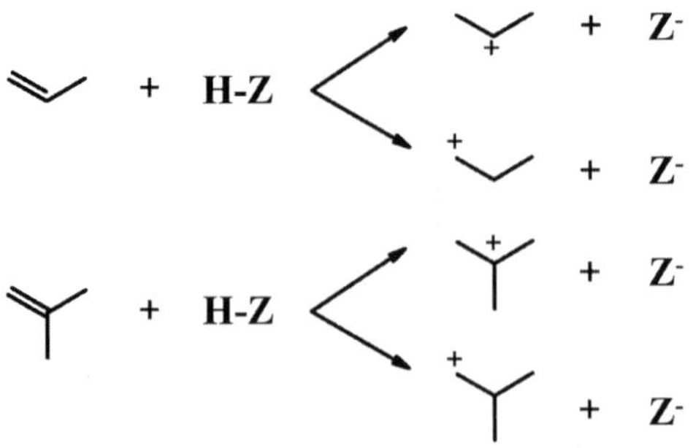

Fig. 4. Schematic reaction mechanisms of propylene and isobutene activation in acidic zeolite.

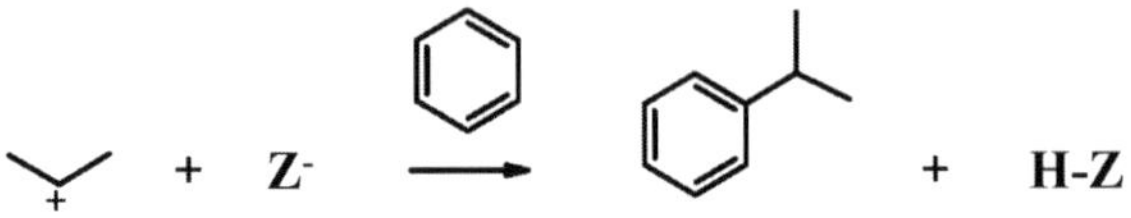

Fig. 5. Schematic reaction mechanism of alkylation of benzene with propylene leading to the formation of cumene catalyzed by acidic zeolite.

of propylene leads to two possible carbocations (see Fig. 4). The activation energy to protonate propylene relates to the stability of the intermediate carbocation. Formation of a primary carbocation is unlikely with the formation of a secondary carbocation being much more easily achieved. Once propylene has been activated through a secondary carbocation intermediate, it can alkylate benzene, and cumene is formed as a product (see Fig. 5), which agrees with the reaction mechanism as commonly depicted in the literature [30,31,45], although the description is oversimplified as revealed when we performed periodic DFT analysis of this reaction pathway. The calculations were obtained using the same method and model as in previous studies [35–39], which allowed a full comparison between the different data. In this reaction pathway, propylene protonation does lead to the formation of carbocationic 'activated' species. Such species do not exist as stable intermediates, since charge separation is energetically too costly [41,45]. On the contrary, within the zeolite, the carbocations are very close to transition states, which can be deduced from their geometries as seen previously but is also apparent when energies of carbocations are compared to those of their related transition-state structures [43], which we illustrate in the next section. The actual reaction mechanisms of benzene alkylation with propylene is shown in Fig. 6.

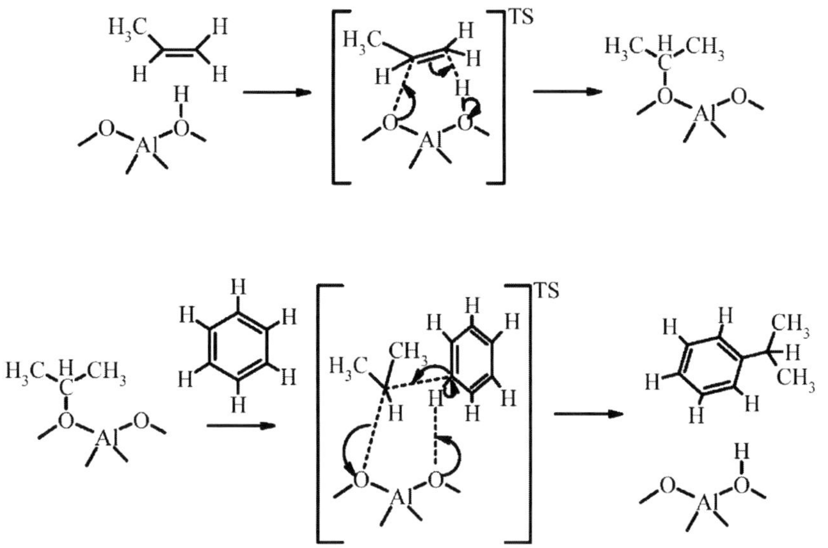

Fig. 6. Reaction mechanisms of alkylation of benzene with propylene leading to the formation of cumene catalyzed by acidic zeolite as they were observed to occur using quantum chemistry tools.

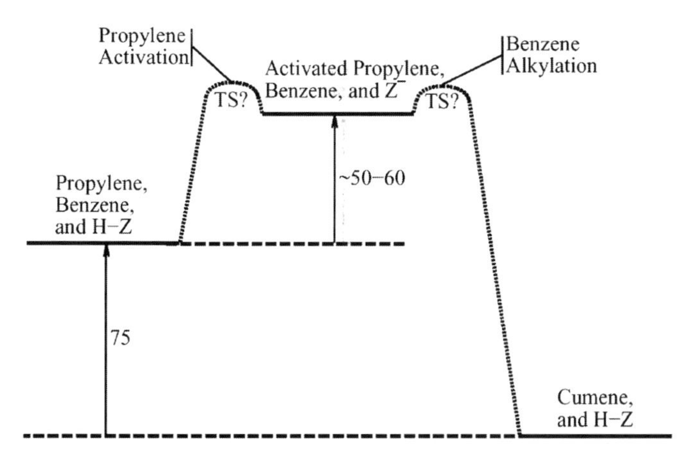

Fig. 7. Schematic reaction energy diagram of the alkylation of benzene with propylene catalyzed by acidic zeolite as it can be estimated using description of the reaction mechanisms as in Figs. 4 and 5. All values in kJ/mol.

Based on the classical reaction mechanism diagrams depicted in Figs. 4 and 5, one can formulate an initial approximate reaction energy diagram (see Fig. 7). The periodic electronic structure calculations lead, however, to a different result (see Fig. 8). Such a difference is important to the kinetic analysis of the reaction [46]. The reaction pathway

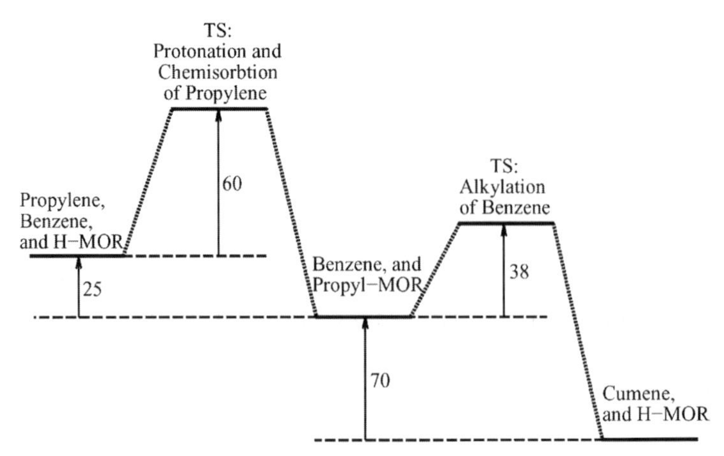

Fig. 8. Reaction energy diagram of alkylation of benzene with propylene catalyzed by acidic mordenite as obtained in the periodic DFT calculations. All values in kJ/mol.

deduced from the theoretical calculations agrees with experimental results [42]: it is reported that the limiting step in this reaction is the formation of the 'activated' olefin. In the reaction pathway as obtained from periodic DFT calculations, the limiting step is protonation/chemisorption of propylene. One notes that a secondary propyl-cation is involved in the alkylation of benzene with propylene transition-state structure. The greater stability of secondary carbocation compared to the primary carbocation explains why alkylation of the transition state cannot lead to linear propyl-benzene.

The activation energy for the alkylation reaction resulting in the formation of cumene can be estimated from the chemisorption activation energy of propylene using the Polanyi–Evans–Brønsted relation [47,48]. This relation states that change(s) in reactant and/or product energy level(s) induce(s) a linear change in the energy of the transition-state structure that link them; usually, this linear dependence is around a half. A closer look at the reaction energy diagram as obtained from the periodic DFT calculations reveals that the reaction energies for these two reactions are not equivalent (see Figs. 7 and 8). The correction of the activation energy in propylene chemisorption with half of the difference in reaction energy between propylene chemisorption and benzene alkylation with propylene (i.e. $\frac{1}{2}$ (70−25) kJ/mol) leads to an activation energy of 38 kJ/mol, which is also the value that is obtained in the periodic calculations. Application of the Polanyi–Evans–Brønsted relation is therefore useful in giving a crude prediction of the activation energy. However, we will see in the next section that application of the relation is not always straightforward and should be applied with caution.

2.2. *Complexity of reaction mechanisms*

Specificity is a key goal of catalysis to accomplish which needs specific zeolite structures, of which we will provide an analysis based on the previously discussed alkylation reaction.

First of all, the activation energy is close to the formation energy of the carbocation. Therefore, it is not expected to find a very different activation energy for competitive propylene oligomerization, or any other reaction in which a secondary carbocation-like transition state is involved. To avoid these undesirable side-chain reactions, a large excess of benzene over propylene is used in practice [42]. However, oligomerization cannot be avoided, leading to a substantial deactivation of the catalyst [31,42].

Any catalytic cycle consists of a sequence of elementary reaction steps, which eventually leads to a restored catalytic site and product(s) (see Fig. 9). As zeolite catalysts are microporous, a special feature with this class of catalyst is the coupling of the reaction at the catalytic site with product and reactant diffusion in the micropores [8]. Before reaching the catalytic site, reactants need to adsorb within the zeolite grain mouth, and to diffuse within the zeolite microporous network. Then, reactants adsorb at the catalytic site, and the reaction can take place. Products desorb from the catalytic site, diffuse away, and finally leave the zeolite catalyst grain. The understanding of a zeolite-catalyzed reaction requires the description of all of these steps [8]. At each of these steps shape selectivity can occur. The dimension of the zeolite micropores becomes important in the control of diffusion of reactant and product. Also, the micropore structure has a predominant role in controlling transition-state selectivity. In case the zeolite micropores are sufficiently large so that products (e.g. cumene) can diffuse easily, this product will be the main product of the reaction. If cumene can hardly diffuse, its residence time in the micropore increases. As a result, it becomes involved in consecutive reactions with benzene. For cumene, this consecutive step may lead to the formation of linear propyl-benzene (see Fig. 10), which can leave more easily the zeolite micropores [42], thereby providing an example of diffusion-controlled selectivity.

The complexity of the overall reaction increases if one recycles the main product and performs another step to achieve dialkylation. Once dialkylated benzene is formed, it needs to diffuse out of the zeolite.

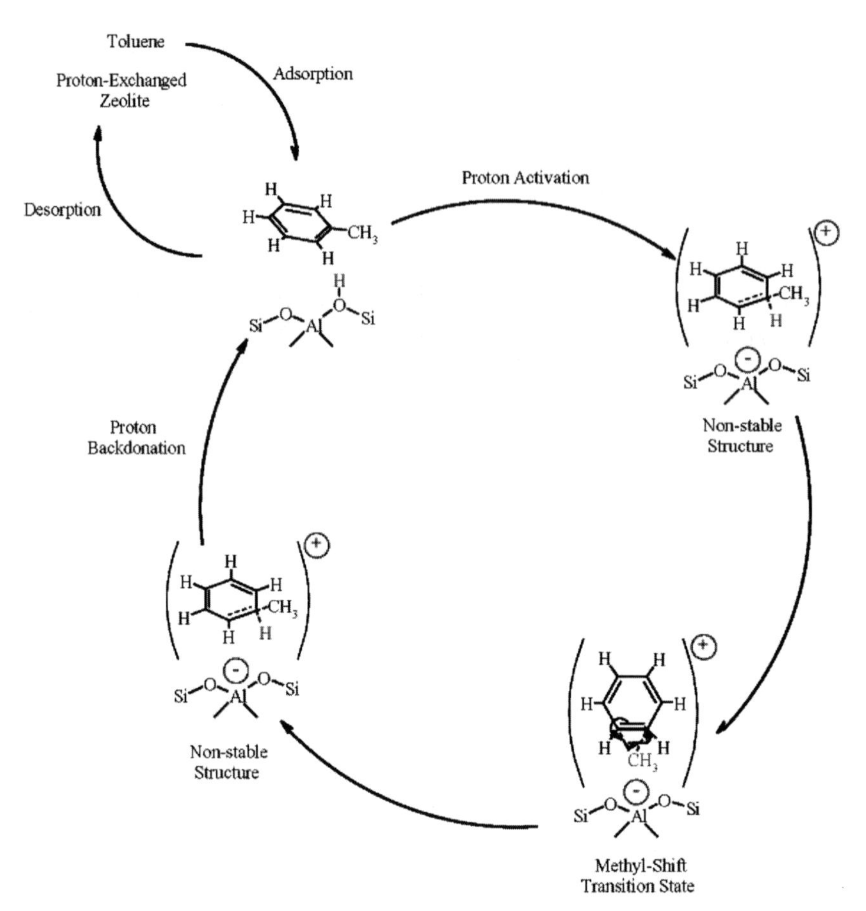

Fig. 9. Zeolite catalytic cycle (the zeolite catalyst reaction cycle is illustrated using toluene methylshift isomerization as support).

The relative diffusion constants of the different isomers will favor selectivity of some products over others. Monomolecular isomerization and transalkylation will occur so as to increase the fraction in the product with the higher-diffusivity constant. Simple consideration of the reaction mechanism as schematized in Figs. 4 and 5 is no longer appropriate, and zeolite micropore effects need to be explicitly considered so as to predict or understand reaction selectivity. For this purpose, it is still possible to use as illustrations some simple schemes such as that shown in Fig. 11. However, implicit assumptions in the scheme depicted in Fig. 11 have to become explicit in order to describe properly the actual relationship between the transition-state and the zeolite micropore structures. It is this relationship that we will discuss in the following sections.

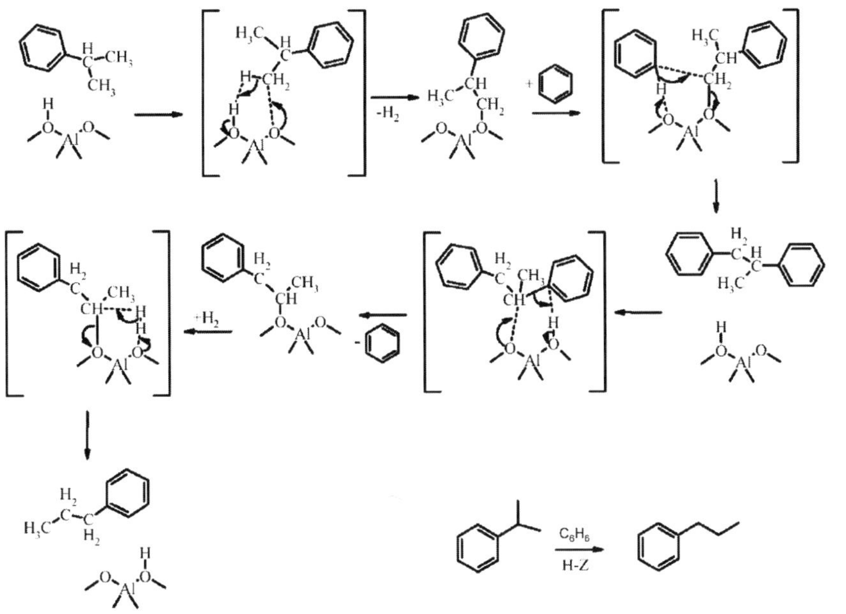

Fig. 10. Reaction mechanisms of the isomerization of cumene to linear propyl-benzene catalyzed by acidic zeolite.

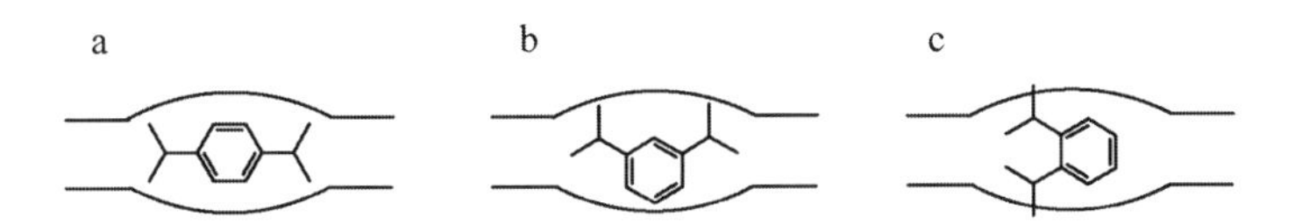

Fig. 11. Schematic illustration of the shape selectivity in zeolite catalyst. (a) *para*-Diisopropyl-benzene can be formed in the micropore. Moreover, it can easily diffuse into the micropore. (b) *meta*-Diisopropyl-benzene can be formed in the micropore. However, it cannot easily diffuse into the micropore, i.e. diffusion selectivity. (c) *ortho*-Diisopropyl-benzene cannot be formed in the micropore because of steric constraints, i.e. transition-state selectivity. These results will be totally different in another zeolite structure.

3. The interaction between activated hydrocarbon and zeolite framework

3.1. Flexibility of the zeolite framework

Unexpected reactivity patterns of a zeolite catalyst have been related to differences in the flexibility of a particular framework compared to that of other zeolite catalysts [49]. In determining the dependence of transition-state structures on the zeolite-framework structure, periodic quantum chemistry methods are of value. All atoms within the periodic unit cell are treated on the same basis, and analysis of the effects of the zeolite framework on reactivity including relaxation can easily be achieved.

To study these effects, we considered the activation of propylene in proton-exchanged chabazite in detail [35]. In this reaction, as previously described, propylene is protonated and becomes covalently bonded to the zeolite wall (see Fig. 2). Starting from a fully optimized unit cell of an acidic chabazite zeolite (H-CHA) in the absence of propylene [35], constraints to the zeolite atoms were gradually removed after propylene was introduced in different states (viz. physisorbed, transition, and chemisorbed states). The results of the calculations are summarized in Table 1.

The activation energy and reaction energy of propylene chemisorption was calculated all along with constraints removal. Careful monitoring of the radial atomic distribution indicated slight changes in zeolite atom positions [35]. However, the effect on computed activation energy was dramatic: a nearly 100% error in activation energy can easily be obtained if in case an approximate zeolite model is used. The activation energy that was obtained in the case of the highest-constrained system was close to that of the cluster approach calculations [50]. For the reaction energy, relaxation turns out also to be essential: the exothermic reaction of chemisorption of propylene becomes endothermic when the zeolite atom positions remained constrained.

These results must lead us to question the value of large cluster (or periodic) calculations where for reasons of computational cost, the

Table 1
Effect of the absence of the zeolite-framework relaxation in the propylene chemisorption in chabazite as observed from the periodic DFT calculations[a]

	1	2	3	4	5
E_{ads}	−21	−20	−19	−17	−16
E_{act}	+56	+60	+62	+70	+91
$\Delta E_{\sigma-\pi}$	−27	−23	−14	+10	+23

[a]All data in kJ/mol. (1) The unit cell and all atomic positions are fully optimized. (2) All atomic positions are optimized in the constant volume and size chabazite unit cell. (3) Si atoms except the first four ones around the aluminum atom are fixed in the constant volume and size chabazite unit cell. (4) The positions of H, C, Al, and four O and four Si around the aluminum atom are optimized in the constant volume and size chabazite unit cell. (5) The positions of H, C, Al, and four O around the aluminum atom are optimized in the constant volume and size chabazite unit cell.

atoms in the zeolite frameworks are held fixed. With a small zeolite cluster model, the influence of the zeolite framework is missing, which can, however, in some circumstances be regarded as an advantage. We will describe the consequences of this approximation more in detail in a later section.

3.2. Induced dipoles

The flexibility of the zeolite framework is important in stabilizing transition-state structures at least in part for electronic reasons. It optimizes polarization of the zeolite oxygen atoms near the carbocationic transition states, as was shown in the toluene methylshift isomerization reaction studied by an analysis of polarization of the electronic density in the system (see Fig. 12). Any modification of the electronic structure of the system induces geometry changes in return [51]; preventing zeolite framework flexibility prohibits optimum polarization of the zeolite atoms. As can be seen in Fig. 12, all zeolitic oxygen atoms surrounding the TS are polarized, and their position is slightly but significantly altered. Therefore, it is extremely important to allow full relaxation of the zeolite framework in the periodic calculations. The flexibility of the zeolite framework and induced dipoles on the zeolite-framework atoms can be considered as the same property as one induces the other, and vice versa.

The proximity of TS to the Brønsted acidic site did not appear to be a significant parameter, which was observed in alkylated aromatic

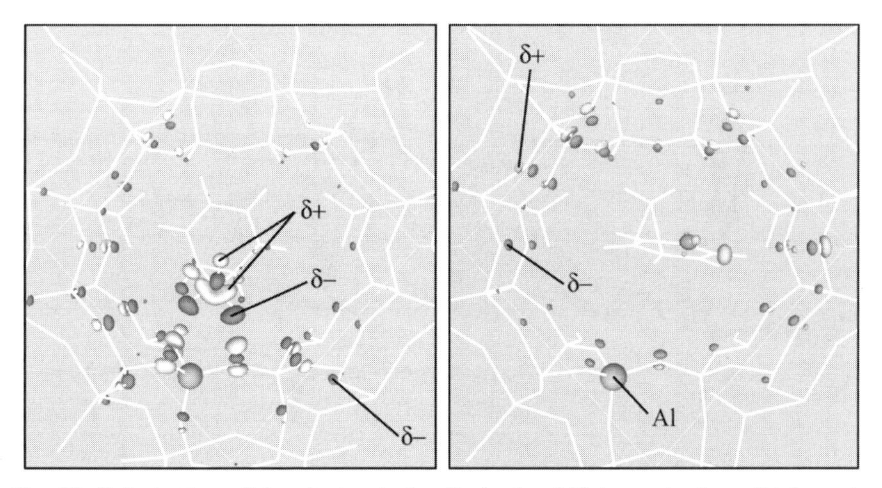

Fig. 12. Polarization of the electronic density in the shift isomerization of toluene in acidic mordenite as obtained in DFT periodic structure calculations.

studies [37], and also in isobutene chemisorption in different proton-exchanged zeolites [43]. However, this observation can hardly be extrapolated to all reactions. For instance, in thiophene cracking, the oxygen atoms of the catalytic site actually initiate the reaction [32]. Then, the reactant has to be located in a very close proximity to the catalytic site.

3.3. Cluster versus periodical approach

We return to this key methodological point which was discussed in the previous chapter. In the previous section, the interaction of the transition-state structures with the zeolite-framework catalyst have been discussed. It appears that elementary reaction steps remain in essence the same when periodic calculation results are compared to the cluster approach calculations. To illustrate this point, the similar transition-state structures for periodic and cluster calculations found in the study of toluene isomerization are of interest (see Fig. 13) [34,37].

Let us analyze the data obtained in more detail. As can be seen in Fig. 13, the geometries of transition states seem to be relatively unaffected by the presence of the zeolite framework. The large difference between these two TS stems from their activation energies: $E_{act} = 180\,kJ/mol$ in the periodic approach, whereas $E_{act} = 280\,kJ/mol$ in the cluster approach. This important ionic stabilization in the periodic structure has important consequences to the reaction pathway of isomerization (see Fig. 14). Generally, one observed few changes for neutral intermediates, but large zeolite solvatation effects for the

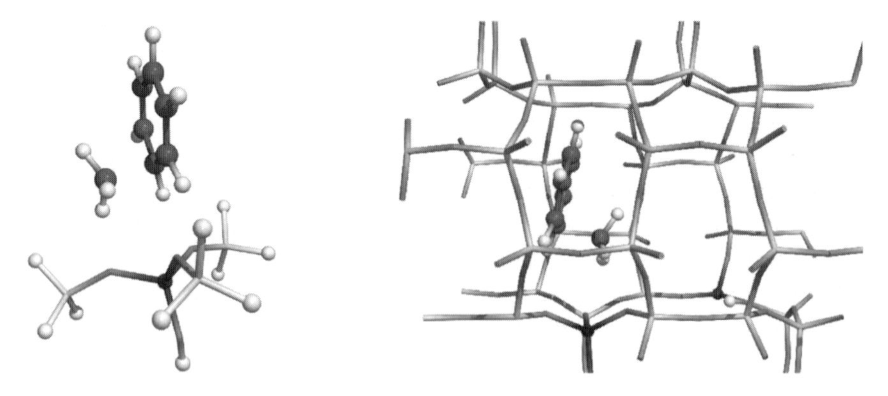

Fig. 13. Geometries of the shift isomerization transition states of toluene catalyzed by acidic zeolite as obtained from the cluster approach method (left) and the periodical structure method (right).

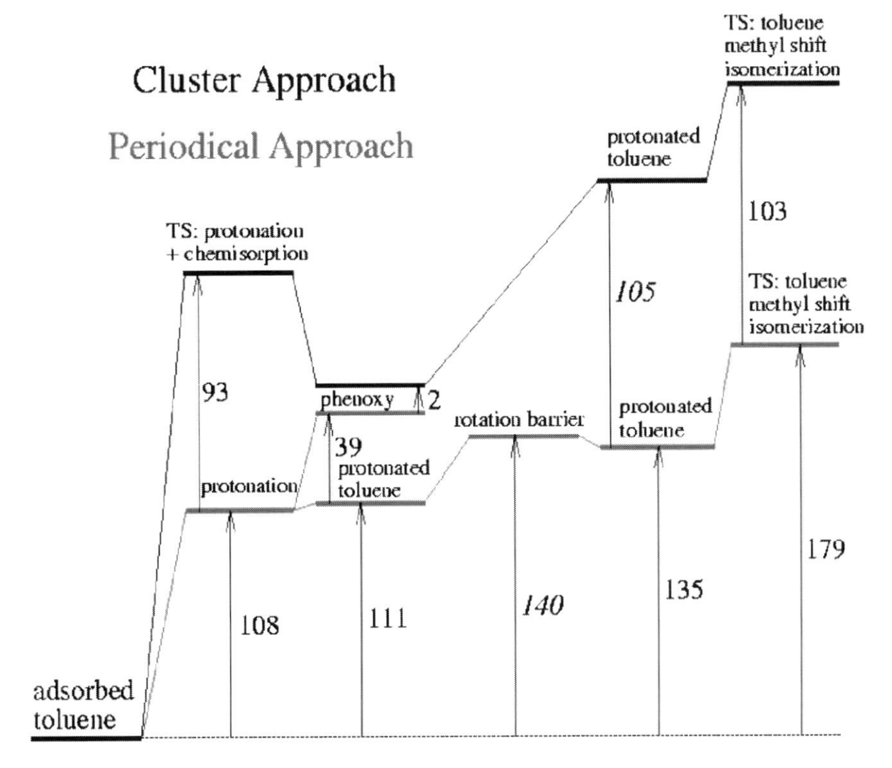

Fig. 14. Reaction energy diagrams of the shift isomerization reaction of toluene catalyzed by acid zeolite. The diagrams using black lines refer to the cluster approach results, and the ones using gray lines to the periodical calculation results (in kJ/mol).

carbocationic transition states. The investigation of the activation of toluene by proton attack reveals a completely different picture from that obtained with the cluster approach. In the periodic calculations, the structure showing the protonation step of toluene becomes an inflection point in the reaction pathway, corresponding to a metastable Wheland complex. Formation of the phenoxy intermediate turns out to be unlikely as the energy of this intermediate remains unchanged compared to that of the cluster approach. The Wheland complex energy level is around +110 kJ/mol with respect to physisorbed toluene.

Zeolite-framework stabilization affects uniformly all transition states and charged transient intermediates. As mentioned earlier, it does not affect the neutral intermediates. This large stabilization of TS structure was also found in other reactions in which toluene is involved [34,37,39]. Similar results have been obtained in other theoretical studies [52,53]. The carbocationic nature of the transition structures was also revealed to be slightly enhanced in these studies [52].

The effect of the zeolite framework on the reaction is not limited to a stabilization of charged species as we described in the previous sections. One of the conclusions that can be derived from Figs. 13 and 14 is that the effect of the framework can be sometimes quite subtle and not straightforward. We will discuss this in the following sections.

3.4. The dependence of transition-state structure on micropore dimension

3.4.1. Micropore size

To determine how the microporous structure affects the substrate transition-state structure, an analysis of isobutene chemisorption in different zeolite crystals has been undertaken [43]. For this purpose, chabazite (CHA), ZSM-22 (TON), and mordenite (MOR) were selected. Chemisorption of isobutene within the protonic zeolite can proceed through two reaction pathways. In the first, an isobutoxy species is formed, and in the second, a linear butoxy species. As explained before, isobutoxide is selectively formed in the experiment as it implies evolution through the most stable tertiary carbocation-like transition state rather than a primary carbocation-like TS structure.

Let us consider the different zeolites that were used in the periodic electronic structure calculations. MOR is a large pore zeolite; it has parallel 12-membered rings in which isobutene is not expected to experience any steric constraints. Furthermore, the Brønsted acidic site that was selected in the calculations was located at the junction of a 12-membered ring and two smaller eight-membered rings. Therefore, alkoxy species are not expected to suffer from steric constraints. TON is a medium-sized micropore zeolite: it has parallel 10-membered ring channels without side pockets or openings of any sort in the channels. The CHA micropore structure consists of supercages connected to each other through eight-membered ring openings. CHA has large cavities, but the curvature of the zeolite wall has a radius that is smaller than that in TON.

The reaction energy diagrams that correspond to the chemisorption in the different zeolites are shown in Fig. 15. As expected, primary carbocation-like transition states show much higher activation energies than tertiary carbocation-like transition states. Protonation to the tertiary carbocation structures occurs readily whatever the zeolite structure, with an activation energy of the order of 25–45 kJ/mol. One notes that the tertiary carbocation energy is very close to that of the corresponding transition state. In contrast, protonation to the

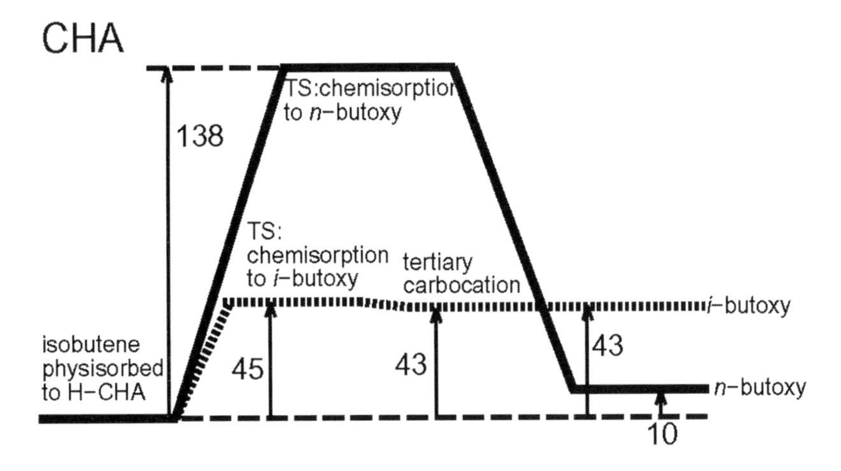

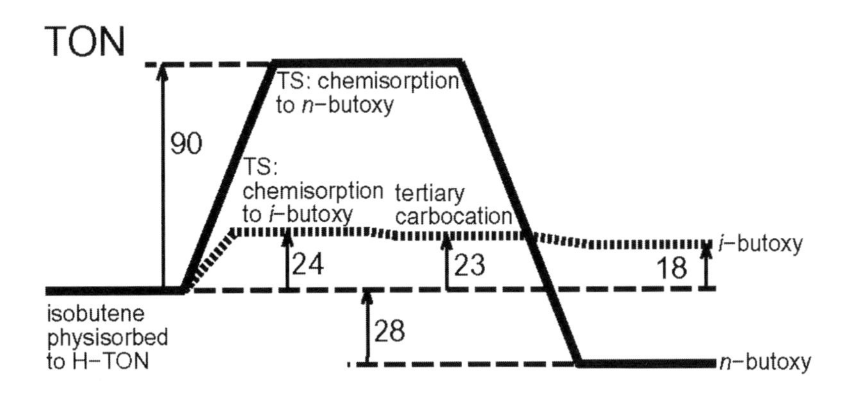

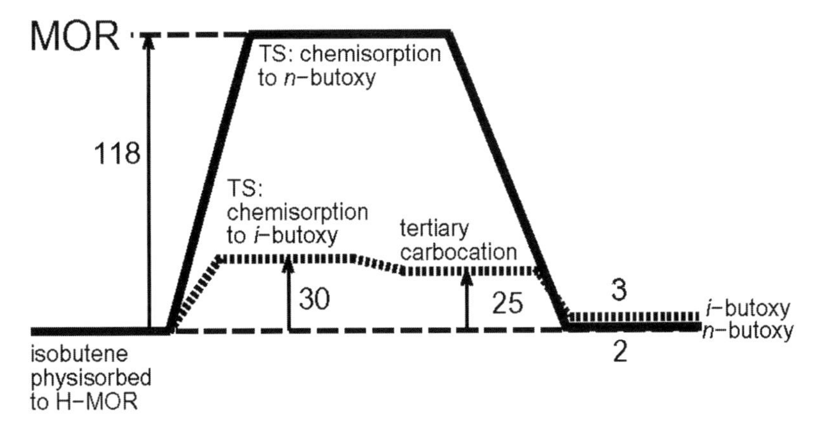

Fig. 15. Reaction energy diagrams of isobutene chemisorption in different acidic zeolites as obtained from the periodic calculations. All data in kJ/mol.

primary carbocation structure and consecutive chemisorption requires an activation energy between 90 and 138 kJ/mol. We will return later to an analysis of these activation energy differences. In case no steric constraints are expected (i.e. in MOR zeolite), the relative energy levels of the alkoxy species are the same in isobutoxy and *n*-butoxy species; they differ, however, in CHA and TON. In these zeolites, the curvature of the zeolite surface induces steric constraints that destabilize the more bulky *i*-butoxy species.

The geometries of the transition states that lead to the formation of *n*-butoxide are shown in Fig. 16. Distances between zeolite oxygen atoms and isobutene hydrogen atoms in the range 0–3 Å are reported for the three zeolite catalysts. One notes that in TON, in which the activation energy is the lowest, the match between the size of the transition-state structure and the zeolite micropore is ideal, which is not the case in CHA and MOR: in these zeolites, the oxygen atoms located to the opposite side of the micropore to which the reaction occurs are less involved in the stabilization of the transition state. Similar results are observed in tertiary carbocation-like transition states, although they are energetically less pronounced.

A striking result is that *i*-butoxide in CHA and TON becomes as stable as the 'free' tertiary carbocation. However, the energies that

CHA

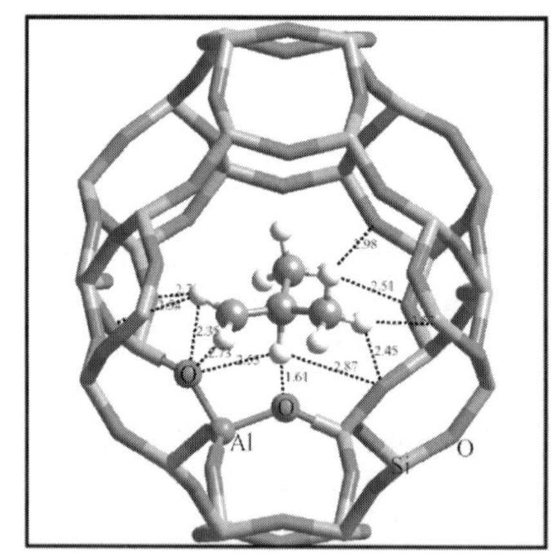

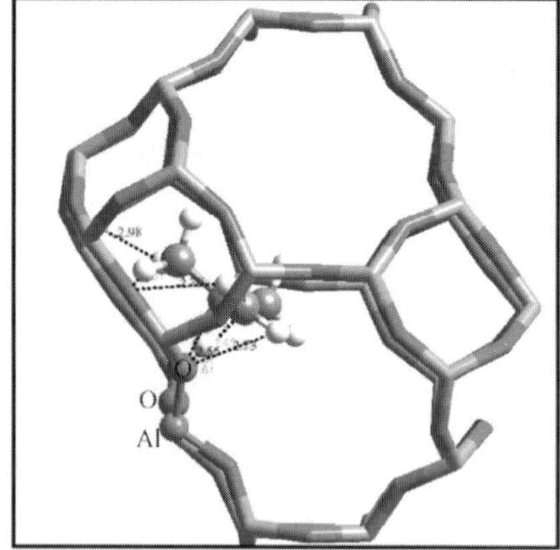

E_{act} = 138 kJ/mol

Fig. 16. Geometries and energies of the transition states of isobutene chemisorption to methyl-2-prop-2-oxide in different acidic zeolites. The broken lines report the interactions below 3 Å between hydrocarbon hydrogen atoms and zeolite oxygen atoms.

TON

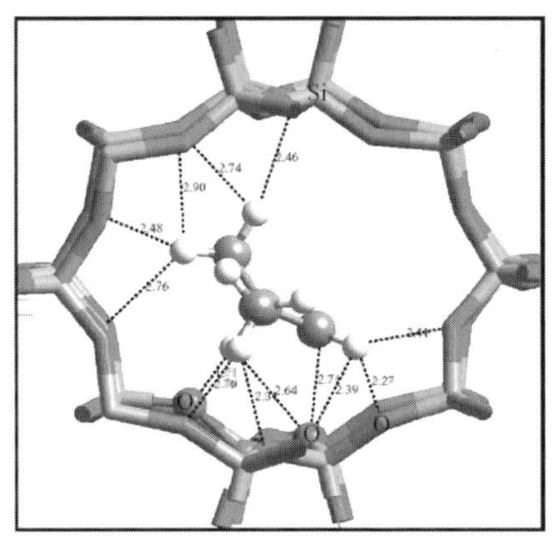

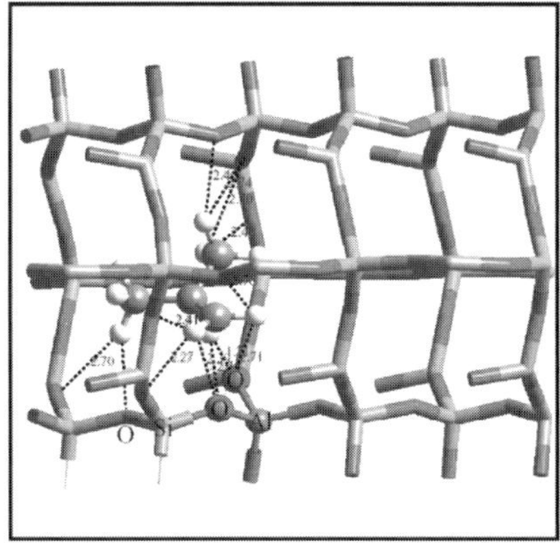

$E_{act} = 90$ kJ/mol

MOR

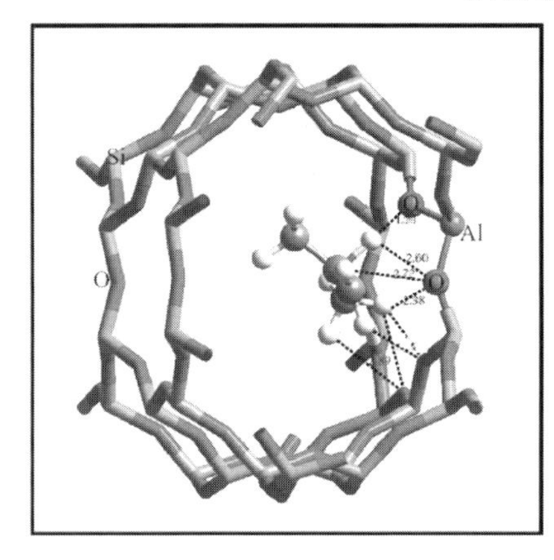

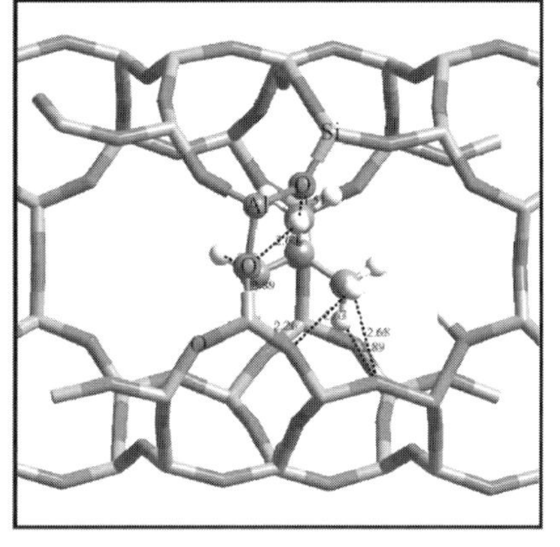

$E_{act} = 118$ kJ/mol

Fig. 16. (Continued)

are involved in the corresponding elementary reaction steps are so small that fast equilibrium between the different systems should be obtained. Such a result agrees with experimental observation [54–60]. In CHA, another contribution is playing an important role. Steric constraints appear to destabilize both transition-state structures and alkoxy species. The steric constraints will be discussed below.

The stabilization of a transition-state structure where there is a good match between the size of the transition state and that of the zeolite framework is unsurprising. The polarization of the zeolite oxygen atoms by the cationic charge is important in the stabilization of carbocationic transition states. Induced dipole-charge electrostatic contributions rapidly decrease with distance. A large micropore gives little TS stabilization. Too small, a micropore zeolite induces steric constraints that destabilizes the TS. Ideal-sized micropores are those for which the TS complex has the best fit. This effect accounts for around 10% of the activation energies in the case of isobutene chemisorption in CHA, TON, and MOR.

From a practical, computational point of view, the most suitable and computationally inexpensive method to determine activation energies is: (i) to determine the transition-state structure geometry using cluster approach calculations, (ii) to evaluate how good is the match of a transition-state complex into micropores of an exhaustive set of representative zeolite crystals using classical dynamic or Monte Carlo simulations. Such an approach is followed by several groups [61,62].

3.4.2. The size and shape of the transition state

We now consider an alternative to the above approach to estimate the effect of the zeolite micropore size and shape on the stabilization of the TS structure. Different size thiophenic derivatives were considered in the thiophenic ring-cracking reaction catalyzed by acidic mordenite. The comparison of the data that were obtained using the periodic DFT method [63] with that of the cluster approach method [32,33] is especially revealing when different-sized transition states are compared.

The thiophenic ring-opening reaction has been shown to be the first step in the hydrodesulfurization of thiophenic derivatives catalyzed by an acidic zeolite (see Fig. 17) [33,64]. Moreover, it is the rate-determining step in this reaction [33,64]. The alkoxy species that is

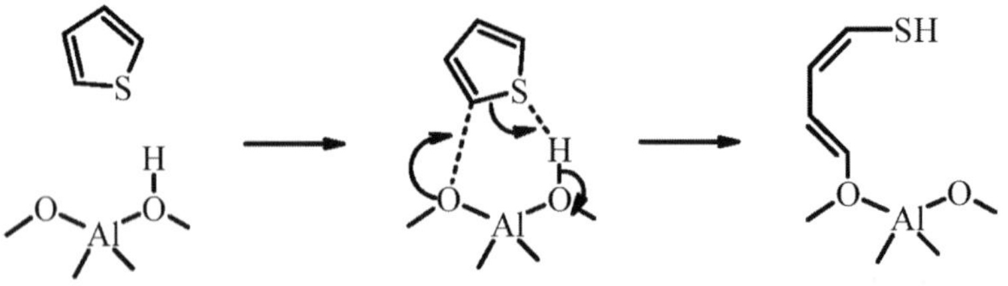

Fig. 17. Reaction mechanisms in the cracking of thiophenic ring catalyzed by acidic zeolite.

formed in this reaction is set free from the zeolite wall when H_2 dissociates. In this step, the Brønsted acidic site is restored. The hydrocarbon molecule becomes a thiol-species from which the sulfur atom can easily be removed [33,64].

The geometries of transition states for cracking are shown in Fig. 18. The thiophene transition-state structure does not closely interact with all parts of the zeolite channel. The dihydrothiophene transition state experiences more contact with the zeolite wall and interacts with all sides of the zeolite micropore. The geometries of the transition states of cracking of thiophene and dihydrothiophene are very similar to those found in cluster calculations. However for dibenzothiophene, a major difference occurs: in the periodic system, the sulfur atom has become protonated in the transition state, contrary to what is observed in the cluster calculations.

A surprising result concerning the calculations on the cracking of thiophene derivatives is that transition-state structures are no longer stabilized by their interactions with the zeolite framework contrary to what is observed with olefin [35,43] or aromatics [34,37] (see Table 2). This feature is apparent when the results of cluster calculations are

Thiophene Cracking

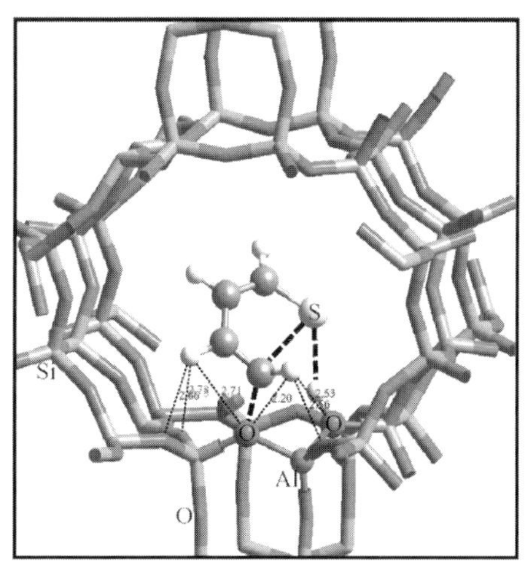

$E_{act} = 318$ kJ/mol

Fig. 18. Geometries and energies of the transition states of thiophene, dihydrothiophene, and DBT thiophenic ring cracking catalyzed by acidic mordenite as obtained from the periodic electronic structure calculations. The broken lines report the interactions below 3 Å between hydrocarbon hydrogen atoms and zeolite oxygen atoms.

Dihydrothiophene
Cracking

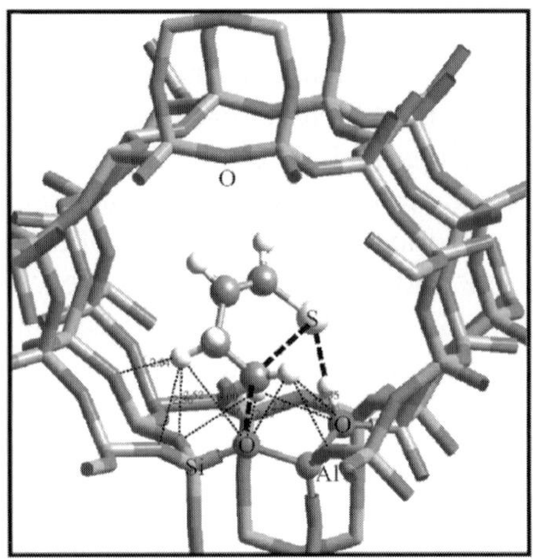

E_{act} = 272 kJ/mol

DBT Cracking

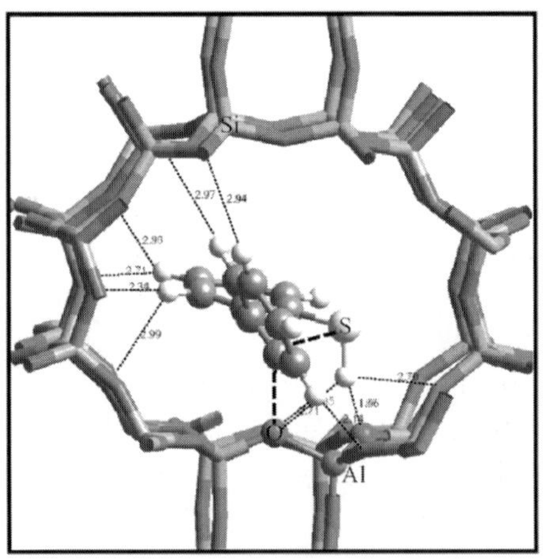

E_{act} = 314 kJ/mol

Fig. 18. (Continued)

compared to results from periodic calculations. Let us consider dibenzothiophene as a specific case: as can be seen from the energy difference between the physisorbed system and the alkoxy species, large steric constraints are involved in this system.

Table 2
Calculated thermodynamic data in cracking of thiophenic derivatives[a]

Reactant	E_{act}		ΔE (chemisorbed–physisorbed)	
	Periodic	Cluster	Periodic	Cluster
Thiophene	318	226	128	108
Dihydrothiophene	272	214	83	73
DBT	314	288	180	90

[a]All data in kJ/mol.

Table 3
Comparative analysis of energy stabilization of some transition states by zeolite framework and of Mulliken charge of the corresponding TS structure

Reaction	Mulliken charge	ΔE_{act} (periodic-cluster)
Thiophene cracking	0.25	92
Dihydrothiophene cracking	0.28	58
Propylene chemisorption	0.60	−34
Toluene isomerization	0.88	−100

The source of the destabilization of the thiophenic transition state is clarified from further analysis of the systems. The stabilization/destabilization of different transition states shows interesting trends when related to the Mulliken charge that is carried by the activated hydrocarbon (see Table 3). It is the positive charge of the carbocation transition state that induces polarization of the zeolite oxygen atoms. Hence, more charge separation in the transition state leads to an increase in the contribution to transition-state energy of the electrostatic term due to the induced dipoles on the zeolite framework. However, when the charge separation in the transition state is insufficient, this stabilization turns into a destabilization.

Comparison of thiophene and dihydrothiophene cracking reveals that the size of the transition state is important. There are two additional hydrogen atoms in dihydrothiophene compared to thiophene. The destabilization of the transition state for dihydrothiophene cracking is 58 kJ/mol, whereas it is 92 kJ/mol for thiophene. This large difference is not explained by the difference in the Mulliken charge.

In dibenzothiophene, we mentioned that steric constraints alter the picture. We will describe in a later section how steric constraints change reaction energies. It is possible to estimate and to correct for this change using the Polanyi–Evans–Brønsted relation [47,48], to be discussed in Sections 4 and 5. Application of this correction leads to an estimated value for the activation energy of dibenzothiophene cracking in the absence of steric constraints of the order ~270 kJ/mol. The transition-state structure for dibenzothiophene cracking is actually stabilized by its contact with the zeolite framework when the zeolite steric constraints contribution can be ignored. We mentioned that contrary to the results for cluster calculations, dibenzothiophene was protonated in the periodic transition state. Therefore, the charge separation is more complete in this TS than in the transition state found by the cluster calculations. Furthermore, the fit of the transition-state structure to the micropore is better than for thiophene or dihydrothiophene.

3.5. Synergy between Brønsted acidic site and Lewis basic site

In all the periodic studies that have been undertaken [35,39,43,63], we systematically designed the systems to avoid the possibility of transition-state structures having contact with two catalytic sites. This procedure helped in comparative analyses of results for the periodic and cluster approaches, as the small zeolite cluster models we used contain only one Brønsted acid site [32–34].

However, it is known from experimental studies [65] that the reactivity is sometimes enhanced when the hydrocarbon interacts with a Brønsted acidic site as well as a Lewis basic site.

As all atoms in the periodic unit cell are described at the same quantum mechanical level, it is quite simple to estimate such an effect, which we considered for the methylshift isomerization of toluene in mordenite. In one case, the transition state interacts with a single catalytic site (the second being quite distant). In others, the transition-state complexes interact with a second catalytic site, and, more specifically, with Lewis basic oxygen. In another case, the transition-state structure and physisorbed toluene interact with the proton of the second Brønsted acid site. The geometries of the TS and activation energies with respect to the corresponding ground state are shown in Fig. 19.

As can be seen in Fig. 19, the associative effect of several catalytic sites can substantially enhance the reactivity. The decrease in activation energy is large enough to favor reaction on the specific location where the Brønsted acidic site and Lewis basic site can interact with the transition-state structure. Interaction of the transition-state complex

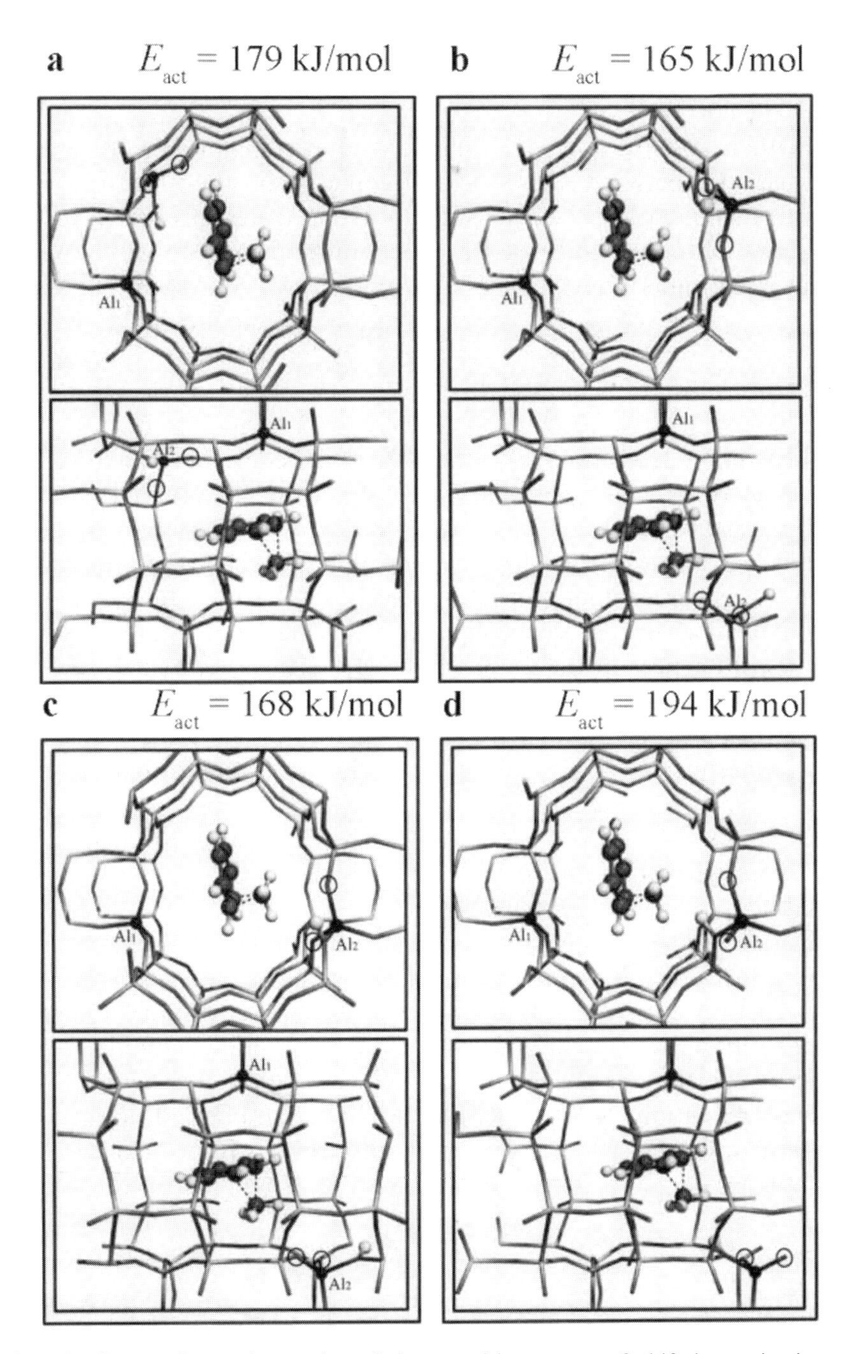

Fig. 19. Geometries and energies of the transition states of shift isomerization of toluene catalyzed by acidic mordenite. The location of the second catalytic changes (see Al_2). The position of the Lewis base oxygen atoms is enhanced by the circles.

with the spectator acidic proton leads to a more unfavorable situation than when the TS complex does not experience contact with the second catalytic site.

Further stabilization of transition-state complexes relate to an adequate fit between the specific geometry of the transition-state complex and the local shape and local electronic properties of the zeolite micropore. Unfavorable configurations can also, however, result as in Fig. 19d.

4. Transition-state shape selectivity

In this section, transition-state selectivity will be discussed based on the results of periodic calculations. The study of the alkylation of benzene with methanol catalyzed by acidic mordenite was the first quantum chemistry study ever of such a phenomenon [36,41]. After this first case, other studies were undertaken to identify the basis of transition-state selectivity in microporous zeolites [37–41]. Transition-state selectivity follows differences in the fit of the substrate transition-state structure with the zeolite micropore space. Steric constraints between the zeolite framework and the TS structure increase the system energy. Then, the reaction tends to follow other pathways in which transition-state structures show less or no steric constraints [37].

4.1. Isomerization of dimethyldibenzothiophene

Isomerization of dimethylated dibenzothiophene provides a very good illustration of transition-state selectivity (see Fig. 20) [38]. The reaction energy diagram of the shift isomerization of dimethyl-dibenzothiophenes catalyzed by acidic mordenite obtained using periodic calculations is shown in Fig. 21.

The activation energy of the 4,6-dimethyldibenzothiophene to 3,6-dimethyldibenzothiophene isomerization with respect to adsorbed

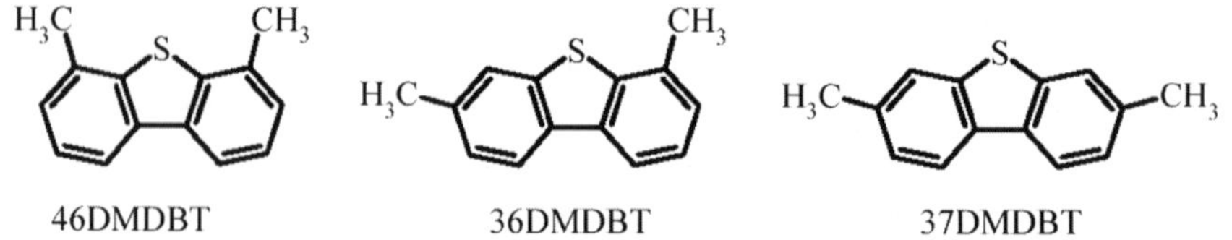

Fig. 20. Isomers of dimethyldibenzothiophene.

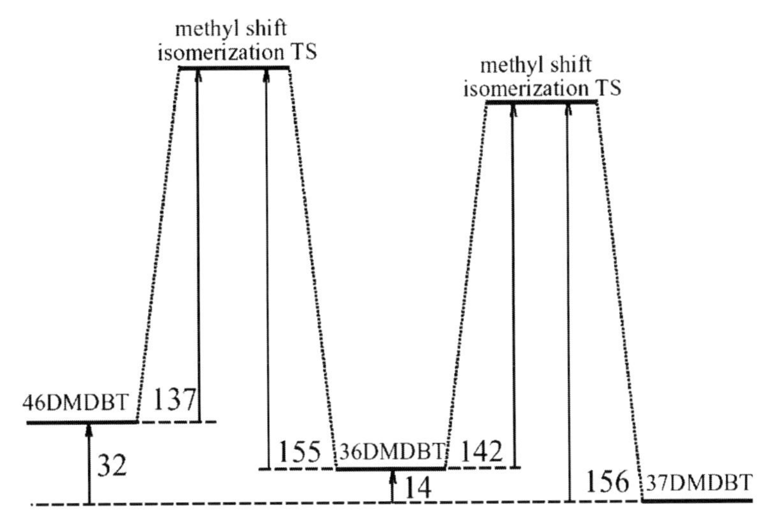

Fig. 21. Reaction energy diagram of the shift isomerization of DMDBT catalyzed by acidic mordenite as obtained from the periodic electronic structure calculations. All data in kJ/mol.

4,6-dimethyldibenzothiophene is $E_{act} = +137$ kJ/mol. It is $E_{act} = +142$ kJ/mol for the 3,6-dimethyldibenzothiophene to 3,7-dimethyldibenzothiophene reaction with respect to adsorbed 3,6-dimethyldibenzothiophene. Interestingly, the activation energies for these two reactions are similar. The same is observed for the isomerizations from 3,6- to 4,6-dimethyldibenzothiophene and 3,7- to 3,6-dimethyldibenzothiophene. In this case, the computed activation energies are $E_{act} = +155$ and $+156$ kJ/mol, respectively. For 3,6-dimethyldibenzothiophene, it is more likely that isomerization leads to the formation of 3,7-dimethyldibenzothiophene than 4,6-dimethyldibenzothiophene as the ΔE_{act} is 13 kJ/mol. In this case, transition-state selectivity favors the isomerization to 3,7-dimethyldibenzothiophene.

The full reaction energy diagram reveals other interesting facts (see Fig. 21). The energy levels of the different physisorbed dimethyldibenzothiophenes are very different from each other because of more or less sterically hampered dimethyldibenzothiophene methyl groups within the narrow mordenite pore.

With respect to the energy of adsorbed 4,6-dimethyldibenzothiophene, the relative energies of adsorbed 3,6-dimethyldibenzothiophene and adsorbed 3,7-dimethyldibenzothiophene are -18 and -32 kJ/mol, respectively. The destabilization of the TS structures follows linearly the destabilization of the physisorbed dimethyldibenzothiophene. One can estimate using the Polanyi–Evans–Brønsted relation [47,48]

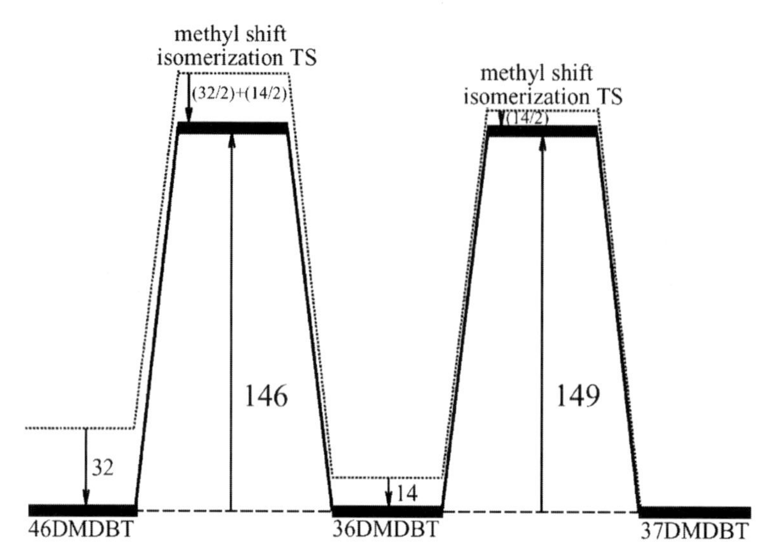

Fig. 22. Application of the Polanyi–Evans–Brønsted relation to the reaction energy diagram of the shift isomerization of DMDBT catalyzed by acidic mordenite.

that the activation energy for a methyl-shift isomerization of dimethyldibenzothiophene is $E_{act} = +148$ kJ/mol in the absence of steric constraints. The energy differences in the physisorbtion of the different dimethyldibenzothiophene isomers are only due to the steric constraints. The Polanyi–Evans–Brønsted relation can be used to deduce a reaction energy diagram in the absence of steric contraints (see Fig. 22).

4.2. Disproportionation of toluene and benzene

In this example, we will show how the Polanyi–Evans–Brønsted relation can be used in the analysis of reaction energy diagrams. From experimental studies in mordenite, the disproportionation between aromatics is known to be suppressed because of steric constraints [66]. Therefore, this zeolite was selected in the periodic quantum chemical calculations of this reaction. Suppression of disproportionation implies that transition-state structures are destabilized and that transition-state selectivity redirects the reaction pathway to monomolecular isomerization. This reaction has been studied in both the cluster approach method [34] and with the periodic method [39]. The reaction mechanism for disproportionation is shown in Fig. 23. No steric constraints exist in the cluster approach method as the zeolite framework is absent. Further analysis of the data was undertaken using previous findings [34,37] and the Polanyi–Evans–Brønsted relation (see Table 4).

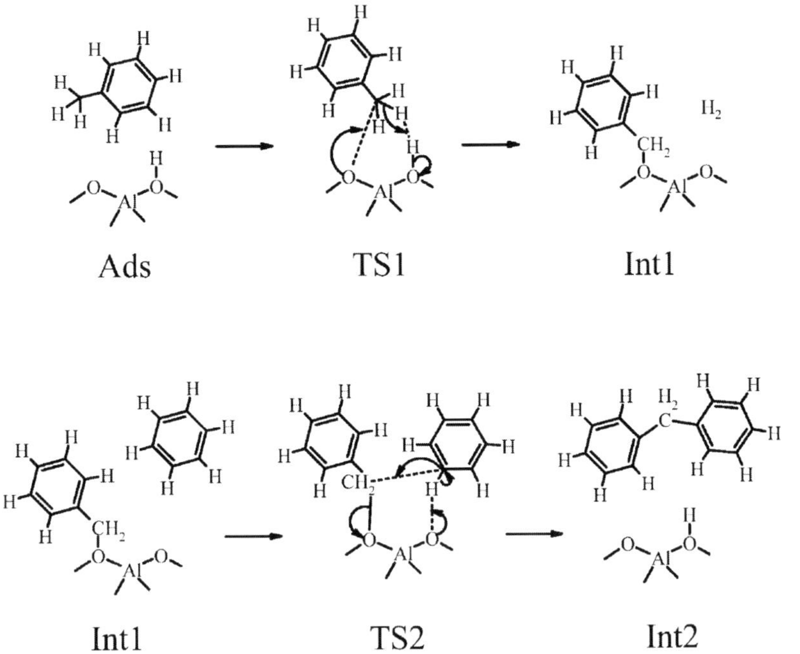

Fig. 23. Reaction mechanisms of the intermolecular isomerization of toluene catalyzed by acidic zeolite.

Table 4
Illustration of the corrective method to estimate thermodynamic data of reaction in zeolite catalyst based on cluster approach data[a]

	1	2	3	4
TS1	297	197	223	221
Int1	106	106	157	157
TS2	253	153	189	193
Int2	47	47	67	67

[a]All data with respect to physisorbed toluene and benzene (in kJ/mol). (1) Cluster approach data [34]. (2) Energy levels of TS are reduced by 100 kJ/mol (see Table 3, and Refs. [34,37,39]). (3) Energy levels of Int1 and Int2 are set to the same than that of periodic calculations, and activation energies are corrected using the Polanyi–Evans–Brønsted relation. (4) Periodic electronic structure calculation data [39].

It appears that the prediction of activation energies can be undertaken with very good accuracy using the cluster approach. This implies that the important computational effort that is required in the transition-state search can be reduced in some circumstances. It makes possible the prediction of activation energies of the disproportionation reaction in the absence of steric constraints. There is good agreement with experimental data: aromatic isomerization follows the disproportionation reaction pathway where there are no steric constraints [33,37,39].

4.3. Monomolecular isomerization of **para**-*xylenes*

Previously, we briefly mentioned that when steric constraints between the zeolite framework and the TS structure increase the system energy, the reaction tends to follow other pathways in which transition-state structures show less or no steric constraints. We will illustrate this feature with the results obtained for monomolecular isomerization of xylenes [37]. Let us consider the methylshift isomerization reaction of *para*-xylene. Two different transition-state structures, both corresponding to methylshift isomerization, were obtained (see Fig. 24). To reach one or the other, different reaction pathways have to be followed. As previously observed for the case of toluene (see Section 3.3 and Ref. [37]), *para*-xylene is being protonated, and the system evolves to a nonstable Wheland complex. From this point, the methylshift transition state can directly be reached (see Fig. 24, left), or the nonstable Wheland complex can change its orientation with respect to the catalytic site prior to the methylshift isomerization (Fig. 24, right). These two alternative reaction pathways were also obtained in periodic calculations in toluene isomerization [37]. The activation energies in toluene isomerization were the same (i.e. $E_{act} \sim 180$ kJ/mol) unlike the case for *para*-xylene isomerization.

As seen in Fig. 24, in the *para*-xylene methylshift transition-state structure (right), the nonparticipating methyl group is close to the zeolite wall. In the other case (left in Fig. 24), the methyl group is oriented along the large 12-membered ring channel of mordenite, and

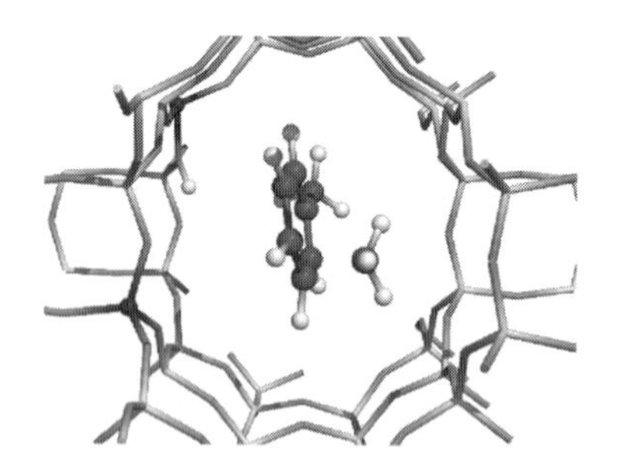

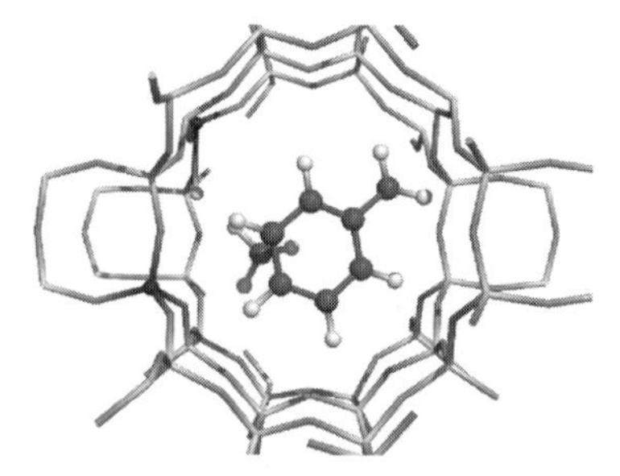

Fig. 24. Front views of the transition states of *para*-xylene methylshift isomerization catalyzed by proton-exchanged mordenite that occur with (right) and without (left) reorientation of the molecule after protonation as obtained from DFT periodic calculations.

hence does not suffer from steric constraints with the zeolite wall. This results in two different activation energies, which prohibit the reaction pathway via transition state (as shown in Fig. 24, right). The difference in activation energy is $\Delta E_{act} = 63$ kJ/mol, and the activation energy in the more favourable case is $E_{act} = 171$ kJ/mol.

5. On the use of the Polanyi–Evans–Brønsted relation

As discussed earlier, we have used the Polanyi–Evans–Brønsted relation in several cases to provide further analysis of the data [35–39,43,63]. This relation is an useful tool in gaining more insight into these systems. However, it is not recommended to use the relation without full calculation of all intermediates and transition states in a system.

Let us describe in more detail what is the basis of the relation [47,48]. In the days of quantum chemistry, calculations of the type which are now routine, were not feasible, although useful semiempirical calculations were achieved [48]. To investigate transition states, the only way was the scanning of a potential energy surface. Of course, this method could only be applied to simple systems (i.e. three center systems), from the results of which it was possible to make an estimate of the activation energy in the reaction (see Fig. 25).

Following this approach, Polanyi and Evans [48] observed that a change in the energy level of reactant and/or product induced a linear change in the activation energy (see Fig. 25). They identified that the parameter in the linear relationship in energy change between reaction energy and activation energy was around a half, although the value of this parameter changes between 0 and 1, depending on whether the transition state has an earlier or later character in the reaction coordinate. The Polanyi–Evans–Brønsted relation was confirmed as valid at DFT level for metal [70], and zeolite-catalyzed reactions [71]. Moreover, application of this relation to zeolite catalysis provides important information. In principle, it can be used only as long as reaction paths do not change. Structural factors may affect, however, this limitation significantly. In addition, in Section 2.1, we noted that the relation is valid as long as one estimates activation energies in systems for which the size of the transition state and the zeolite catalyst structure are the same.

However, as soon as one of these conditions is not fulfilled, the Polanyi–Evans–Brønsted relation is unsuccessful in establishing a dependence between reaction energy and activation energy (see Sections 3.4 and 3.5).

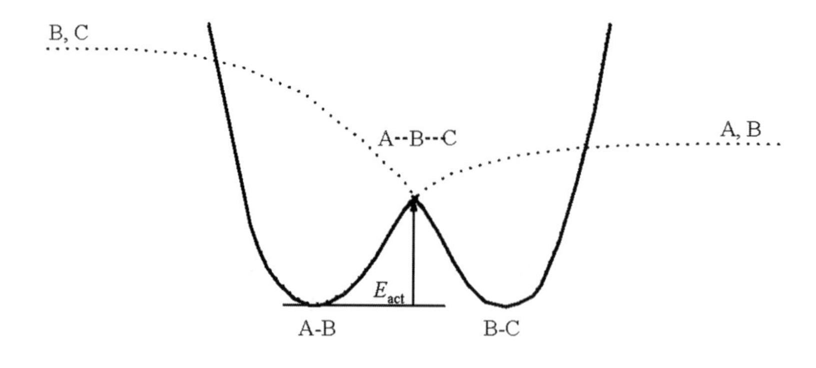

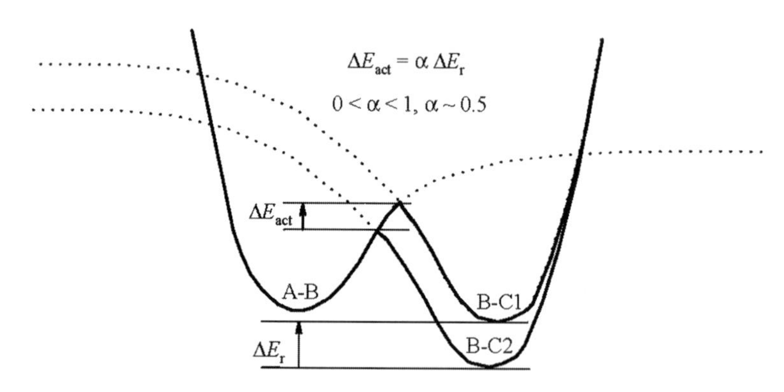

Fig. 25. The Polanyi–Evans–Brønsted relation as historically obtained [48].

As discussed previously, the zeolite framework stabilizes carbocationic transition-state structures. This stabilization is mainly caused by short-range electrostatic interactions between the zeolitic oxygen atoms and the transition-state structures. Let us return to the case of isobutene chemisorption in proton-exchanged zeolites (see Section 3.4.1). In this study, the zeolite frameworks are very different for CHA, TON, and MOR zeolites. Therefore, the interactions between the zeolite framework and TS structures differ, as also do the stabilization energies of TS structures by the zeolite frameworks. However, the interaction of neutral species (viz. physisorbed isobutene and covalently bonded alkoxides) with the zeolite framework is not of same nature as that which dominates the transition-state structures. In the first case, van der Waals energy contributions dominate, whereas in the latter case it is the short-range electrostatic contributions. Differences in the stabilization energy of transition-state complexes that result from size differences of the zeolite cage cannot be estimated from the energy variation of reactant or product. Hence, the Polanyi–Evans–Brønsted relation is unsuccessful in this case.

6. Concluding remarks

This chapter has focused on important mechanistic concepts derived from recent results. Details have been published elsewhere [32–41,43,67–69]. A major advance has been the introduction of full periodic structure calculations in quantum chemical studies of elementary reaction steps, in which some cases have significantly refined results obtained using approximate cluster models [41].

The flexibility of the zeolite framework is important in accurately describing reactivity. This flexibility is largely a local property. However, macro-rearrangements of zeolite crystal can also be important in catalysis [49,72]. These effects cannot be practically described using periodic electronic structure calculations as the size of the unit cell that can be reasonably calculated remains limited (i.e. a maximum of 300–500 zeolite atoms, which gives a maximum size of unit cell around $20 \times 20 \times 20$ Å^3).

Short-range polarization of the zeolite framework oxygen atoms is essential for the stabilization of transition-state complexes. The importance of matching transition-state structure size and shape with the dimensions of the zeolite micropores supports the idea that improved zeolite catalysts can be developed with greater specificity. The lock and key principle [73] also controls transition-state selectivity in zeolite catalysis.

Studies of effect of the influence of a second catalytic site has revealed the importance of such an interaction with TS complexes. The periodic electronic structure method is the more suitable tool to analyze this effect as all atoms in the unit cell are treated at an equal quantum chemistry level.

Finally, transition-state selectivity has been studied. Polanyi–Evans–Brønsted relation has been shown to be extremely useful to analyze the importance of steric constraints effect to a reaction pathway.

Acknowledgements

The studies described in this work have been supported by Totalfinaelf in the framework of the European Research Group 'Ab Initio Molecular Dynamics Applied to Catalysis', supported by the Centre National de la Recherche Scientifique (CNRS), the Institut Français du Pétrole (IFP), and Totalfinaelf, in collaboration with the University of Vienna and the Technical University of Eindhoven. The authors thank Dr. F. Hutschka from Totalfinaelf company, and Professor J. Hafner from Vienna University.

References

1. Van Santen, R.A. and Kramer, G., *J. Chem. Rev.*, **95**, 637–660 (1995).
2. Meier, W.M., Olson, D.H. and Baerloch, C., *Zeolites*, **17**, 1–230 (1996).
3. Thomas, J.M. and Thomas, W.J., *Principles and Practice of Heterogeneous Catalysis*. VCH Publishers, New York, 1997.
4. Fraenkel, D. and Levy, M., *J. Catal.*, **118**, 10–21 (1989).
5. Chen, N.Y., Degnan Jr., T.F. and Smith, C.M., *Molecular Transport and Reaction in Zeolites, Design and Application of Shape Selective Catalysts*. VCH Publishers, New York, 1994, pp. 195–289.
6. Tsai, T.-C., Liu, S.-B. and Wang, I., *Appl. Catal. A*, **184**, 355–398 (1999).
7. Van Santen, R.A. and Niemantsverdriedt, J.W., *Chemical Kinetics and Catalysis*. Plenum Press, New York, 1995, pp. 243–247.
8. Van Santen, R.A. and Rozanska, X., In: Chakraborty, A. (Ed.), *Molecular Modeling and Theory in Chemical Engineering*; In: Wei, J., Seinfeld, J.H., Denn, M.M., Stephanopoulos, G., Chakraborty, A., Ying, J. and Peppas, N. (Eds.), *Adv. Chem. Eng.* Academic Press, New York, 2001, Vol. 28, pp. 399–437.
9. Fricke, R., Kosslick, H., Lischke, G. and Richter, G., *Chem. Rev.*, **100**, 2303–2405 (2000).
10. De Man, A.J.M. and Van Santen, R.A., *Zeolites*, **12**, 269–279 (1992).
11. Lee, C., Parrillo, D.J., Gorte, R.J. and Farneth, W.E., *J. Am. Chem. Soc.*, **118**, 3262–3268 (1996).
12. Gorte, R.J., *Catal. Lett.*, **62**, 1–13 (1999).
13. De Man, A.J.M., Van Beest, B.W.H., Leslie, M. and Van Santen, R.A., *J. Phys. Chem.*, **94**, 2524–2543 (1990).
14. Schröder, K.P. and Sauer, J., *J. Phys. Chem.*, **100**, 11043–11049 (1996).
15. Yang, L., Trafford, K., Kresnawahjuena, O., Sepa, J. and Gorte, R.J., *J. Phys. Chem. B*, **105**, 1935–1942 (2001).
16. Zhen, S. and Seff, K., *Microporous Mesoporous Mater.*, **39**, 1–18 (2000).
17. Guisnet, M. and Magnoux, P., *Catal. Today*, **36**, 477–483 (1997).
18. Barrer, R.M., *Zeolite and Clay Minerals as Sorbents and Molecular Sieves*. Academic Press, New York, 1978, pp. 18–19.
19. Nicholas, J.B., *Topics Catal.*, **4**, 157–171 (1997).
20. Rigby, A.M., Kramer, G.J. and Van Santen, R.A., *J. Catal.*, **170**, 1–10 (1997).
21. Van Santen, R.A., *Catal. Today*, **38**, 377–390 (1997).
22. Haw, J.F., Richardson, B.R., Oshiro, I.S., Lazo, N.D. and Speed, J.A., *J. Am. Chem. Soc.*, **111**, 2052–2058 (1989).
23. Haw, J.F., Nicholas, J.B., Xu, T., Beck, L.W. and Ferguson, D.B., *Acc. Chem. Res.*, **29**, 259–267 (1996).
24. Haw, J.F., Xu, T., Nicholas, J.B. and Goguen, P.W., *Nature*, **389**, 832–835 (1997).
25. Kazansky, V.B. and Senchenya, I.N., *J. Catal.*, **119**, 108–120 (1989).

26. Sinclair, P.E., De Vries, A., Sherwood, P., Catlow, C.R.A. and Van Santen, R.A., *J. Chem. Soc. Faraday Trans.*, **94**, 3401–3408 (1998).
27. Rigby, A.M. and Frash, M.V., *J. Mol. Catal. A*, **126**, 61–72 (1997).
28. Olah, G.A. and Donovan, D.J., *J Am. Chem. Soc.*, **99**, 5026–5039 (1997).
29. Olah, G.A., Prakash, G.K.S. and Sommer, J., *Superacids*. Wiley Interscience, New York, 1985.
30. Venuto, P.B., *Microporous Mesoporous Mater.*, **2**, 297–411 (1994).
31. Hölderich, W.F. and Van Bekkum, H., In: Van Bekkum, H., Flaningen, E.M. and Jansen, J.C. (Eds.), *Introduction to Zeolite Science and Practice*. Elsevier, Amsterdam, The Netherlands, 1991, Vol. 58, pp. 631–726.
32. Rozanska, X., Van Santen, R.A. and Hutschka, F., *J. Catal.*, **200**, 79–90 (2001).
33. Rozanska, X., Saintigny, X., Van Santen, R.A., Clémendot, S. and Hutschka, F., *J. Catal.*, **208**, 89–99 (2002).
34. Rozanska, X., Saintigny, X., Van Santen, R.A. and Hutschka, F., *J. Catal.*, **202**, 141–155 (2001).
35. Rozanska, X., Demuth, T., Hutschka, F., Hafner, J. and Van Santen, R.A., *J. Phys. Chem. B*, **106**, 3248–3254 (2002).
36. Vos, A.M., Rozanska, X., Schoonheydt, R.A., Van Santen, R.A., Hutschka, H. and Hafner, J., *J. Am. Chem. Soc.*, **123**, 2799–2809 (2001).
37. Rozanska, X., Van Santen, R.A., Hutschka, F. and Hafner, J., *J. Am. Chem. Soc.*, **123**, 7655–7667 (2001).
38. Rozanska, X., Van Santen, R.A., Hutschka, F. and Hafner, J., *J. Catal.*, **205**, 388–397 (2002).
39. Rozanska, X., Van Santen, R.A. and Hutschka, F., *J. Phys. Chem. B*, **106**, 4652–4657 (2002).
40. Rozanska, X., Van Santen, R.A. and Hutschka, F., In: Nascimento, M.A.C. (Ed.), *Theoretical Aspects of Heterogeneous Catalysis*. Kluwer Academic Press, Dordrecht, The Netherlands, 2001, pp. 1–28.
41. Rozanska, X. and Van Santen, R.A., In: Auerbach, S.M., Carrado, K.A. and Dutta, P.K. (Eds.), *Handbook of Zeolite Catalysts and Microporous Materials*. Marcel Dekker, Inc., New York, 2002, pp. 785–832.
42. Derouane, E.G., He, H., Hamid, S.B.D.-A. and Ivanova, I.I., *Catal. Lett.*, **58**, 1–18 (1999).
43. Rozanska, X., Van Santen, R.A., Demuth, T., Hafner, J. and Hutschka, F., *J. Phys. Chem. B*, **107**, 1309–1315 (2003).
44. Paukshtis, E.A., Malysheva, L.V. and Stepanov, V.G., *React. Kinet. Catal. Lett.*, **65**, 145–152 (1998).
45. Martens, J.A. and Jacobs, P.A., In: Ertl, G., Knözinger, H. and Weitkamp, J. (Eds.), *Handbook of Heterogeneous Catalysis*. VCH, Weinheim, Germany, 1997, pp. 1137–1149.
46. De Gauw, F.J.M.M., Van Grondelle, J. and Van Santen, R.A., *J. Catal.*, **204**, 53–63 (2001).
47. Brønsted, J.N., *Chem. Rev.*, **5**, 231–338 (1928).
48. Evans, M.G. and Polanyi, N.P., *Trans. Faraday Soc.*, **34**, 11–29 (1938).
49. Hammonds, K.D., Deng, H., Heine, V. and Dove, M.T., *Phys. Rev. Lett.*, **78**, 3701–3704 (1997).
50. Rozanska, X., Van Santen, R.A. and Hutschka, F., unpublished results.
51. Ugliengo, P., Ferrari, A.M., Zecchina, A. and Garrone, E., *J. Phys. Chem.*, **100**, 3632–3645 (1996).

52. Vollmer, J.M. and Truong, T.N., *J. Phys. Chem. B*, **104**, 6308–6312 (2000).
53. Boronat, M., Zicovich-Wilson, C.M., Corma, A. and Viruela, P., *Phys. Chem. Chem. Phys.*, **1**, 537–543 (1999).
54. Kondo, J.N., Yoda, E., Wakabayashi, F. and Domen, K., *Catal. Today*, **63**, 305–308 (2000).
55. Domokos, L., Lefferts, L., Seshan, K. and Lercher, J.A., *J. Mol. Catal. A*, **162**, 147–157 (2000).
56. Rutenbeck, D., Papp, H., Freude, D. and Schwieger, W., *Appl. Catal. A*, **206**, 57–66 (2000).
57. Rutenbeck, D., Papp, H., Ernst, H. and Schwieger, W., *Appl. Catal. A*, **208**, 153–161 (2001).
58. Patrigeon, A., Benazzi, E., Travers, Ch. and Bernhard, J.Y., *Catal. Today*, **65**, 149–155 (2001).
59. Philippou, A., Dwyer, J., Ghanbari, A., Pazè, C. and Anderson, M.W., *J. Mol. Catal. A*, **174**, 223–230 (2001).
60. Patrylak, K.I., Bobonych, F.M., Voloshyna, Y.G., Levchuck, M.M., Solomakha, V.M., Patrylak, L.K., Manza, I.A. and Taranookha, O.M., *Catal. Today*, **65**, 129–135 (2001).
61. Clark, L.A., Ellis, D.E. and Snurr, R.Q., *J. Chem. Phys.*, **114**, 2580–2591 (2001).
62. Schenk, M., Berend, S., Vlugt, T.J.H. and Maesen, T.L.M., *Angew Chem. Int. Ed.*, **40**, 736–739 (2001).
63. Rozanska, X., Van Santen, R.A., Hutschka, F. and Hafner, J., *J. Catal.*, **214**, 68–77 (2003).
64. Saintigny, X., Van Santen, R.A., Clémendot, S. and Hutscka, F., *J. Catal.*, **183**, 107–118 (1999).
65. Adeeva, V., Liu, H.Y., Xu, B.-Q. and Sachtler, W.M.H., *Topics Catal.*, **6**, 61–76 (1998).
66. Halgeri, A.B. and Das, J., *Appl. Catal. A*, **181**, 347–354 (1999).
67. Rozanska, X., Van Santen, R.A. and Hutschka, F., In: Galarneau, A., Di Renzo, F., Fajula, F. and Vedrine, J. (Eds.), *Zeolites and Mesoporous Materials at the Dawn of the 21st Century*. Elsevier, Amsterdam, The Netherlands, 2001, *Stud. Surf. Sci. Catal.*, Vol. 135, pp. 2611–2617.
68. Vos, A.M., Rozanska, X., Schoonheydt, R.A., Van Santen, R.A., Hutschka, F. and Hafner, J., In: Galarneau, A., Di Renzo, F., Fajula, F. and Vedrine, J. (Eds.), *Zeolites and Mesoporous Materials at the Dawn of the 21st Century*. Elsevier, Amsterdam, The Netherlands, 2001, *Stud. Surf. Sci. Catal.*, Vol. 135, pp. 2461–2468.
69. Rozanska, X., Van Santen, R.A., Hutschka, F. and Hafner, J., In: Galarneau, A., Di Renzo, F., Fajula, F. and Vedrine, J. (Eds.), *Zeolites and Mesoporous Materials at the Dawn of the 21st Century*. Elsevier, Amsterdam, The Netherlands, 2001, *Stud. Surf. Sci. Catal.*, Vol. 135, pp. 2596–2603.
70. Logadottir, A., Rod, T.H., Nørskov, J.K., Hammer, B., Dahl, S. and Jacobsen, C.J.H., *J. Catal.*, **197**, 229–231 (2001).
71. Kramer, G.J., Van Santen, R.A., Emeis, C.A. and Nowak, A.K., *Nature*, **363**, 529–531 (1993).
72. Kramer, G.J., De Man, A.J.M. and Van Santen, R.A., *J. Am. Chem. Soc.*, **113**, 6435–6441 (1991).
73. Behr, J.P., *The Lock and Key Principle. The State of the Art-100 Years On*. Wiley, Chichester, 1994.

Computer Modelling of Microporous Materials
C.R.A. Catlow, R.A. van Santen and B. Smit (editors)

Chapter 7

Structure and reactivity of metal ion species in high-silica zeolites

G.M. Zhidomirov* and A.A. Shubin**

Boreskov Institute of Catalysis, Siberian Branch of the Russian Academy of Sciences, Pr. Akad. Lavrentieva 5, Novosibirsk 630090, Russia

R.A. van Santen***

Schuit Institute of Catalysis, Eindhoven University of Technology, P.O. Box 513, 5600 MB, Eindhoven, The Netherlands

1. Introduction

In this chapter, the emphasis moves to reactivity at metal ion sites in zeolites. One of the recent trends in the development of heterogeneous catalysts is connected with metal-containing zeolites [1]. Metal cations can be introduced into a zeolite at the synthesis stage as well as via various methods of zeolite modification (such as liquid- or solid-state ion-exchange technique, chemical vapor deposition). There are now a number of unique catalysts (especially MFI zeolites-based ones) showing high activity and selectivity in the diverse processes. Typical examples of such catalysts and processes are:

- Ti-containing zeolites (hydrocarbon oxidation by hydrogen peroxide);
- Fe zeolites (direct oxidation of benzene into phenol by nitrous monoxide);
- Zn and Ga zeolite (dehydrogenation and aromatization of alkanes);
- Cu, Fe, and Co zeolites (SCR of NO_x).

*E-mail: zhi@catalysis.nsk.su
**E-mail: a.a.shubin@catalysis.nsk.su
***E-mail: r.a.v.santen@tue.nl

The key question in understanding how these catalysts work is the determination of the structure of the cation species. The latter depends in general on the lattice type, the zeolite composition (Si/Al ratio), the method of cation introduction and the redox treatments. One can consider various basic structural types of cation localization within the zeolite lattice: lattice positions, cation positions, so-called charged oxo-ions containing extra-lattice oxygen (ELO), as well as channel- and cavity-stabilized small neutral oxo-clusters.

Fine dispersion of the active component seems to be one of the main factors that make molecular sieves very promising for the development of efficient catalysts. Indeed, the metal ions localized both in the lattice and cationic positions are essentially the stabilized isolated ions embedded into silica surroundings. The latter may prevent the leaching of the active component during the reaction, as, for example, in the case of Ti-silicalites. It is known also that agglomeration of Ti ions can activate the side reactions and thus deteriorate the process selectivity [1]. When binuclear sites are needed for efficient catalytic action, high-silica zeolites are also promising since they allow the stabilization of binuclear oxo-ions. The zeolite matrix can enforce a high-energy geometry at the metal site and thus increase its chemical activity. Considerable attention has been focused in recent years on this problem in the current literature, and it will be one of the main topics discussed in the present review.

Another aspect of the correlation between the stability and reactivity of the cation species relates to possible competition between different structural forms of the cations located in various positions of the zeolite matrix, such as the above-mentioned lattice positions, cationic positions, oxo-ions, and oxo-clusters. The relative stability of these forms depends on numerous factors: zeolite structural type, the Si/Al ratio, the content of active component and the method of its insertion into zeolite matrix, subsequent oxidizing or reducing treatments, and conditions of the catalytic reaction. All these factors develop possibilities of controlling the formation of the optimum catalytic structure; but, on the other hand, lead to very complex phenomena that are difficult to reproduce and that complicate structural, spectroscopic, and kinetic studies.

The definition of the task for computer modeling of zeolite structures with embedded metal ions raises the problem of the choice of the calculation method (force-field approach, semiempirical or nonempirical quantum chemical methods, or density functional theory), as well as the problem of model choice (cluster models, embedding cluster models or periodical calculations). The two previous chapters have considered these issues in detail. We will return when appropriate to those methodological aspects, but our emphasis will be on the qualitative physicochemical significance of the calculated results.

2. Metal cations in lattice positions

As noted, cations can be occluded in the silicious lattice during zeolite synthesis or due to postsynthesis chemical treatment. Such isomorphic substitution can be comparatively easily provided for Al, Ga, B, Fe, Ti. Intensive search for other isomorphically substituted cations (V, Cr, Mn, Co, Cu, Zr, Sn, etc.) has been continued with a view to design promising materials with advanced physicochemical properties or novel catalysts (see Ref. [1]). Encouraging results have been reported, but subsequent structural studies often failed to prove the lattice localization of the cations. Embedding of three-valent cations (Al, Ga, B, Fe) into silicious lattice leads to effective negative charges localized in these sites and thus forms the cationic position for the metal ion. The latter acts as a counterion, which compensates the negative charge of the lattice. The proton counterions will give rise to Bronsted acidity, which is the most strong in the case of Al and Ga. Catalytic activity of these sites relates essentially to the counterions activity. The intriguing question as to whether there is direct participation of regular lattice cations such as Al(III), Ga(III), Fe(III) in adsorption and catalytic interactions accompanied with increasing of the coordination number by means of the reagents involved seems to have a negative answer. In this regard, the effect of high catalytic activity of the lattice Ti(IV) ions to selective oxidation of hydrocarbons by hydroperoxide is of particular interest.

The first task for theoretical analysis of such systems related to the studies of structural stability of titanium ions localized in the lattice positions of silicalite with respect to possible formation of small extra-lattice oxide/hydroxide clusters in the silicalite channels and cavities. Both the force-field method and comparative nonempirical quantum-chemical calculations of model lattice and extra-lattice titanium oxide clusters [2–4] proved an energetic preference for the substitution of Si by Ti at lattice sites. Since the ultimate concentration of titanium ions entrapped in the lattice positions is relatively low (2–3%), the question arises as to the preferential localization of Ti^{4+} ions in various crystallographic positions of the MFI structure. A set of theoretical and experimental studies has been devoted to this problem.

Reviewing the current state of the problem, Lamberti et al. [5] note that both calculation results reported in different works and interpretation of experiments vary considerably. Nevertheless, a key characteristic of local geometry — an extension of the Ti–O bond to ~ 1.8 Å in comparison to average length of Si–O bond equal to ~ 1.6 Å in ZSM-5 zeolite — is well proved by various calculation methods. At the same time, the conclusion on preferential localization of Ti ions in specific sites of the silicalite lattice depends strongly on the model choice and on the level of quantum-chemical simulation. Millini et al. [6] performed a comparative evaluation of the stability of clusters $Si[OSi(OH^*)_3]_4$, which kept the crystalline geometry of different lattice sites, and concluded that differences in local stabilization energies are spread over about ~ 30 kcal mol^{-1}. Here and subsequently H* means boundary H atom saturating dangling bond of a cluster. The T12, T10, T7, T1, and T11 sites of the silicalite structures were found to be the less-stable sites. One would have assumed that various structural defects including isomorphic substitution of Si by Ti could localize exactly in these lattice sites.

Similar simulation of clusters $Ti[OSi(OH^*)_3]_4$, having Ti–O = 1.8 Å, reported in the same paper, supported the above conclusion only partially and proved some preference of T12, T10, and T3 for the substitution. In general, the estimated difference in energetic stability of Ti ions localized in different silicalite positions did not exceed 4 kcal mol^{-1} that suggests an almost homogeneous distribution of Ti ions throughout the zeolite lattice. Essentially the same conclusion was reached by Jentys and Catlow [2]. The calculated stabilization energies differed by no more than 5 kcal mol^{-1}; T6 and T19 sites were slightly more favorable. Vayssilov and van Santen [7] found that energy of the clusters $Ti(OSIH_3^*)_4$, calculated with full and constrained energy optimization for T6, T8, and T12 lattice sites varied within less than 0.3 kcal mol^{-1}. Oumi et al. [8] applied the method of molecular dynamics (MD) to study the question how the Ti ions embedded into different lattice positions of the MFI matrix affect the anisotropy of the lattice expansion. Comparative analysis of the MD results and the data of the TS-1 structural studies suggested the T8 position to be the most probable for Ti isomorphous substitution. The combination of the Monte Carlo method with MD calculations [9] showed clear preference of T2 and T12 positions for the substitution. However, as noted by Riccardi et al. [10], the distribution of Ti ions through the lattice positions of zeolite should be controlled to some extent by the conditions of TS-1 synthesis, in particular by the presence of an aqueous solution, since hydration energies may increase (up to 10 kcal mol^{-1}) the energetic benefits of particular positions.

Some discrepancy between experimental results reported in various works is also observed. Indeed, the powder X-ray diffraction studies suggested the preference of the T10 and T11 positions [11], the powder neutron diffraction studies — of the T6, T7, and T11 [5] and of T3, T7, T8, T10, and T12 [12] positions. The latter work is of particular interest, because the authors used a semiempirical quantum-chemical method PM3(d) and force-field method to calculate the Ti distribution throughout the TS-1 lattice positions, and attempted to account directly for the template effect on that distribution. However, neither calculation result was consistent with experimental data reported in the same work.

The Ti distribution throughout the lattice positions of silicalite may be of key importance with respect to catalytic activity of TS-1 for selective oxidation of hydrocarbons by hydroperoxide. The activity may be related to the availability of an active site for reagents, as well as to a possible effect of the site structure on its reactivity; this latter question will be considered in the end of this section. It has generally been assumed that the key step of the process was the activation of hydroperoxide, accompanied by the hydrolysis of one or two of the Ti–O–Si bridges (see Fig. 1).

The initial structure (**1a**) includes Ti(IV) ion in effective tetrahedral coordination of a lattice position of silicalite. Structure (**1b**) may appear as a result of an interaction with a water molecule, or as a result of postsynthesis treatment [13]. As shown by Gleeson et al. [14], the contribution of that structure may be essential when the Ti concentration is ca. 1 wt.%.

The existence of titanyl Ti=O species (structure **1f**) in Ti-silicalites has been extensively discussed in the literature. Recently, defect site (**1f**) was suggested as the result of heat treatment of the catalyst and as the active site in the selective oxidation of cyclohexane with hydroperoxide [15]; this explains the higher catalytic activity resulting from catalyst treatment at increasing temperature. In systematic theoretical study of the mechanism of partial oxidation in Ti-silicalite catalysis [16], structure (**1f**) was discussed as a possible important intermediate in the process. It was noted that protic solvents should promote formation of titanyl species.

Currently, it is generally agreed both in experimental and theoretical studies that structure (**1c**) is critically important for the transfer of oxygen atom from catalyst to substrate molecule. This point will be discussed in more detail later. Structure (**1d**) is assumed to be less active for the oxidation reaction; moreover, it may be responsible for

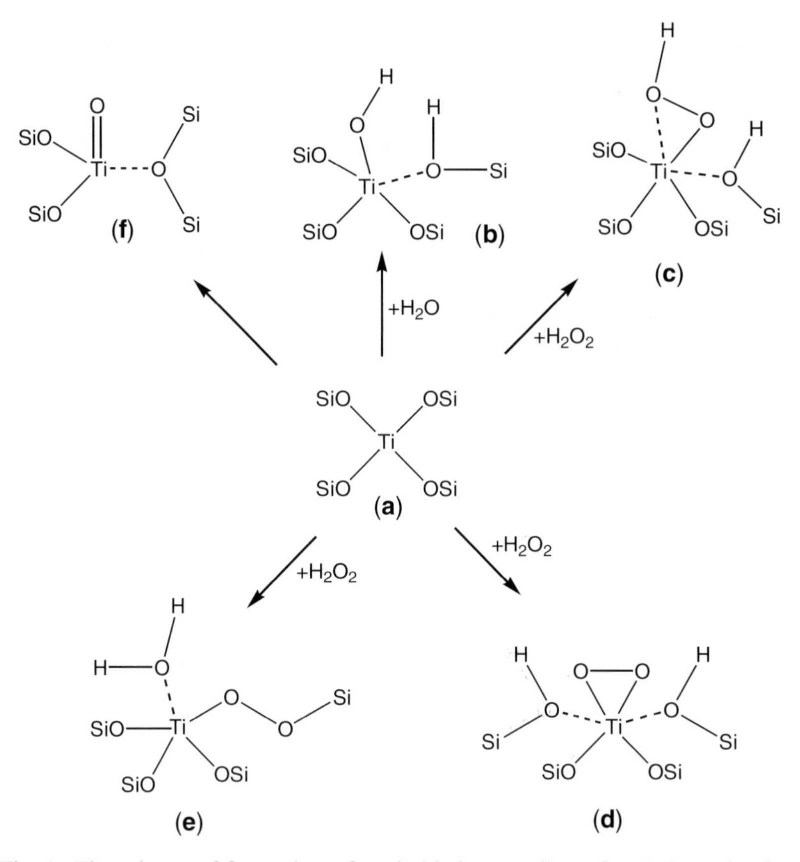

Fig. 1. The scheme of formation of probable intermediates for H_2O_2 activation.

catalyst deactivation [17]. Finally, structure (**1e**) was suggested in theoretical studies [18].

The main problem of computer modeling of structures (**1b**–**1e**) relates to accounting the structural relaxation of the lattice in the formation of hydroperoxide activation intermediate, since the latter process is accompanied by the hydrolysis of the lattice bonds. This problem is even more profound at the formation of defect structure **1f**. Formally, the most adequate approach would be to impose periodic boundary conditions on the calculation model; probably the embedded cluster model will be efficient for this problem. In essence, pure cluster calculations can model only the limiting cases of very flexible or rather rigid lattices that requires either fully optimized cluster geometry or constrained optimization; the latter can be realized, for example, by fixing the terminal hydrogen atoms (H*) saturating the dangling bonds of the cluster.

Calculations of the structures (**1a**) and (**1d**) [19] with the modeling of those by fully optimized clusters $Ti(OSiH_3^*)_4$ and $(OO)Ti(OSiH_3^*)_2(HOSiH_3^*)_2$ proved considerable geometric changes to take place at the hydrolysis of the Ti–O bonds; the newly formed fragments $HOSiH_3^*$ move away from the Ti coordination sphere and the Ti–O distance elongates by more than 0.5 Å. The corresponding increase in the deformation energy estimated in the framework of special simulation with using constrained optimization was appeared to be equal to 11.2 $kcal\,mol^{-1}$ per one hydrolized group. Similar geometric changes resulting from the hydrolysis of the Ti–O bond and formation of structure (**1c**) were found by Vayssilov and van Santen [7]. The authors found that hydrolysis is endothermic (21 $kcal\,mol^{-1}$), assuming the lattice constraints imposed on the cluster geometry. Note also that calculation of the substitution of silicon–oxygen surrounding of Ti ion with hydroxyl groups, for example, by the reaction:

$$Ti(OSi(OH)_3)_4 + nH_2O \rightarrow Ti(OSi(OH)_3)_{4-n}(OH)_n + nSi(OH) \qquad (1)$$

is endothermic as well [20].

Periodical calculations using CRYSTAL program by Zicovich-Wilson et al. [21] also revealed the endothermic effect of the Ti–O bond hydrolysis in TS-1 (8.4 $kcal\,mol^{-1}$) in relation to monohydrate adsorption complex; the activation energy of the process was 20 $kcal\,mol^{-1}$. Similar conclusion was suggested in periodical embedded calculations by Riccardi et al. [10]. The aqua-complexes appeared to be more favorable than the hydrolyzed structures by 11–27 $kcal\,mol^{-1}$ depending on the lattice position. In recent work of Munakata et al. [18], the cluster $Ti(OSi(OH^*)_3)_4$ was chosen as the basic model and the calculations were performed with constrained optimization imposed by the lattice position T1. The estimated endothermic effect of the reaction

$$(\mathbf{1a}) + HOOH \rightarrow \geqq Ti{-}OOH + HO^- \qquad (2)$$

was 10 $kcal\,mol^{-1}$ for the dissociation of the adsorbed hydroperoxide; the adsorption energy was 7.4 $kcal\,mol^{-1}$. The process activation energy was 16.5 $kcal\,mol^{-1}$. In earlier calculations of a 'more rigid' cluster $Ti(OSiH_3)_4$ [7], the hydroperoxide adsorption energy was found to be 7.9 and 0.8 $kcal\,mol^{-1}$ for fully optimization of the cluster geometry and constrained optimization, respectively. Cluster calculations assuming $Si(OH^*)_3$ to be the boundary groups rather than SiH_3^* seem to be more useful for the modeling of local structural relaxation of the active site. At the same time, the question arises as to whether the adsorption and catalytic activities of the Ti ions vary depending on the lattice positions. As mentioned above, periodical embedded calculations by Riccardi et al. [10] showed that the adsorption energy of water molecule varied considerably depending on the lattice position of Ti ions. Of particular interest in this regard is the work of Damin et al. [22], in which an embedded cluster scheme was used in order to take into account the effect of structural lattice constraints on the relaxation geometry of the active site in calculations of the adsorption of H_2O and NH_3 molecules. Structural restrictions were imposed by taking into account the zeolite rings in the outer parts of the calculated structures, and it was concluded that the resulting geometrical structures of the local surrounding of the Ti ion could influence remarkably the adsorption energy. Depending on the constrained structural model, the calculated adsorption energies of the water molecule varied within 3.3–4 $kcal\,mol^{-1}$, in contrast to 0.5 $kcal\,mol^{-1}$ calculated for the unconstrained cluster. Qualitatively the same results were obtained for the adsorption of the ammonia molecule, although the adsorption energy in this case was higher. The calculated effect can be attributed to the influence of the above-mentioned structural destabilization on the adsorption activity; this influence will be discussed in more detail in the following section. Note that these values of water adsorption energy reported by Damin et al. [22] are considerably lower than those cited above. The reason is that these energies were calculated taking into account the basis superposition errors.

To summarize the above results, it should be noted that the process of active-site formation owing to hydrolysis of the Ti–O–Si bridge by adsorbed water or reagent (HOOH) needs definite energetic inputs. In this regard, structure (**1b**) can be considered as the more probable active site. As mentioned above, such structures are formed in TS-1 during synthesis and postsynthesis treatment. Most of the published theoretical studies of the formation of hydroperoxide-activated intermediates are based exactly on this model of the active site.

The transition to the active intermediate (**2b**) → (**2c**) can proceed most readily by the reaction of the hydroxyl substitution with the –OOH hydroperoxide fragment (see Fig. 2). The energetic effect of the reaction is low: 2.9 or −3.6 $kcal\,mol^{-1}$, as estimated by Zhidomirov et al. [23] and by Sinclair and Catlow [16], respectively. The estimated values slightly differ because the former calculations assumed the η^1-coordination of –OOH fragment to the Ti ion, while the latter ones — the η^2-coordination. Energetic difference of structures η^1 and η^2

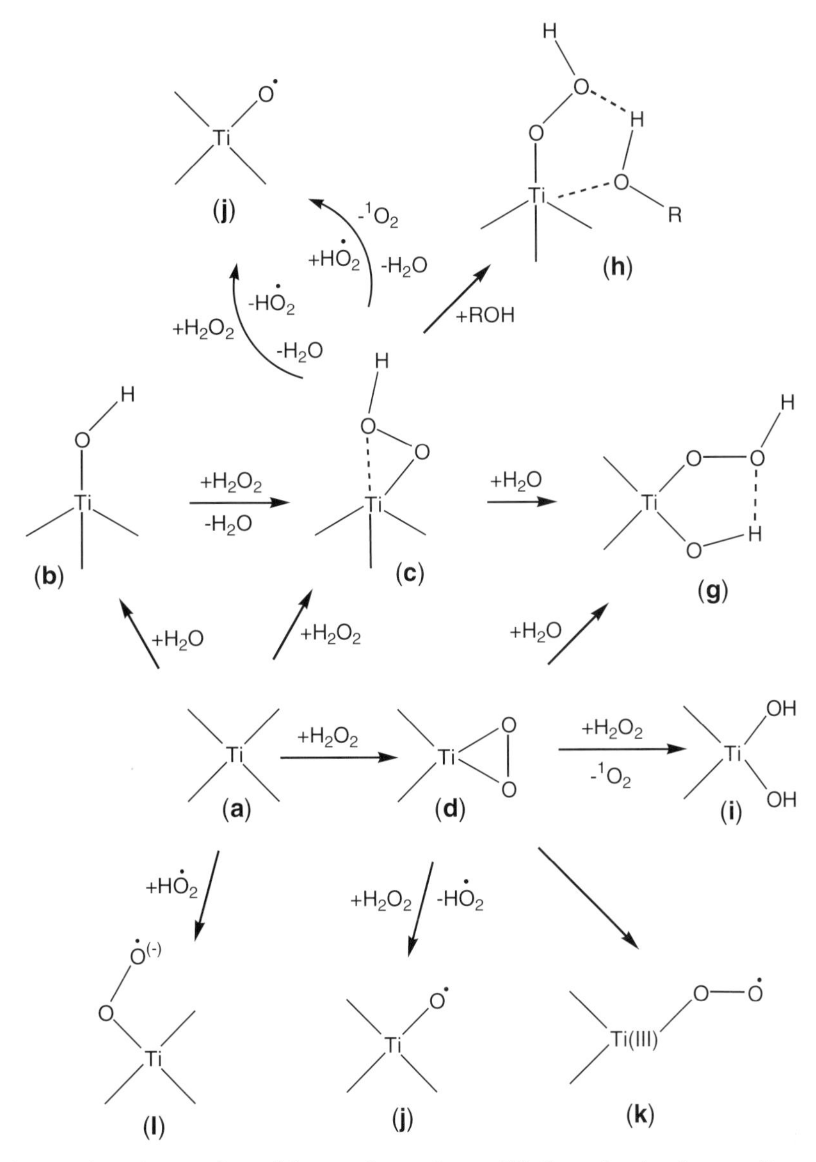

Fig. 2. The scheme of possible transformations of H_2O_2 activation intermediates.

amounts to 8.0 $kcal\,mol^{-1}$, as proved by Karlsen and Schöffel [24]. According to calculations of Sinclair and Catlow [16], the activation energy of the reaction with the participation of the adsorbed hydroperoxide was equal to 13 $kcal\,mol^{-1}$. The solvent effect mentioned in some studies (see, for example, Ref. [25]) was attributed to the formation of the active titanium hydroperoxo complex in the form of five-membered ring (see structure **2h**). Note that if R=H, this structure corresponds to the case of the adsorbed state of the water molecule resulting from the substitution reaction. This cyclic structure, on the one hand, stabilizes the η^1-coordination, which seems to be structurally favorable for the reaction; and on the other hand, it can appear promising to reduce the activation energy of the oxidation reaction. In theoretical studies, the oxidation reaction is usually modeled by the ethylene epoxidation reaction. Generally speaking, theoretical calculations have not confirmed unambiguously the necessity of the formation of the cyclic active intermediate for an efficient reaction [16,18,23].

In some studies, comparative analysis of the stability of various intermediates of the hydroperoxide activation and their activity in the considered reaction was based on the minimum cluster $Ti(OH)_4$. Karlsen and Schöffel [24] calculated the reaction transition state based on cluster $(HO)_3Ti(\eta^2\text{-}OOH)$ and obtained the value of the activation energy equal to 21 kcal mol^{-1}. The calculation was made assuming the transfer of the oxygen atom nearest to the Ti ion in the hydroperoxo group. Neurock and Manzer [26] proved this reaction pathway to be preferable to the transfer of a distant oxygen atom. However, calculations based on the cluster $(HO)_3Ti(\eta^1\text{-}OOH)$ by Zhidomirov et al. [23] resulted in the respective activation energies of 14.4 and 16.5 kcal mol^{-1}, i.e. showed insignificant energetic preference of the transfer of the nearest oxygen atom. In the latter work the influence of hydrogen bonding of one of the hydroxyl groups with hydroperoxide fragment on the reaction was also considered. The hydrogen bond stabilized the η^1-structure of the hydroperoxide fragment. In this case, the calculations of the activation energy of ethylene epoxidation assuming both oxygen transfer pathways showed that the energetic barrier for the transfer of the distant oxygen atom remained almost the same (15.1 kcal mol^{-1}), whereas the transfer of the nearest oxygen atom was practically prohibited by the barrier of 31.7 kcal mol^{-1}. The latter result can be explained by the difficulty of formation of the η^2-structure.

It seems interesting to consider the formation of the structure (**2d**) and its potential role in the oxidation reactions. Evaluation of the energy of the reaction

$$(HO)_3TiOOH \rightarrow (HO)_2TiOO + H_2O \qquad (3)$$

resulted in values of ca. 21.5 kcal mol^{-1} [24,27]. The high endothermic effect of the reaction is attributed to elimination of the water molecule from the Ti coordination sphere. Calculations on the complex

$(HO)_2TiOO$ [28] showed that the energy of H_2O adsorption was equal to 25.9 kcal mol^{-1}. Calculations of the stabilization of the Ti-peroxide structure for the transformation of cluster $HOOTi(OH)(OSiH_3^*)_2$ into cluster $OOTi(OH_2)(OSiH_3^*)_2$ gave similar results. Therefore, it seems inadequate to conclude that the formation of the Ti-peroxide structure is energetically very unfavorable, see also Ref. [23]. This suggestion has indirect experimental confirmation provided by the ESR studies of coordinated ion-radical O_2^-, formed by the adding of H_2O_2 to TS-1 [29], see structure (**2l**). At the same time, comparative analysis of the activation energies of ethylene epoxidation through complexes $(HO)_3TiOOH$ (11.0 kcal mol^{-1}) and $(HO)_2Ti(OH_2)OO$ (17.4 kcal mol^{-1}) made by Yudanov et al. [28] proved the preference of the Ti-hydroperoxide structure for the reaction.

As mentioned above, the role of structure (**1f**), containing titanyl oxygen, in the hydroperoxide oxidation reactions catalyzed by TS-1 is discussible. Sinclair and Catlow [16] suggested that structure (**1f**) could appear as a key intermediate of the catalytic cycle. Reacting with hydroperoxide, it easily transforms into structure (**2g**) releasing 24 kcal mol^{-1}. This energetic effect is even higher in the case of strained defect siloxane structures [27].

We should note two works, which suggest unconventional oxidation reaction pathways. Vayssiiov and van Santen [7] considered the oxygen atom transfer from the coordinated hydroperoxide molecule to the ethylene molecule and found the process activation energy equal to 19.8 kcal mol^{-1}. Munakata et al. [18] considered a modified activation of H_2O_2 that includes the hydrolysis of the Ti–O–Si bridge, formation of Ti-hydroperoxide structure with subsequent transfer of the oxygen atom form the latter structure to the silicious lattice, leading to the formation of the peroxide structure Si–O–O–Si (structure **1e**). The authors predicted the latter structure to be more stable (by 7.6 kcal mol^{-1}) and estimated the transfer activation energy to equal 16 kcal mol^{-1}. Calculation of the transition state of the ethylene epoxidation reaction for structure (**1e**) resulted in the activation energy value of 18 kcal mol^{-1}.

Possible participation of the radical hydroperoxide activation intermediates (**2j**), (**2k**), and (**2l**) in the oxidation reactions on TS-1 is also under discussion in the literature. Structures (**2j**) and (**2k**) were considered by Clerici [30] with respect to peculiarities of the hydroxylation reaction on TS-1, and structure (**2j**) was suggested as the preferential one. Indeed, energy evaluations of transformations (**2d**) $\rightarrow$ (**2j**) and (**2d**) $\rightarrow$ (**2k**) (see Fig. 2) for structure $(HO)_2TiOO$ by Ruzankin et al. [31] give the values of 7.7 and 13.2 kcal mol^{-1}, respectively.

Possible generation of singlet oxygen $^1O_2$ at the activation of H_2O_2 on TS-1 and mesoporous silicates TiMCM [32] is another problem attracting urgent interest. Meanwhile, only hypotheses could be suggested on the molecular mechanism of the $^1O_2$ formation. One of them suggests the participation of the radical intermediates [30] (see transfer **(2c)** → **(2j)**). Another concept could consider the substituting of the peroxide **(2d)** fragment, for example, at transition **(2d)** → **(2i)** (see Fig. 2), as was suggested by Milov and Zhidomirov [27] in the calculations of the activated hydroperoxide on the TiMCM-41 active site modeled by the $Si_7O_{12}H_7TiOH$ cluster based on structural motif of silsesquioxane species.

Zeolite structural defects can facilitate the embedding of metal ions into the zeolite silicious lattice. It was suggested, for example, that vanadium incorporation into the lattice occurs at the defect sites, see [33] and references therein. One of the inherent defect structures in zeolites is: $\equiv SiO^- – HOSi \equiv$ as proved by NMR ^{29}Si [34] and ^1H MAS NMR [35] studies. Other defects can be associated with the hydrolyzed Si–O–Si bridges or with the T vacancies. Neutron powder diffraction studies [36] identified T6, T7, T10, and T11 as the preferable lattice sites for the T vacancies.

Such vacancies can appear during the process of zeolite dealumination. Some clustering of the vacancies is also probable. The defect centers can stabilize the polyvalent metal ions, for example, at the reaction of chlorinated metal compounds with the hydroxyl groups of the defects. Preliminary zeolite dealumination may appear to be an efficient tool for implanting polyvalent ions into siliceous wall of the zeolite matrix, as was demonstrated by immobilizing Ti ions [37] and V ions [33] in zeolites. Structures of various defects existing in the zeolite lattice including those resulting from the hydrolysis of the Si–O–Si bridges and formation of the T vacancies (nest defects) were calculated within the periodical DFT model by Sokol et al. [38]. Bordiga et al. [39] considered the ordered system of hydrogen-bonded hydroxyl groups in the nest defects and calculated its vibration spectrum in the framework of embedded model. The cluster $Si(OSiH_3^*)_4$ was used by Ruzankin et al. [40] to model the nest defects. For this purpose, the cluster geometry had been optimized, then the terminal H* atoms were fixed, and the defect vacancy was calculated assuming the elimination of central Si atom and formation of four hydroxyl groups. The resulting structure is presented in Fig. 3.

The structure found was used for further calculations of the immobilized ions Fe(III) and Fe(II). El-Malki et al. [41] observed the immobilization of Fe ions in ZSM-5 zeolites through the interaction of

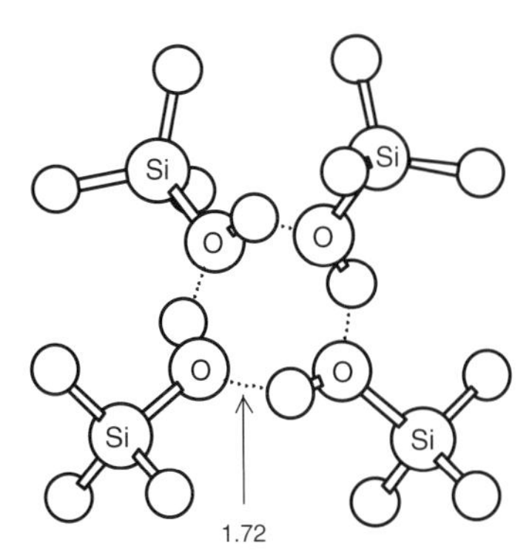

Fig. 3. Calculated structure of nest defect in zeolites.

$FeCl_3$ with internal hydroxyl groups. Simulation of local lattice relaxation was of a primary importance for the molecular modeling of iron ions immobilized in the vacancy defect. Two limiting cases of the 'rigid' and 'flexible' lattice were considered that required the constrained geometry optimization such as that described early and the full optimization of the cluster geometry. As would be expected, the stabilization energies of Fe ions were essentially higher in the latter case. The lower stabilization seems to correspond to the higher adsorption and catalytic activity. In fact, the adsorption energy of atomic oxygen on Fe(II) was higher by 11 kcal mol^{-1} in the case of constrained geometry optimization. The energy of the Fe(II)–O bond is especially interesting with respect to the formation of highly active oxidation center at the first step of N_2O molecule decomposition on Fe(II) in ZSM-5 zeolite [42] with subsequent elimination of dinitrogen.

3. Metal ions in cationic positions

The formation of cationic positions of monovalent metal ions in zeolites are determined by distribution of aluminum through zeolite lattice. It is generally considered that there are preferential lattice sites for Al localization exist in a high-silica MFI zeolite structure, but the list of favored sites and the energies of Al stabilization depend strongly on the computational approach. The force-field approach for the lattice, taking into account geometry relaxation of the lattice after substitution of Si by Al showed difference in substitution energy for various lattice sites within the range of about 4.6 kcal mol^{-1} [43]. A slightly wider range of 9 kcal mol^{-1} with the preference of T1, T7, T8, and T9 sites was found by Nachtigallova et al. [44]. Sastre et al. [45]

has attracted attention to the importance of taking into account the influence of the template on the formation of the Al preferential location during the zeolite synthesis. The force-field model calculation of Al distribution in ITQ-7 the zeolite predicted a span of about 24 kcal mol^{-1} over 64 different acid sites. Recently, synchrotron powder diffraction study of CsZSM-5 zeolite gave information on the Al distribution and the T4, T7, T10, T11, and T12 sites were found as probable for Al atoms [46].

One possible stabilization form of bivalent metal ions in cationic positions in zeolites is presented by stabilization as a hydroxyl compound, (HO)M(II)/Z, that turns us again to the problem of monovalent metal ions in cationic positions. Two other generally discussed stabilization forms, M(II)/Z and M(II)–O–M(II)/Z, require a pair of relatively closely-located Al ions. Taking into account Löwenstein's ru e, the distance between two Al ions ($R_{Al–Al}$) in the next nearest positions in ZSM-zeolite is equal to 4.6–6 Å. According to Rice et al. [47], the limit $R_{Al–Al}$ for the M(II)/Z stabilization is 5.75 Å. As the $R_{Al–Al}$ exceeds that value, bivalent ion would be most probably stabilized in forms M(II)–O–M(II)/Z or (HO)M(II)/Z.

For a rough estimation of the content of the M(II)/Z stabilization sites at the given Si/Al ratio, we can use the fraction of Al ions with the next nearest neighbors (NNN). Bell [48] considered the dependence of NNN from Si/Al ratio using various calculation approaches and found out that at Si/Al = 50 only 20% of Al atoms comply with the above condition. More accurate calculation [47] with the use of ensemble averaging of the Al distribution decreases that value to 10%; further accounting for energetic nonequivalence of Al distribution through the lattice positions diminishes that value to 5%. Besides, this approach neglected the interactions between Al atoms, which obviously affects the energetic preference for a particular Al distribution throughout the zeolite lattice. Dempsey et al. [49] and Sonnemans et al. [50] considered the electrostatic interactions that stabilize the long-distance distribution of Al atoms. On the other hand, the tendency for local relaxation subsequent to isomorphic substitution of Si by Al [51] will cause the reverse effect. Similar influences will produce the tendency for condensation of aluminum species in H-zeolites [52]. Direct comparative cluster calculations of the localization of two Al ions in the H-form of faujasite [51] showed the preference of their next-nearest-neighbor sitting in four-membered rings. Comparative periodical VASP calculations of the localization of two Al ions in the H-chabazite lattice [53] suggested preferential localization of both ions in one of four-membered rings of the zeolite structure.

The generally accepted molecular models for the cationic position of bivalent metal ion in the MFI zeolites are the six-, five-, and four-membered rings containing two Al ions. Rice et al. [54] carried out a systematic comparative calculation of the stability of bivalent ions in the ZSM-5 zeolite depending on the size and shape of the zeolite rings. It was found that Cu(II), Co(II), Fe(II), and Ni(II) tended to localize in five-membered rings, while Pd(II), Pt(II), Ru(II), Rh(II), Zn(II) — in six-membered rings. Recently, general attention has been attracted to the approach used by Wichterlova et al. [55,104] for the simulation of preferential localization of Co(II) in pentasil zeolites with MOR, FER, and MFI topology by comparative analyzing of the UV–VIS–NIR spectra of Co(II). Three preferential positions have been separated, referred to as α, β, and γ. The α position — an effective six-membered ring on the wall of direct channel, formed by two interconnecting five-membered rings, and readily available to reagents — seems to be of special interest with respect to catalytic action.

In any case, in zeolites with a rather high Si/Al ratio the structures with a larger distance between Al ions than that inside one zeolite ring could be considered as the cationic positions. Well-known variants of cation stabilization in such positions is associated with the formation of oxygen-bridged binuclear structures M(II)–O–M(II)/Z. Extra-lattice oxygen in such a structure gives rise to specific properties showing obvious catalytic potentials. This point will be considered in more detail in the next section.

Comparison of the energetic stability of various cation species is a key task for theoretical simulation of cation localization in zeolite lattice. Comparing the stability of M(II)/Z located in different cationic positions Z_i, it is reasonable to consider respective H-forms:

$$M(II)/Z_i + 2H^+/Z_i \rightarrow M(II)/Z_i + 2H^+/Z_i \quad (4)$$

In some cases it may be interesting to estimate the energy of the M(II)/Z species reduction by hydrogen to produce the H-form cationic position and metal species:

$$M(II)/Z + H_2 \rightarrow 2H^+/Z + M \quad (5)$$

or intra-zeolite oxide species:

$$M(II)/Z + H_2O \rightarrow 2H^+/Z + MO \quad (6)$$

The estimation by reactions (5) and (6) strongly depend on the M and MO models. Rice et al. [54] used for this purpose the formation energy of bulk metals and oxides and found that both reactions for most metals and cationic positions had considerable exothermic effect. The resistance of M(II)/Z species to reduction decreased in the order $Co > Zn > Fe > Ni > Cu > Ru \sim Rh > Pd > Pt$. Nevertheless, this approach seems to be inaccurate for the evaluation of the stability of small metal or oxide clusters in zeolite channels, since in this case the energy effect would be considerably lower. We shall consider this problem in more detail below, in the section devoted to stabilization of Zn(II) in ZnZSM-5 zeolites. We shall also consider there the dissociative adsorption of water in zinc zeolite:

$$M(II)/Z + H_2O \rightarrow (HO)M(II)/Z + H^+/Z \tag{7}$$

and demonstrate that the process energy is controlled by the stability of the M(II)/Z species in a particular cationic position.

Another interesting problem is a comparative estimate of metal species stabilized in the hydroxide and binuclear forms:

$$(HO)M(II)/Z + (HO)M(II)/Z \rightarrow M(II){-}O{-}M(II)/Z + H_2O \tag{8}$$

The calculations made by Rice et al. [54] showed a considerable endothermic effect of reaction (8) with the participation of Cu, Fe, Ni, Zn, and Pd. This problem will be discussed in more detail below.

Let us consider some aspects of calculations of the chemisorption and catalytic activity of metal ions in cationic positions. The high requirements for the calculation level (expanded basis sets, accounting for electronic correlation) enforce often the use of the minimal cluster structures to model the active centers. Generally speaking, each calculation becomes a compromise between the quantum-chemical level and the molecular level of active structure modeling. The modeling of monovalent metal ions or the hydroxy form of bivalent ions in the cationic positions of zeolite lattice is based often on the well-known 3T: $(H^*O)_3SiO{-}Al(OH^*)_2{-}OSi(OH^*)_3$ and 5T: $(H^*O)_3SiO{-}Al(OSi(OH^*)_3)_2{-}OSi(OH^*)_3$ models containing one Al ion in tetrahedral coordination. The model can be modified, based on different choices of the boundary fragments. The majority of the calculations concerned the problem of SCR of NO_x on CuZSM-5 or CoZSM-5 zeolites. Trout et al. [56] performed systematical studies of the structures and thermodynamic characteristics of various intermediates of NO, N_2O, and NO_2 decomposition on Cu^+/Z, using the 5T cluster as the cationic position model. A reaction pathway was proposed based on calculations considering the adsorption of two NO molecules, desorption of N_2O with the formation of CuO/Z species and subsequent interaction of N_2O and CuO/Z with the elimination of N_2. Desorption of O_2 completed the catalytic cycle. Solans-Monfort et al. [57] considered alternative mechanism of the process through a nitroso-nitrosyl intermediate and concluded in favor of the former mechanism. In these calculations, the 3T cluster modeled the Cu^+/Z species in cationic position. Rice et al. [58] considered the deactivation of CuZSM-5 and CoZSM-5 in the reaction with water and found that the latter zeolite was more stable. Systematic calculations of the NO_2 adsorption on Zn(I), Cu(I), Ni(I), Co(I), and Fe(I) within the 3T cluster model showed that coordination through two oxygen atoms was more favorable. The adsorption energy followed the order Zn > Ni > Cu > Fe > Co. Kim et al. [59] used the 3T cluster model in the form of $[(H^*O)_3SiO–Al(OH^*)_2–OSi(OH^*)_3]^-$ for the modeling of cationic position for Co(II) and the calculation of NO adsorption.

To illustrate the calculation of the adsorption and catalytic activities of bivalent metal ions embedded in high-silica zeolites, we shall discuss some recent results for ZnZSM-5 and FeZSM-5 zeolites. Two five-membered rings in the straight and sinusoidal channels, respectively, as well as four-membered ring and α-site were chosen for molecular modeling of Zn^{2+} cationic sites in ZSM-5 zeolite [60]. Additionally, a six-membered ring in faujasites was considered for comparison. Some of these sites are shown for ZSM-5 in Fig. 4.

Hydrogen boundary atoms (H*) were used to saturate dangling bonds of the Si and Al atoms in the cluster models of these structures. Special restrictions were imposed on the optimization of positions of these boundary H* atoms. At the first step of the optimization procedure the entire zeolite cluster frame was chosen according to experimental X-ray diffraction data [61]. Only Si–H* and Al–H* bond distances were optimized, while the positions of other atoms (except Zn), as well as directions of O–H* bonds, were kept fixed according to crystallographic data. The Zn^{2+} ion was allowed to move freely in the structure. Subsequently the second step of optimization was performed with fixed positions of H* atoms. Positions of these atoms obtained after the first step of this procedure were saved and used further in the restricted geometry optimizations of H-forms and other zeolite structures.

As discussed in previous chapters, procedures using the experimental X-ray diffraction data and subsequent restricted geometry optimization

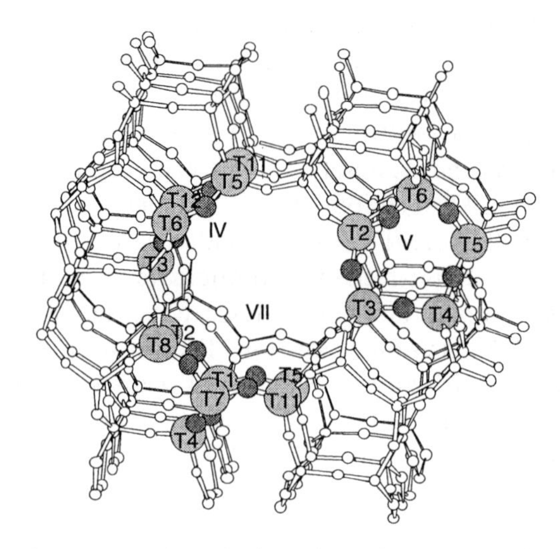

Fig. 4. Location of some possible cationic positions in ZSM-5 zeolite crystal structure: a 5-ring in the straight channel (**IV**), a 5-ring in the sinusoidal channel (**V**), and α-site (**VII**).

(with fixed boundary H* atoms) are rather artificial method and can lead to some inaccuracy. In the present case, the most important factors contributing to the inaccuracy are the following:

(1) the use of an 'average' experimental X-ray diffraction structures, because there is no possibility to know from experiment at which positions of the ring Si atoms are isomorphically substituted by Al, which can result in some artificial tensions and deformations in the vicinity of Al atoms;
(2) basis set dependence of the difference between experimental and optimized bond lengths and angles;
(3) the use of fixed positions of boundary atoms H* obtained from the Zn-form of the ring may influence the calculated value of the energy difference between cationic and H-form, since this can result in significant stress when such positions are used for the corresponding H-form.

Nevertheless, this scheme can help in estimating the role of the real lattice geometry of zeolite rings. Moreover, for test purposes in order to trace the influence of the applied geometrical constraints on the stabilization of Zn^{2+} ions additional full geometry optimization calculations of clusters imitating Zn^{2+} containing six-, five-, and four-membered rings were also performed. These calculations correspond to the case of reasonably high flexibility of the zeolite lattice and reveal some common tendency of the Zn^{2+} ion stabilization in different rings.

Most calculations were performed at the SCF Hartree-Fock (HF) level with LANL2DZ pseudopotential using GAUSSIAN-98 program [62]. For optimized geometries single-point calculations were also performed using DFT with gradient-corrected Becke exchange and Pedrew correlation functionals. It was verified that the precision of the calculation at the LANL2DZ level was sufficient for the purposes of investigation. In the recent calculations of vibrational frequencies [63–65], the B3LYP hybrid functional and a more sophisticated set of basis functions were applied. The 6-31G* basis set was used for zinc, while the 6-311G** basis set was used for the zeolite oxygen atoms; hydrogen atoms of bridging hydroxyl groups, H_2, CH_4, and C_2H_6. Zeolite Al and Si atoms, as well as hydrogen atoms (H*) were treated in D95-Dunning/Huzinaga basis set. This compromise choice allowed us to save computational time and, at the same time, to use extended models for frequency calculations without significant loss of accuracy.

The stabilization energies (E_{st}) of Zn^{2+} ions in various probable cationic positions of ZSM-5 zeolite were calculated according to the reaction:

$$Zn^{2+}/Z + H_2 \rightarrow 2H^+/Z + Zn^0 \qquad (9)$$

The results of the calculation of E_{st}, as well as the energy of the ethane heterolytic dissociation reaction through the 'alkyl' route

$$Zn^{2+}/Z + C_2H_6 \rightarrow H^+/Z + C_2H_5^-Zn^{2+}/Z \qquad (10)$$

and the elimination of H_2

$$H^-Zn^{2+}/Z + H^+/Z \rightarrow Zn^{2+}/Z + H_2 \qquad (11)$$

after the dehydrogenation reaction are shown in Table 1. In order to discuss the reactivity of these sites it is reasonable to compare the calculated energies with those for the Zn^{2+} ion in the four-membered ring. It was shown earlier by Frash and van Santen [66] that this site could be active in the ethane dehydrogenation reaction. Intermediates and transition states for 'alkyl' (10) and 'carbenium' ($Zn^{2+}/Z + C_2H_6 \rightarrow H^-Zn^{2+}/Z + C_2H_5^+/Z$) mechanism were calculated at the MP2/6-31G**//B3LYP/6-31G* level. The activation energies for the heterolytic dissociation of ethane in these mechanisms were evaluated as 18.6 and 53 kcal mol^{-1}, respectively. The conclusion was made that the 'alkyl' route is preferential for the ethane activation.

Table 1
Calculated energies (kcal mol^{-1}) of the reactions: $Zn^{2+}/Z + H_2 \rightarrow (2H^+)/Z + Zn^0(E_{st})$, $Zn^{2+}/Z + C_2H_6 \rightarrow (ZnC_2H_5^+ + H^+)/Z(E_{ads}(C_2H_6))$, and $(ZnH^+ + H^+)/Z \rightarrow Zn^{2+}/Z + H_2(E_{des}(H_2))$ for different zeolitic rings (energy is positive for endothermic reactions)

	E_{st}	$\Delta E_{ads}(C_2H_6)$	$\Delta E_{des}(H_2)$
4T-ring	16.3	8.7	4.1
5T-ring (straight channel)	25.5	14.2	−0.8
5T-ring (zig-zag channel)	35.7	12.9	1.1
α-Site	19.4	9.9	1.2

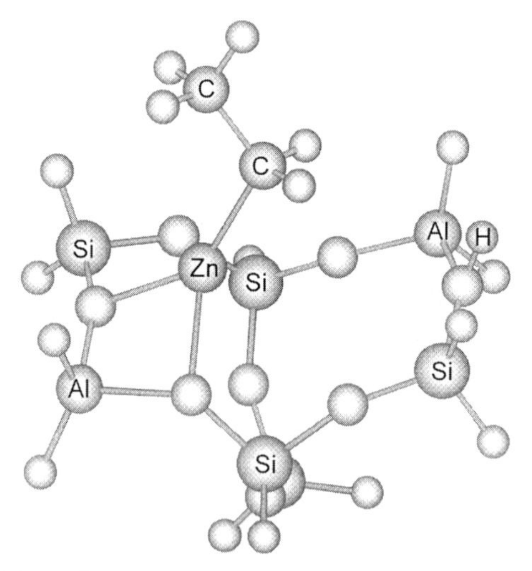

Fig. 5. Decomposition of ethane on α-site in ZSM-5 zeolite (restricted geometry optimization).

The calculations showed that stabilization of Zn^{2+} in five-membered rings is significantly stronger than in four-membered rings, which correlates with their activity in the ethane heterolytic dissociation. It is interesting that E_{st} and $E_{ads}(C_2H_6)$ for Zn^{2+} in the α-site (Fig. 5) are quite similar to those obtained for the four-membered ring. The α-site was proposed [55] as one of the favorable structures for localization of two-valent exchangeable cations in MFI systems. In general, the energies of intermediates of ethane dehydrogenation are similar for the different sites considered above for ZnZSM-5 and the largest differences are obtained for the first step of the reaction. The energies of H_2 recombination are comparatively small for these sites (see Table 1).

An attempt to stabilize zinc oxide particles in ZnHY zeolites [67] showed that these species are rather unstable and disappear during the dehydration of the zeolite (vacuum treatment at temperature above 300°C). This process is accompanied by a decrease in the intensity of IR bands from the bridged hydroxyl group. Calculations on the Zn^{2+} zeolite cationic sites permitted consideration of the question of the comparative stability of small oxide particles in zeolite cavities. We should first note that the reduction of very small oxide species by hydrogen is an exothermic process. At the DFT level of calculation [60] the gain in energy for the exothermic reaction of ZnO molecules with dihydrogen:

$$ZnO + H_2 \rightarrow Zn^0 + H_2O \tag{12}$$

is 69.2 kcal mol^{-1}. Note that the increase of the size of oxide particles substantially decreases and even reverses the overall energy effect of this reaction. In particular, a calculation for the hypothetical 'cubane' oxide structure of $(ZnO)_4$ gives only 8.9 kcal mol^{-1}. Nevertheless, it is evident that conditions for formation of oxide species $(ZnO)_n$ in faujasites are, as a rule, unfavorable. For example, for the faujasite-like 6-ring structure with three-coordinated Zn^{2+}, the heat of the reaction (at DFT level)

$$x\,Zn^{2+}/Z\ (6\text{-ring}) + x\,H_2O \rightarrow (ZnO)_x + 2x\,H^+/Z\ (6\text{-ring}) \tag{13}$$

is 142.1 and 64.0 kcal mol^{-1} per one Zn^{2+} ion for $x = 1, 4$, which corresponds to the formation of an isolated ZnO molecule and a cubic $(ZnO)_4$ cluster, respectively.

It is not surprising that the low stabilization energy for Zn^{2+} in ZSM-5 compared to faujasites is favorable for formation of small zinc oxide clusters. For example, for the 5-ring in the sinusoidal channel of ZSM-5, the energy of the reaction

$$x\,Zn^{2+}/Z\ (5\text{-ring}) + x\,H_2O \rightarrow (ZnO)_x + 2x\,H^+/Z\ (5\text{-ring}) \tag{14}$$

is 104.9 and 26.8 kcal mol^{-1} per one Zn^{2+} ion for formation of isolated ZnO molecule or $(ZnO)_4$ cluster, respectively. Additional accounting of the stabilization of the zinc oxide particles by the zeolite lattice could decrease these values further.

It is of interest that for faujasites, calculations of the interaction of Zn^{2+}/Z with water showed that the energy of molecular adsorption is quite significant. Depending on the cluster model it is calculated as

24–30 kcal mol^{-1} for 6-ring structures, which is in qualitative agreement with experimental data [68]. As for the reaction

$$Zn^{2+}/Z\ (6\text{-ring}) + H_2O \rightarrow ((ZnOH)^{+} + H^{+})/Z\ (6\text{-ring}) \quad (15)$$

it was impossible to find for faujasites any structure corresponding to the heterolytic dissociation of water. Irrespectively of the starting point the optimization resulted, for both full and restricted optimization, in the molecular water adsorption.

In contrast to faujasite the structure corresponding to water heterolytic dissociation can be found for all five-membered ring structures from ZSM-5. The reaction

$$Zn^{2+}/Z\ (5\text{-ring}) + H_2O \rightarrow ((ZnOH)^{+} + H^{+})/Z\ (5\text{-ring}) \quad (16)$$

is exothermic with the energy gain 20–30 kcal mol^{-1} depending on the site structure.

The adsorption and activation of water molecules on metal cationic species in zeolites is commonly suggested as an important step in many catalytic processes. Water can participate in a reaction as a reagent or assist in the passage of elementary steps. Both variants of the activity of water molecules were demonstrated by Barbosa and van Santen [69] in their theoretical study of nitrile hydrolysis reaction on Zn(II) ion-exchanged zeolites. The previously mentioned four-membered ring was chosen as the cationic position for the Zn(II) ion and the acetonitrile hydrolysis reaction at this site was analyzed. The reactivity of the metal cation was found to be rather similar to that for enzyme carbonic anhydrase active site. Three steps in the reaction were considered: hydration, isomerization, and product desorption; a number of various reaction ways were calculated. It was shown that both the roles of water molecules as proton donors in hydration and tautomerization and their copromotion influence in several elementary acts were important.

Generally speaking, there are two main factors in the destabilization of cation position and increase of Lewis acidity of the Zn^{2+}: sterical hindrance of the cation adsorption due to structural peculiarities of the site and the decrease of adsorption interaction which could be associated with distant mutual siting of two aluminum charge-compensating ions in the zeolite lattice. The theoretical study of methane adsorption at Zn^{2+}, stabilized in four- and five-membered rings, showed that together with reduction of the stabilization energy of

Zn^{2+} in the four-membered ring (the Zn–O bond length is 2.07 Å for the four-membered ring, while for the five-membered ring it is 1.98 Å [60,70]) there is an increase of the CH_4 adsorption energy (5 kcal mol^{-1} for 4-ring and 0.3 kcal mol^{-1} for 5-ring) and, respectively with a shift of the symmetric stretching frequency of the adsorbed CH_4 molecule (-38 cm^{-1} for 4-ring and -10 cm^{-1} for 5-ring). An example of the second effect was studied by Yakovlev et al. [71] where stabilization of Zn^{2+} in the zeolite fragment consisting of two connected five-membered rings with distant placing of two Al ions was considered. Reduction of stability and high reactivity of these cationic positions were revealed, but the study of CH_4 adsorption and perturbation of the CH_4 IR frequency bands was not reported. An abnormal frequency shift ($\sim$110 cm^{-1}) of the symmetric (A_1) stretching band for the methane molecule adsorbed on ZnZSM-5 zeolite was observed in Ref. [72]. It was impossible to explain such a shift by interaction of CH_4 with traditional forms of zinc ion species [70]. Below we illustrate how Zn^{2+} in the cation position can really cause a large perturbation of IR vibration frequencies [65].

The two connected 5-rings on the wall of straight channel of ZSM-5 zeolite have been suggested as a possible cationic site for the zinc ion, with Al ions placed in T_{12} and T_8 lattice positions. The cluster $ZnAl_2Si_6O_9H^*_{14}$ (called as Z_dZn in Ref. [71]) was chosen for model calculations. Quantum chemical computations have been carried out within the gradient-corrected density functional theory with B3LYP functional using the GAUSSIAN-98 program, partial geometry optimization procedure, and basis sets described above.

The cluster structure of the Zn^{2+} ion stabilized in the cationic position of Z_dZn cluster with the distant placing of two Al ions is shown in Fig. 6.

As expected the zinc cation is localized strongly near one of the Al ions (the T_{12} lattice position in this computation) which have a larger separation. Though the effective coordination number (Fig. 6) of Zn^{2+} ion by oxygen ions of zeolite lattice is 4, the bond lengths with the two oxygen ions neighboring to aluminum are somewhat shorter than the other two (see Table 2). The C–H bond lengths in adsorbed CH_4 (Table 2, Fig. 7) are changed slightly in comparison with the bond length in the free methane molecule (-1.091 Å).

The adsorption of CH_4 initiates also some increase in the Zn–O bond length (Table 2). Changes of the atomic charges under methane adsorption indicate electron transfer (-0.1) from the methane molecule to the zinc cation and to the neighboring oxygen ions according to the Lewis acidity of zinc in the cationic position. Calculated vibrational

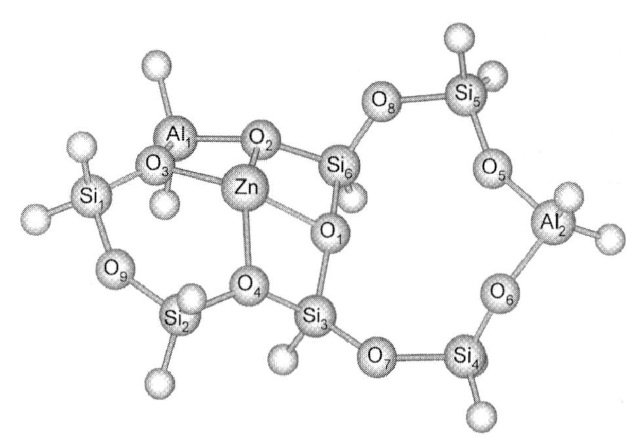

Fig. 6. Cluster structure (Z_dZn) of Zn^{2+} ion stabilized in the cationic position with the distant placing of two Al ions.

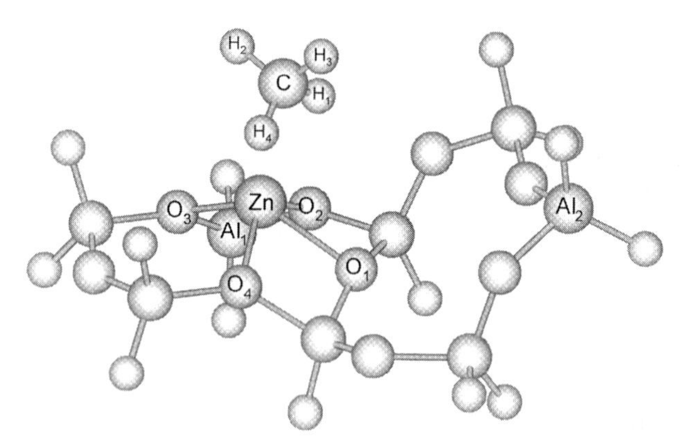

Fig. 7. Methane adsorption on Z_dZn cluster in ZnZSM-5 zeolite.

Table 2
Bond lengths and distances (in Å) for Z_dZn active site

	Z_dZn	CH_4/Z_dZn
$Zn–O_1$	2.081	2.122
$Zn–O_2$	1.973	2.013
$Zn–O_3$	1.916	1.947
$Zn–O_4$	2.019	2.104
Zn–C		2.366
$C–H_1$		1.101
$C–H_2$		1.088
$C–H_3$		1.090
$C–H_4$		1.106

Table 3
Calculated and experimental IR frequencies (cm^{-1}) for methane in vacuum and methane adsorbed on Z_dZn cluster

CH_4 (exp.)	CH_4 (theor.)	CH_4/Z_dZn
1306 (T_2)	1342	1251 (−91) 1355 (+13) 1407 (+65)
1533 (E)	1561	1549 (−12) 1579 (+18)
2917 (A_1)	3026	2916 (−110)
3020 (T_2)	3132	3015 (−117) 3104 (−28) 3170 (+38)

frequencies for free and adsorbed CH_4 are presented in Table 3. The most intense (originating from IR forbidden A_1 band of free methane molecule) C–H stretching band of adsorbed CH_4 is strongly shifted to lower frequencies (~117 cm^{-1}), which is in reasonable agreement with experiment. The high-frequency T_2 asymmetric C–H stretching band of free CH_4 is split due to adsorption, and its components are also shifted. The intensity of this band is low in contrast to free methane where this band is the most intense.

It is also interesting to consider adsorption and chemical properties of Z_dZn cluster and calculate the molecular and dissociative adsorption of H_2 [65] in connection with experimental investigations of low-temperature H_2 adsorption on ZnZSM-5 zeolite [72,73]. The adsorption energy of H_2 (Fig. 8) is 7.7 kcal mol^{-1}, which is a reasonable value for the interaction with strong Lewis acid sites [74].

Some activation of H_2 is also apparent in the elongation of the H–H bond length (0.762 Å) in comparison with the bond length for the free molecule (0.744 Å). Such an activation stimulates further dissociation of H_2 on the site. The first step of the dissociation process involves migration of one hydrogen atom to one of the nearest oxygen ion of the zeolite structure and formation of a Zn–H chemical bond with the second hydrogen atom. According to computational analysis one possible target for hydrogen atom migration is the zeolite lattice $O_{(8)}$ oxygen ion. The structure of the corresponding transition state is shown in Fig. 9.

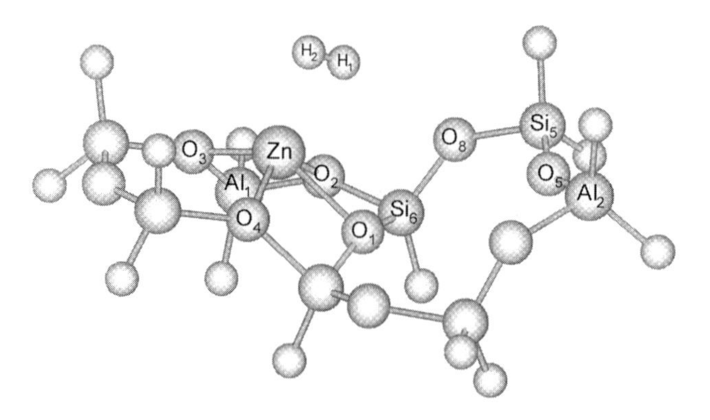

Fig. 8. H_2 adsorption on Z_dZn cluster in ZnZSM-5 zeolite.

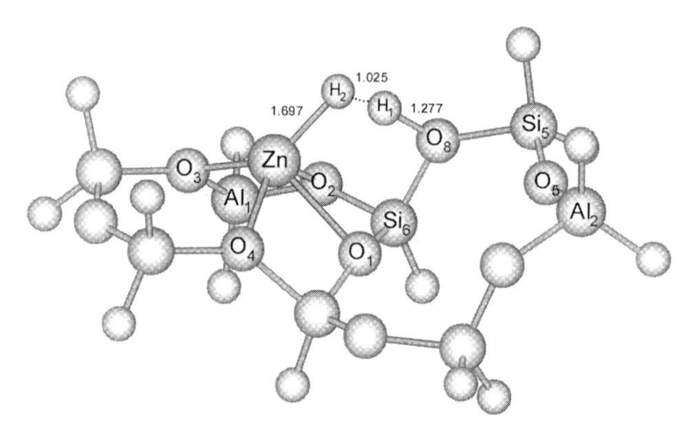

Fig. 9. Transition state for H_2 dissociation on Z_dZn cluster.

The H–H distance was increased to 1.025 Å. The Mulliken atomic charges on hydrogen atoms are $+0.332$ ($H_{(1)}$) and -0.229 ($H_{(2)}$), which is in agreement with the chemical meaning of the path of dissociation. The resulting structure (not shown) is characterized by the increase of positive atomic charge on $H_{(1)}$ and the considerable decrease of the negative atomic charge on $H_{(2)}$. It is certainly not the final position for the migration of hydrogen. The most stable localization of $H_{(1)}$ should be near one of two oxygen ions connected with Al. The next possible step of $H_{(1)}$ migration is complex. It includes the 'rotation' of $O_{(8)}$–$H_{(1)}$ bond around Si_5–$O_{(8)}$–Si_6 quasilinear fragment in order to form intermediate transition state for the subsequent step of $H_{(1)}$ proton migration from $O_{(8)}$ to $O_{(5)}$. The transition state for this proton transfer is shown in Fig. 10, while the resulting structure for the second step of migration is shown in Fig. 11.

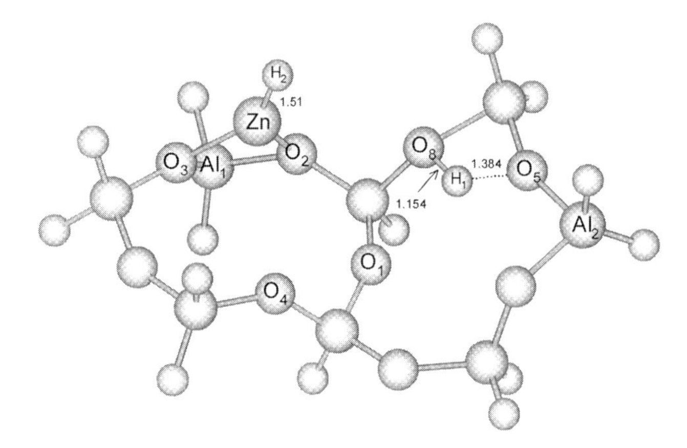

Fig. 10. Transition state for $H_{(1)}$ proton transfer from $O_{(8)}$ to $O_{(5)}$ during the process of its migration through Z_dZn cluster.

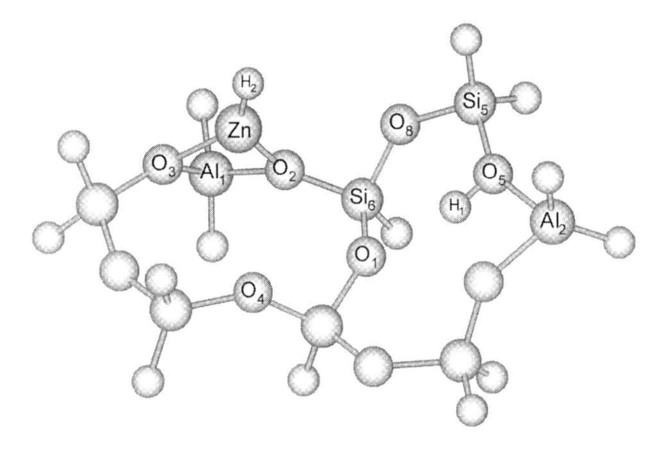

Fig. 11. One of the most stable (final) structures for $H_{(1)}$ proton migration through Z_dZn cluster.

The Zn–$H_{(2)}$ and O–$H_{(1)}$ stretching vibrational frequencies are presented in Table 4. The energy path for $H_{(1)}$ migration is shown in Fig. 12.

It can be concluded that H_2 dissociation followed by migration of H atom to the second Al ion does not meet with serious energy hindrance.

Fe-exchange MFI zeolites have been reported to display high activity for selective benzene oxidation to phenol with N_2O as oxidant [75,76]. The main question in the understanding of the origin of the catalytic

Table 4
Calculated IR vibrational frequencies (cm^{-1}) for free H_2, adsorbed H_2, and H_2 dissociated on Z_dZn cluster

	H_2	H_2/Z_dZn (Fig. 8)	$(ZnH^+ + H^+)/Z_dZn$ (Fig. 11)
$\nu_{H\text{–}H}$	4419	4149 (−270)	
$\nu_{O\text{–}H}$			3449
$\nu_{Zn\text{–}H}$			1958

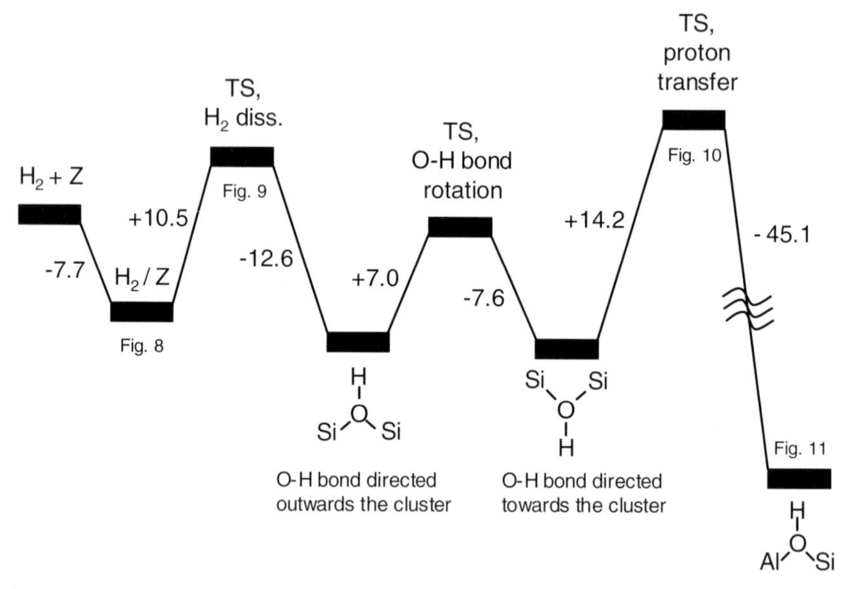

Fig. 12. Energy path for $H_{(1)}$ proton migration through Z_dZn cluster. LM, local minimum; TS, transition state. Two proposed, but not computed local minima are shown with dashed lines.

activity of these systems concerned the nature of the active sites, which were called α-sites or Fe_α. High-temperature treatments were found to be very important for the formation of α-sites and steaming technique was especially effective in increasing of α-sites concentration. During high-temperature pretreatment, practically all iron migrates from the lattice into extra-lattice position [77]. It has been proposed that the extra-framework iron oxide particles could be partially destroyed by steaming with conversion of iron to cationic positions [78]. After the treatment an essential part of the iron cations appears to be in the Fe^{2+} state [79]. These sites can be easily oxidized to Fe^{3+} through decomposition of N_2O and formation of α-oxygen.

There is a correlation between the number of α-oxygen sites and that of the amount of reduced Fe^{2+} [80]. The nuclearity of these α-sites is strongly debated [76]. Recently, it has been concluded that probably the α-sites are binuclear complexes. However, in the oxidation cycle, each iron ion reacts independently [80]. This was the main reason for choice of a single Fe^{2+} cation for the active site by Kachurovskaya et al. [81]. In principle, isolated cations can be localized in a cation position of the zeolite or in some defect structure with a pure silicate surrounding. The variant with a cation position was chosen. Unfortunately, there are only limited experimental data on the nature of α-oxygen and on the reaction intermediates. The bond energy of the oxygen atom in the α-oxygen structure was estimated to be about 60 $kcal\,mol^{-1}$; in addition, the kinetic isotope H/D effect (KIE) in benzene oxidation by 'α-oxygen' was found to be absent [42]. An intriguing vibrational frequency band at 2874 cm^{-1} was observed during the adsorption of benzene in a zeolite where α-oxygen was deposited. This band disappeared after completing the reaction [82].

Crystallographic data for the α-position of MFI has been chosen as the initial model for cluster calculations. This site is constructed from two intersecting five-membered zeolite rings and forms an effective six-membered ring. DFT cluster model calculations were carried out using the B3LYP hybrid functional and 6-31G* basis set for all types of atoms. The optimized structure of the Fe^{2+} ion model for the α-site is shown in Fig. 13A.

The cation is stabilized by the interaction with four oxygen ions from Si–O–Al fragments of the effective six-membered ring of the α-cationic position. The next step is to generate the α-oxygen species responsible for the catalytic activity. This center is formed due to the decomposition of a nitrous oxide molecule on the α-site. The reaction energy associated with the reaction

$$Fe/Z + N_2O \rightarrow OFe/Z + N_2 \tag{17}$$

is ~ -10 $kcal\,mol^{-1}$. The resulting Fe=O bond distance of 1.59 Å (Fig. 13B) corresponds to the bond distance in ferryl oxygen species. The Fe=O bond energy was calculated by evaluating the reaction energy for:

$$2OFe/Z \Rightarrow 2Fe/Z + O_2 \tag{18}$$

The value is 55.5 $kcal\,mol^{-1}$, which is in good agreement with the value of 60 $kcal\,mol^{-1}$ obtained from calorimetric measurements [42].

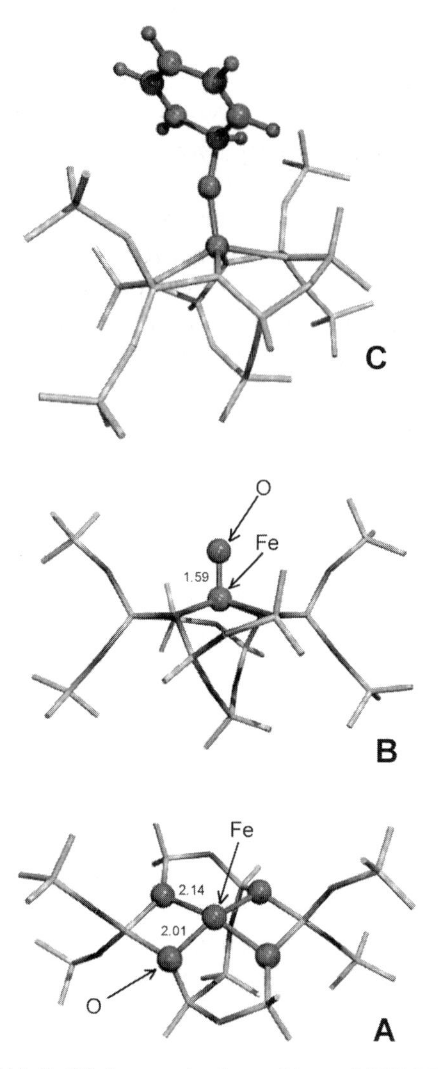

Fig. 13. Structures: (A) Fe(II) in α-cationic position of ZSM-5; (B) α-oxygen after decomposition of a nitrous oxide molecule; (C) benzene oxide adsorbed on Fe/ZSM-5 model cluster.

Panov et al. [42] proposed that the reaction of benzene oxidation proceeds via the initial formation of unstable arene oxides, which can spontaneously isomerize into a phenolic product:

$$Fe_2O + C_6H_6 \xrightarrow{(1)} Fe_2\text{----}O(C_6H_6) \xrightarrow{(2)} Fe_2\text{----}O(H)\text{—}C_6H_5$$

An experimental observation of the absence of a kinetic H/D isotope effect led to the proposal of step (1) being rate limiting.

To study the catalytic mechanism, attempts were made to find the adsorption complex of the benzene molecule with model cluster. Depending on the initial geometry, either benzene oxide was formed or no adsorption of benzene took place, which points to a low activation energy of benzene oxide formation or its absence. Also geometry optimization was performed for benzene oxide adsorbed on the Fe/ZSM-5 model cluster (Fig. 13C). The energy of benzene oxide formation from free benzene and the initial OFe/Z cluster is 23.2 $kcal\,mol^{-1}$. The ease of benzene oxide formation during the adsorption of benzene on the oxygen containing α-center allows us to conclude that step 1 is not rate limiting. The process of phenol formation (2) proceeds via breaking of a C–H bond; thus it cannot be rate limiting. The experimental evidence of the absence of KIE can be explained since the process consists of several steps. The evolution of benzene oxide by breaking one of the two C–O bonds was studied. This intermediate is less stable by 7.0 $kcal\,mol^{-1}$ than the adsorbed benzene oxide. According to the calculations, formation of this intermediate is a potential candidate for the rate-limiting step of benzene to phenol conversion. This would also comply with the experimental observation of the absence of the kinetic isotope effect (KIE). Although this intermediate was not observed, its formation was mentioned in a number of papers (Ref. [83,103]. One possible pathway for the transformation of this intermediate is the formation of the keto-tautomer of phenol. The idea of keto-tautomer formation during photolysis of benzene oxide was stated in the literature [84,85] and is in agreement with this study. The transformation of the intermediate to keto-tautomer is favorable by about −42.1 $kcal\,mol^{-1}$. Further indications for the formation of this keto-tautomer are found in IR studies during conversion of benzene to phenol [82]. IR spectra for all intermediate molecules were calculated. The band shift (about 200 cm^{-1}) is found only for the keto-tautomer of phenol. Thus, the experimentally observed band at 2874 cm^{-1} [82] has to be attributed to the C–H vibrations of the CH_2 group of keto-tautomer of phenol.

To clarify the reaction path for benzene hydroxylation, the adsorption complex of phenol with Fe/ZSM-5 was further studied. The energy of transformation of benzene oxide to phenol on Fe/ZSM-5 is −39.2 $kcal\,mol^{-1}$ and from the keto-tautomer to phenol is −4.2 $kcal\,mol^{-1}$. This value includes the difference in the energies of free phenol and benzene oxide or its keto-tautomer, and the difference in adsorption energies on the zeolite cluster. The adsorption energy of phenol on Fe/ZSM-5 is $-23\ \mathrm{kcal\,mol^{-1}}$.

The final complex from which phenol is released involves Fe^{2+} that needs to be regenerated back to Fe–O upon decomposition of N_2O. Summarizing, the calculated data indicate that the reaction path for the hydroxylation of benzene proceeds as shown in the scheme.

For the total mechanism, this implies that desorption of phenol is the key step in the reaction process, that probably explains the peculiarities of the kinetics of the reaction [86].

It is interesting to compare the described results of the energy path calculations of benzene hydroxylation with those for another model of active sites [87]. 3T cluster $H_3^*SiO–Al(OH^*)_2–OSiH_3^*$ was chosen as cationic site for Fe^+ ion, which transformed to OFe(III)/Z species after N_2O decomposition. This structure was suggested as the model of α-oxygen. DFT calculations of the reaction pathways of hydroxylation of methane and benzene molecules were performed. The same mechanism of the reaction was found in both cases, consisting in C–H dissociation and the formation of the H_3C–Fe–OH/Z or C_6H_5–Fe–OH/Z intermediates. The activation energies for the formation of intermediates were evaluated as 12.8 and $15.4\ \mathrm{kcal\,mol^{-1}}$, respectively and the next step of the reaction was found as limiting step in the hydroxylation process. Probably the main problem in such an approach is the experimental evidence for a strongly different kinetic H/D isotope effect in the case of methane and benzene hydroxylation by α-oxygen [42]. Another question is associated with using of Fe(I) state instead of Fe(II) for or Fe_α. Nevertheless, the discussion of the catalytic activity of OM(III)/Z species in zeolites is interesting and important for three-valent cations: Al(III), Ga(III), Fe(III), Co(III), etc. These species are suggested to be very active as was shown by calculations of CH_4 dissociation on OGa(III)/Z [88] and of N_2O decomposition on OAl(III)/Z [89].

4. Metal oxo-ions

Stabilization of multivalent metal ions in zeolites can be achieved by the incorporation of ELO into the structure and formation of bi- or multinuclear complexes, wherein each cation interacts efficiently only with the zeolite fragment containing a single Al ion. Although such complexes can form even in low-silica zeolites, they seem to be of particular importance in high-silica zeolites, where the overexchange phenomenon is often related to them. The tendency for stochastic distribution of Al throughout the zeolite lattice, especially in the case of high zeolite modules (Si/Al > 50), leads to zeolite structures with relatively large distances between neighboring Al ions. As a result, the metal cation can interact efficiently with the zeolite fragment containing a single Al ion. In this case, the stability of multivalent cations is probably provided by the formation of structures of the types [M–O–M]2, containing ELO. Boudart et al. [90] observed redox pairs of iron ions in Fe/Y zeolite. The bridging position of oxygen was confirmed by Mössbauer [91] and IR [92] spectroscopic studies. The catalytic activity of such systems is associated both with their binuclear structure and peculiar chemical properties of the ELO, differing from those of the lattice oxygen. In particular, much attention has been given to the redox properties of the binuclear structures [93] that were assumed to be responsible for some distinctive features of the kinetics of catalytic reactions, for example, generation of oscillations of N_2O decomposition [94,95].

The majority of theoretical studies of oxo-binuclear structures in zeolites were devoted to the structures and chemical properties of oxygen-bridged Cu^{2+} ion pairs. Lattice energy minimization technique was used by Sayle et al. [96] to identify a model for the active-site configuration within the NO decomposition CuZSM-5 catalyst. Binuclear hydroxide cluster [Cu(II)–OH–Cu(I)] was found as a part of six-membered cycle {–O–Cu(II)–OH–Cu(I)–O–Al–} in the vicinity of single Al ion in zeolite lattice. It was concluded that this cluster is strongly anchored by interaction with framework aluminum. The most likely configuration of the six-membered cycle is a strained

(Cu–Cu distance is 3.1 Å) which probably can be important for the chemical activity of these species. The formation of binuclear hydroxide cluster in CuY zeolite was also studied [97] and a comparison with CuZSM-5 was provided.

Molecular dynamics simulations and MO calculations were employed to study the coordination structures of Cu ionic species in CuZSM-5 zeolite by Teraishi et al. [98]. The six-membered ring with two Al ions, occupying T8 sites, was found as the most probable cationic position for Cu(II) and the calculated structure was in agreement with experimental data. Further, for this site the binuclear oxocation Cu(II)–O–Cu(II) and pair of Cu(I) ions were simulated. The Cu–Cu distance in former species was in agreement with EXAFS data. But the existence of the Cu(I) dimer was questioned according to MD simulation and calculation of photoluminescence spectra.

Oxygen-bridged Cu^{2+} ion pairs [ZCu(II)–O–Cu(II)Z] and [ZCu(II)–O_2–Cu(II)Z] were studied by Goodman et al. [99] in the frame of DFT approach; here Z is the minimal model of zeolite fragment, containing single Al ion: $Al(OH)_4$. The conclusion was made that the calculations support the existence of both types of oxygen-bridged oxocations in the CuZSM-5 zeolite. The reduction of the bridged structures by H_2 and CO was found to be exothermic. For example,

$$\mathrm{ZCu{-}O{-}CuZ} + \mathrm{H_2} \rightarrow \mathrm{ZCuOH_2} + \mathrm{CuZ} \quad -13\ \mathrm{kcal\,mol^{-1}} \tag{19}$$

$$\mathrm{ZCu{-}O{-}CuZ} + \mathrm{CO} \rightarrow \mathrm{ZCuCO_2} + \mathrm{CuZ} \quad -22\ \mathrm{kcal\,mol^{-1}} \tag{20}$$

And according to the reactions

$$2\mathrm{ZCu} + \mathrm{N_2O} \rightarrow \mathrm{ZCu{-}O{-}CuZ} + \mathrm{N_2} \quad -69\ \mathrm{kcal\,mol^{-1}} \tag{21}$$

$$\mathrm{ZCu{-}O{-}CuZ} + \mathrm{N_2O} \rightarrow \mathrm{ZCu{-}O_2{-}CuZ} + \mathrm{N_2} \quad -14\ \mathrm{kcal\,mol^{-1}} \tag{22}$$

the conversion of N_2O on the ZCu is favorable but the elimination of O_2 from the last structure is strongly endothermic ~ 70 kcal mol^{-1}.

The oxo-bridged binuclear structure $(Zn–O–Zn)^{+2}$ was considered by Yakovlev et al. [71]. The same type of Z_d cluster model with two Al in connected 5-rings was chosen as the structure for Zn_2O localization and DFT calculations. It is interesting to compare some properties of Z_dZn_2O and Z_dZn cation species. First of all, we note the lower Lewis acidity of the former. The calculated LUMO energies for these sites were −2.8 and −4.1 eV, respectively. According to the reaction:

$$Z_dH_2 + 2ZnO \rightarrow Z_dZn_2O + H_2O \quad -165.6\ \text{kcal mol}^{-1} \tag{23}$$

the H-form of the structure is less favorable than Zn-form, which agrees with the possibility of preparing ZnZSM-5 by deposition of gaseous ZnO. At the same time, the interaction with water gave rise to more favorable structure.

$$Z_dZn_2O + H_2O \rightarrow Z_d(ZnOH)(ZnOH) \quad -23.9\ \text{kcal mol}^{-1} \tag{24}$$

As a test of reactivity of the sites the calculations of intermediates of the 'alkyl' route of the ethane dehydrogenation reaction were performed:

$$Z_dZn_2O + C_2H_6 \rightarrow Z_d(ZnOH)(ZnC_2H_5) \quad -17.9\ \text{kcal mol}^{-1} \tag{25}$$

$$Z_dZn + C_2H_6 \rightarrow Z_d(H)(ZnC_2H_5) \quad -26.0\ \text{kcal mol}^{-1} \tag{26}$$

which was in agreement with previously discussed 'instability–reactivity' correlation. In the case of more strong stability of Zn(II) in zeolite rings with two Al's in the structure, as was considered in the previous section, these reactions were notable endothermic but the energies of H_2 removal (11) were comparatively small. Here we have opposite situation and recombination reactions

$$Z_d(ZnOH)(ZnH) \rightarrow Z_dZn_2O + H_2 \quad +37.2\ \text{kcal mol}^{-1} \tag{27}$$

$$Z_d(H)(ZnH) \rightarrow Z_dZn + H_2 \quad +32\ \text{kcal mol}^{-1} \tag{28}$$

are strongly unfavorable.

The same cluster model was applied for the DFT studies of Fe(II)–O–Fe(II) and (HO)Fe(III)–O–Fe(III)(OH) oxo-binuclear structures in FeZSM-5 zeolite by Yakovlev et al. [100]. The activity of these species in N_2O decomposition was considered. The first step of the reaction is commonly suggested as the subsequent decomposition of two N_2O molecules. For Fe(II) species:

$$Z_dZn_2O + N_2O \rightarrow Z_d(OFe{-}O{-}Fe) + N_2 \quad -31.4\ \text{kcal mol} \tag{29}$$

$$Z_d(OFe{-}O{-}Fe) + N_2O \rightarrow Z_d(OFe{-}O{-}FeO) + N_2 \quad -9\ \text{kcal mol}^{-1} \tag{30}$$

and analogously for Fe(III) species:

$$Z_d(HOFe{-}O{-}FeOH) + N_2O \rightarrow Z_d[(HO)OFe{-}O{-}Fe)] + N_2 \quad -12\ \mathrm{kcal\,mol^{-1}} \tag{31}$$

$$Z_d[(HO)OFe{-}O{-}FeOH)] + N_2O \rightarrow Z_d[(HO)OFe{-}O{-}FeO(OH)] + N_2 \quad -1\ \mathrm{kcal\,mol^{-1}} \tag{32}$$

As can be expected, the decomposition on Fe(II) species is more favorable. The removal of O_2 is evidently the energetically limiting step of the process. It was evaluated as 34 kcal mol^{-1} for Z_d(OFe–O–FeO) and 37 kcal mol^{-1} for Z_d[(HO)OFe–O–FeO(OH)]. It was concluded that below 200°C the binuclear cluster is further hydroxylated and is probably inactive in N_2O decomposition. Above this temperature and up to 500°C the catalytic site has (HO)Fe(III)–O–Fe(III)(OH) structure and above 500°C this site is predominantly Fe(II)–O–Fe(II). The reaction paths in both cases are rather similar.

In concluding this section we note that oxo-iron binuclear species were also considered as $(HO)_2Fe(OH)_2Fe(OH)_2$ extra-lattice forms by Filatov et al. [101]. They studied N_2O decomposition on partially dehydroxylated cluster resulting in the formation of terminal Fe=O group. Arbuznikov and Zhidomirov [102] found an isomer of this structure in the form of $[(HO)_2Fe{-}O{-}O{-}Fe(OH)_2]$ peroxo group. This site was shown to be very reactive toward CH_4 dissociative adsorption.

5. Outlook

Metal cation species in zeolite matrix are a very significant family of catalytic systems. Recently, high-silica zeolites attracted great attention as one of the most effective supports for active components. The possibility of changing over a wide range the conditions for stabilization of active species is an important factor promoting the use of these systems as catalysts. One of the key points in understanding metal–zeolite catalysis is the pattern of Al distribution over the zeolite lattice as a function of synthesis conditions and Si/Al ratio. Clearly, the peculiarities of synthesis techniques should be taken into account during theoretical modeling the aluminum distribution. The cluster model approach with the use of various constraints imposed by the lattice appeared to be effective in the studies of the structure and reactivity of active sites in these cases. Nevertheless, the need to apply embedding scheme and especially periodical calculations is quite evident. A number of problems in the theoretical interpretation of catalytic activity of polyvalent cations in high-silica zeolites are of particular interest now and the field is open to further study using theoretical methods.

Acknowledgements

The authors are very thankful for the help in preparation of this review to Drs. N.A. Kachurovskaya, A.L. Yakovlev and L.A.M.M. Barbosa. The Dutch Science Foundation is greatfully acknowledged for the financial support of the collaborative Russian-Dutch Project NWO-19-0411999.

References

1. Centi, G., Cavani, F. and Trifiro, F. (Eds.), *Selective Oxidation by Heterogeneous Catalysis*. Kluwer/Plenum Publ., New York, 2001, 505 pp.
2. Jentys, A. and Catlow, C.R.A., *Catal. Lett.*, **22**, 251–257 (1993).
3. Yudanov, I.V., Avdeev, V.I. and Zhidomirov, G.M., *Phys. Low-Dim. Struct.*, **4/5**, 43–46 (1994).
4. De Man, A.J.M. and Sauer, J., *J. Phys. Chem.*, **100**, 5025–5034 (1996).
5. Lamberti, C., Bordiga, S., Zecchina, A., Artioli, G., Marra, G. and Spano, G., *J. Am. Chem. Soc.*, **123**, 2204–2212 (2001).
6. Millini, R., Peregro, G. and Seiti, K., *Stud. Surf. Sci. Catal.*, **84**, 2123–2129 (1994).
7. Vayssilov, G. and van Santen, R.A., *J. Catal.*, **175**, 170–174 (1998).
8. Oumi, Y., Matsuba, K., Kubo, M., Inui, T. and Miyamoto, A., *Microporous Mater.*, **4**, 53–57 (1995).
9. Njo, S.L., van Koningsveld, H. and de Graaf, B., *J. Phys. Chem. B*, **101**, 10065–10068 (1997).
10. Riccardi, G., de Man, A. and Sauer, J., *Phys. Chem. Chem. Phys.*, **2**, 2195–2204 (2000).
11. Marra, G.I., Artioli, G., Fitch, A.N., Milanesio, M. and Lamberti, C., *Microporous Mesoporous Mater.*, **40**, 85–94 (2000).
12. Hijar, C.A., Jacubinas, R.M., Eckert, J., Henson, N.J., Hay, P.J. and Ott, K.C., *J. Phys. Chem. B*, **104**, 12157–12164 (2000).
13. Lamberti, C., Bordiga, S., Arduino, D., Zecchina, A., Geobaldo, F., Spano, G., Genoni, F., Petrini, G., Carati, A., Villain, F. and Vlaic, G., *J. Phys. Chem. B*, **102**, 6382–6390 (1998).

14. Gleeson, D., Sankar, C., Catlow, C.R.A., Thomas, J.M., Spano, G., Bordiga, S., Zecchina, A. and Lamberti, C., *Phys. Chem. Chem. Phys.*, **2**, 4812–4817 (2000).
15. Fejes, P., Nagy, J.B., Lazar, K. and Halasz, J., *Appl. Catal. A: General*, **190**, 117–135 (2000).
16. Sinclair, P.E. and Catlow, C.R.A., *J. Phys. Chem. B*, **103**, 1084–1095 (1999).
17. Notari, B., *Adv. Catal.*, **41**, 253–334 (1996).
18. Munakata, H., Oumi, Y. and Miyamoto, A., *J. Phys. Chem. B*, **105**, 3493–3501 (2001).
19. Zhidomirov, G.M., Arbuznikov, A.V., Yudanov, I.V. and Kachurovskaya, N.A., *React. Kinet. Catal. Lett.*, **57**, 263–274 (1996).
20. Zhanpeisov, N.U., Matsuoka, M., Yamashita, H. and Anpo, M., *J. Phys. Chem. B*, **102**, 6915–6920 (1998).
21. Zicovich-Wilson, C., Dovesi, R. and Corma, A., *J. Phys. Chem. B*, **103**, 988–994 (1999).
22. Damin, A., Bordiga, S., Zecchina, A. and Lamberti, C., *J. Chem. Phys.*, **117**, 226–237 (2002).
23. Zhidomirov, G.M., Yakovlev, A.L., Milov, M.A., Kachurovskaya, N.A. and Yudanov, I.V., *Catal. Today*, **51**, 397–410 (1999).
24. Karlsen, E. and Schöffel, K., *Catal. Today*, **32**, 107–114 (1996).
25. Clerici, M.G. and Ingallina, P., *J. Catal.*, **140**, 71–83 (1993).
26. Neurock, M. and Manzer, L.E., *Chem. Commun.*, 1133–1134 (1996).
27. Milov, M.A. and Zhidomirov, G.M., *React. Kinet. Catal. Lett.*, **75**, 147–155 (2002).
28. Yudanov, I.V., Gisdakis, P., Di Valentin, C. and Rösch, N., *Eur. J. Inorg. Chem.*, 2135–2145 (1999).
29. Anpo, M., Che, M., Fubini, B., Garrone, E., Giamello, E. and Paganini, M.C., *Topics Catal.*, **8**, 189–198 (1999).
30. Clerici, M.G., *Topics Catal.*, **15**, 257–263 (2001).
31. Ruzankin, S.F., Milov, M.A. and Zhidomirov, G.M., To be submitted to *React. Kinet. Catal. Lett.*, (2004).
32. van Laar, F.M.P.R., de Vos, D.E., Vanoppen, D.L., Pierard, F., Brodkorb, A., Mesmaeker, A.K.-D. and Jacobs, P.A., In: Treacy, M.M.J. (Ed.), *Proc. 12th International Zeolite Conference*. Materials Research Society, Warrendale, 1999, pp. 1213–1219.
33. Dzwigaj, S., Massiani, P., Davidson, A. and Che, M., *J. Mol. Catal. A: Chemical*, **155**, 169–182 (2000).
34. Boxhoorn, G., Kortbeek, A.G.T.G., Hays, G.R. and Alma, N.C.M., *Zeolites*, **4**, 15–21 (1984).
35. Koller, H., Lobo, R.F., Burkett, S.L. and Davis, M.E., *J. Phys. Chem.*, **99**, 12588–12596 (1995).
36. Artioli, G., Lamberti, C. and Marra, G.L., *Acta Crystallogr. B*, **56**, 2–10 (2000).
37. Kraushaar, B. and van Hooff, J.H.C., *Catal. Lett.*, **31**, 91–102 (1995).
38. Sokol, A.A., Catlow, C.R.A., Garces, J.M. and Kuperman, A., In: Treacy, M.M.J. (Ed.), *Proc. 12th International Zeolite Conference*. Materials Research Society, Warrendale, 1999, pp. 457–464.
39. Bordiga, S., Ugliengo, P., Damin, A., Lamberti, C., Spoto, G., Zecchina, A., Spano, G., Buzzoni, R., Dalloro, L. and Rivetti, F., *Topics Catal.*, **15**, 43–52 (2001).

40. Ruzankin, S.F., Shveigert, I.V. and Zhidomirov, G.M., *J. Struct. Chem.*, **43**, 229–233 (2002).
41. El-Malki, El-M., van Santen, R.A. and Sachtler, W.M.H., *J. Phys. Chem. B*, **103**, 4611–4622 (1999).
42. Panov, G.I., Uriarte, A.K., Rodkin, M.A. and Sobolev, V.I., *Catal. Today*, **41**, 365–385 (1998).
43. Schröder, K.-P., Sauer, J., Leslie, M. and Catlow, C.R.A., *Zeolites*, **12**, 20–23 (1992).
44. Nachtigallova, D., Nachtigall, P., Sierka, M. and Sauer, J., *Phys. Chem. Chem. Phys.*, **1**, 2019–2026 (1999).
45. Sastre, G., Fornes, V. and Corma, A., *J. Phys. Chem. B*, **106**, 701–708 (2002).
46. Olson, D.H., Khosrovani, N., Peters, A.W. and Toby, B.H., *J. Phys. Chem. B*, **104**, 4844–4848 (2000).
47. Rice, M.J., Chakraborty, A.K. and Bell, A.T., *J. Catal.*, **194**, 278–285 (2000a).
48. Bell, A.T., In: Centi, G., et al. (Eds.), *Catalysis by Unique Metal Ion Structures in Solid Matrices*. Kluwer Academic Publishers, 2001, NATO Science Series II, Vol. 13, pp. 55–73.
49. Dempsey, E., Kühl, G.H. and Olson, D.H., *J. Phys. Chem.*, **73**, 387–390 (1969).
50. Sonnemans, M.H.W., den Heijer, C. and Crocker, M., *J. Phys. Chem.*, **97**, 440–445 (1993).
51. Schröder, K.-P. and Sauer, J., *J. Phys. Chem.*, **97**, 6579–6581 (1993).
52. Pelmenschikov, A.G., Paukshtis, E.A., Edisherashvili, M.O. and Zhidomirov, G.M., *J. Phys. Chem.*, **86**, 7051–7055 (1992).
53. Barbosa, L.A.M.M., van Santen, R.A. and Hafner, J., *J. Phys. Chem.*, **123**, 4530–4540 (2001a).
54. Rice, M.J., Chakraborty, A.K. and Bell, A.T., *J. Phys. Chem. B*, **104**, 9987–9992 (2000b).
55. Wichterlova, B., Dedecek, J. and Sobalic, Z., In: Treacy, M.M.J. (Ed.), *Proc. 12th International Zeolite Conference*. Materials Research Society, Warrendale, 1999, pp. 941–973.
56. Trout, B.L., Chakraborty, A.K. and Bell, A.T., *J. Phys. Chem.*, **100**, 17582–17592 (1996).
57. Solans-Monfort, X., Branchadell, V. and Sodupe, M., *J. Phys. Chem. B*, **106**, 1372–1379 (2002).
58. Rice, M.J., Chakraborty, A.K. and Bell, A.T., *J. Phys. Chem. A*, **102**, 7498–7504 (1998).
59. Kim, J.S., Park, S.-E., Chang, J.-S., Park, Y.K. and Lee, C.W., In: Treacy, M.M.J. (Ed.), *Proc. 12th International Zeolite Conference*. Materials Research Society, Warrendale, 1999, pp. 437–443.
60. Shubin, A.A., Zhidomirov, G.M., Yakovlev, A.L. and van Santen, R.A., *J. Phys. Chem. B*, **105**, 4928–4935 (2001).
61. Lermer, H., Draeger, M., Steffen, J. and Unger, K.K., *Zeolites*, **5**, 131–134 (1985).
62. Frisch, M.J., Trucks, G.W., Schlegel, H.B., Scuseria, G.E., Robb, M.A., Cheeseman, J.R., Zakrzewski, V.G., Montgomery Jr., J.A., Stratmann, R.E., Burant, J.C., Dapprich, S., Millam, J.M., Daniels, A.D., Kudin, K.N., Strain, M.C., Farkas, O., Tomasi, J., Barone, V., Cossi, M., Cammi, R., Mennucci, B., Pomelli, C., Adamo, C., Clifford, S., Ochterski, J., Petersson, G.A., Ayala, P.Y., Cui, Q., Morokuma, K., Salvador, P., Dannenberg, J.J., Malick, D.K., Rabuck, A.D., Raghavachari, K., Foresman, J.B., Cioslowski, J.,

Ortiz, J.V., Baboul, A.G., Stefanov, B.B., Liu, G., Liashenko, A., Piskorz, P., Komaromi, I., Gomperts, R., Martin, R.L., Fox, D.J., Keith, T., Al-Laham, M.A., Peng, C.Y., Nanayakkara, A., Challacombe, M., Gill, P.M.W., Johnson, B., Chen, W., Wong, M.W., Andres, J.L., Gonzalez, C., Head-Gordon, M., Replogle, E.S. and Pople, J.A., Gaussian 98, Revision A.11, Gaussian, Inc., Pittsburgh, PA, 2001.

63. Zhidomirov, G.M., Shubin, A.A., Kazansky, V.B., Solkan, V.N., van Santen, R.A., Yakovlev, A.L. and Barbosa, L.A.M.M., In: *Catalysis for Sustainable Development*. Russian-Dutch Workshop, Novosibirsk, pp. 14–17.
64. Zhidomirov, G.M., Shubin, A.A., Kazansky, V.B. and van Santen, R.A., *Int. J. Quantum Chem.*, in press (2004).
65. Shubin, A.A., Zhidomirov, G.M., Kazansky, V.B. and van Santen, R.A., *Catal. Lett.*, in press (2003).
66. Frash, M. and van Santen, R.A., *Phys. Chem. Chem. Phys.*, **2**, 1085–1089 (2000).
67. Kazansky, V.B., Borovkov, V.Yu., Serykh, A.I., van Santen, R.A. and Stobbelaar, P.J., *Phys. Chem. Chem. Phys.*, **1**, 2881–2886 (1999).
68. McCusker, L.B. and Seff, K., *J. Phys. Chem.*, **85**, 405–410 (1981).
69. Barbosa, L.A.M.M. and van Santen, R.A., *J. Mol. Catal. A: Chemical*, **166**, 101–121 (2001).
70. Barbosa, L.A.M.M., Zhidomirov, G.M. and van Santen, R.A., *Phys. Chem. Chem. Phys.*, **2**, 3909–3918 (2000).
71. Yakovlev, A.L., Shubin, A.A., Zhidomirov, G.M. and van Santen, R.A., *Catal. Lett.*, **70**, 175–181 (2000).
72. Kazansky, V.B., Kustov, L.M. and Khodakov, A.M., In: Jacobs, P.A. and van Santen, R.A. (Eds.), *Zeolites: Facts, Figures, Future*. Elsevier, Amsterdam, 1989, pp. 1179–1182.
73. Kazansky, V.B., Borovkov, V.Yu., Serikh, A.I., van Santen, R.A. and Anderson, B.G., *Catal. Lett.*, **66**, 39–47 (2000).
74. Barbosa, L.A.M.M., Zhidomirov, G.M. and van Santen, R.A., *Catal. Lett.*, **77**, 55–62 (2001b).
75. Panov, G.I., Sobolev, V.I. and Kharitonov, A.S., *J. Mol. Catal.*, **61**, 85–97 (1990).
76. Panov, G.I., *CATTECH.*, **4**, 18–32 (2000).
77. Bordiga, S., Buzzoni, R., Geobaldo, F., Lamberti, C., Giamello, E., Zecchina, A., Leofanti, G., Petrini, G., Tozzola, G. and Vlaic, G., *J. Catal.*, **158**, 486–501 (1996).
78. Zhu, Q., Mojet, B.L., Janssen, R.A.J., Hensen, E.J.M., van Grondelle, J., Magusin, P.C.M.M. and van Santen, R.A., *Catal. Lett.*, **81**, 205–212 (2002).
79. Ovanesyan, N.S., Shteinman, A.A., Dubkov, K.A., Sobolev, V.I. and Panov, G.I., *Kinet. Catal.*, **39**, 792–797 (1998).
80. Dubkov, K.A., Ovanesyan, N.S., Shteinman, A.A., Starokon, E.V. and Panov, G.I., *J. Catal.*, **207**, 341–352 (2002).
81. Kachurovskaya, N.A., Zhidomirov, G.M., Hensen, E.J.M. and van Santen, R.A., submitted to *Catal. Lett.*, **86**, 25–31 (2002).
82. Panov, G.I., Dubkov, K.A. and Paukshtis, Y.A., In: Centi, G., et al. (Eds.), *Catalysis by Unique Metal Ion Structures in Solid Matrices*. Kluwer Academic Publishers, 2001, NATO Science Series II, Vol. 13, pp. 149–163.
83. Kasperek, G.J. and Bruce, T.C., *J. Am. Chem. Soc.*, **94**, 198–202 (1972).
84. Klotz, B., Barnes, I., Becker, K.H. and Golding, B.T., *J. Chem. Soc. Faraday Trans.*, **93**(8), 1507–1516 (1997).

85. Jerina, D.M., Witkop, B., McIntosh, C.L. and Chapman, O.L., *J. Am. Chem. Soc.*, **96**, 5578–5580 (1974).
86. Ivanov, A.A., Chernavsky, V.S., Gross, M.J., Kharitonov, A.S. and Panov, G.I., prepared to *Appl. Catal. A: General*, **249**, 327–343 (2002).
87. Yoshizawa, K., Shiota, Y., Yumura, T. and Yamabe, T., *J. Phys. Chem. B*, **104**, 734–740 (2000).
88. Broclawik, E., Himei, H., Yamadaya, M., Kubo, M. and Miyamoto, A., *J. Chem. Phys.*, **103**, 2102–2108 (1995).
89. Yakovlev, A.L. and Zhidomirov, G.M., *Catal. Lett.*, **63**, 91–95 (1999).
90. Boudart, M., Garten, R.L. and Delgas, W.N., *J. Phys. Chem.*, **73**, 2970–2979 (1969).
91. Garten, R.L., Delgass, W.N. and Boudart, M., *J. Catal.*, **18**, 90–107 (1970).
92. Dalla Betta, R.A., Garten, R.L. and Boudart, M., *J. Catal.*, **41**, 40–45 (1976).
93. Chen, H.-Y., El-Malki, El-M., Wang, X. and Sachtler, W.M.H., In: Centi, G., et al. (Eds.), *Catalysis by Unique Metal Ion Structures in Solid Matrices*. Kluwer Academic Publishers, 2001, NATO Science Series II, Vol. 13, pp. 75–84.
94. Lei, G.-D., Adelman, B.J., Sarkany, J. and Sachtler, W.M.H., *Appl. Catal. B*, **5**, 245–256 (1995).
95. El-Malki, El-M., van Santen, R.A. and Sachtler, W.M.H., *J. Catal.*, **196**, 212–223 (2000).
96. Sayle, D.C., Catlow, C.R.A., Gale, J.D., Perrin, M.A. and Nortier, P., *J. Phys. Chem. A*, **101**, 3331–3337 (1997).
97. Catlow, C.R.A., Bell, R.G., Gale, J.D., Lewis, D.W., Sayle, D.C. and Sinclair, P.E., In: Deruane, E.G., et al. (Eds.), *Catalytic Activation and Functionalisation of Light Alkanes*. Kluwer: Dordrecht, Neth. Academic Publishers, 1998, pp. 189–214.
98. Teraishi, K., Ishida, M., Irisava, J., Kume, M., Takahashi, Y., Nakano, T., Nakamura, H. and Miyamoto, A., *J. Phys. Chem. B*, **101**, 8079–8085 (1997).
99. Goodman, B.R., Schneider, W.F., Hass, K.C. and Adams, J.B., *Catal. Lett.*, **56**, 183–188 (1998).
100. Yakovlev, A.L., Zhidomirov, G.M. and van Santen, R.A., *J. Phys. Chem. B*, **105**, 12297–12302 (2001).
101. Filatov, M.J., Pelmenschikov, A.G. and Zhidomirov, G.M., *J. Mol. Catal.*, **80**, 243–251 (1993).
102. Arbuznikov, A.V. and Zhidomirov, G.M., *Catal. Lett.*, **40**, 17–23 (1996).
103. Hodgson, D., Zhang, H.-Y., Nimlos, M.R. and McKinnon, J.T., *J. Phys. Chem. A*, **105**, 4316–4327 (2001).
104. Wichterlova, B. and Bell, A.T., In: Centi, G., et al. (Eds.), *Catalysis by Unique Metal Ion Structures in Solid Matrices*. Kluwer Academic Publishers, NATO Science Series II, Vol. 13, 2001, pp. 187–204.

Computer Modelling of Microporous Materials
C.R.A. Catlow, R.A. van Santen and B. Smit (editors)

Chapter 8

Template–host interaction and template design

Dewi W. Lewis

Department of Chemistry, University College London, 20 Gordon St., London WC1H 0AJ, UK

1. Introduction

In this chapter, we turn our attention to the challenges posed by understanding and guiding zeolite synthesis. The synthesis of many, if not most, synthetic microporous and mesoporous materials is facilitated by the use of organic so-called template molecules. Until the early 1960s, formation of synthetic zeolites mimicked that of natural zeolites in having alkali or alkaline earth metal cations as the extra-framework component. Barrer and Denny [1] pioneered the use of organic species as a component in the synthesis mixture, a development that saw a rapid expansion in the number of zeolitic materials that could be synthesised. The aim of 'templating' is to form a material that has specific structural features, which are usually complementary to the structure of the organic template molecule. If, therefore, we wish to synthesise a zeolite with cages, we would probably use a small and essentially spherical molecule. On the contrary, if a 1-D channel system is desired then a longer linear molecule may be selected. Since such molecules are incorporated into the structure of the solid during synthesis, they have a direct bearing on the structure and concomitant properties of the solid formed. However, the degree to which this 'templating effect' is expressed and even the mechanism of the action is often a matter of debate. Subsequently, the templating effect is also often referred to as a 'structure-directing' effect and the template molecules as structure-directing agents (SDAs). Both terms will be used interchangeably in this chapter.

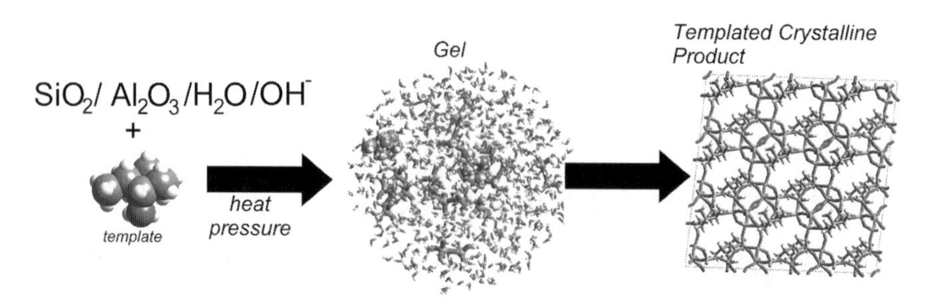

Fig. 1. Amorphous gels condense around template molecules to form crystalline microporous solids.

During the synthesis of microporous solids — mesoporous solids will not be discussed here — the synthesis medium is typically a sol or a gel to which the template molecules are added. Synthesis then occurs under hydrothermal conditions, the framework condensing from the gel, encapsulating the template and producing the solid. The concept is illustrated in Fig. 1. The choice of template molecule is but one of a wide range of synthesis conditions — temperature, time, pressure, concentration, agitation, etc. — and the influence of all the remaining parameters is significant. Nevertheless, it is evident that careful selection of a template can have a significant effect on the product formed: witness the large number of structures which can only be synthesised through the use of templates. Thus, template choice can be considered as one of the primary routes to the formation of specific (and targeted) microporous solids.

The efficacy of the templating or structure-directing action can be simply considered a measure of the closeness of 'fit' between the template and the solid structure that is formed [2–5]. Many structures clearly reflect the structure of the template, for example, the intersecting channels of the zeolite ZSM-5 which is templated by tetrapropylammonium, the small cage of sodalite as templated by tetramethylammonium or more dramatically, the three-fold symmetry of the ZSM-18 structure [6] when templated by a complex trisquaternary ammonium ion (Fig. 2). However, there are many other cases where the correlation is less satisfactory. For example, ZSM-5 can also be synthesised using α,ω-diamines, which at first sight do not have a structure which we may expect to form an intersecting channel system. However, they do possess a similar molecular cross-section that is reflected in the *diameter* of the pores formed. The literature reveals that many microporous structures are formed by a number of organic templates (or none) and similarly many organic molecules can

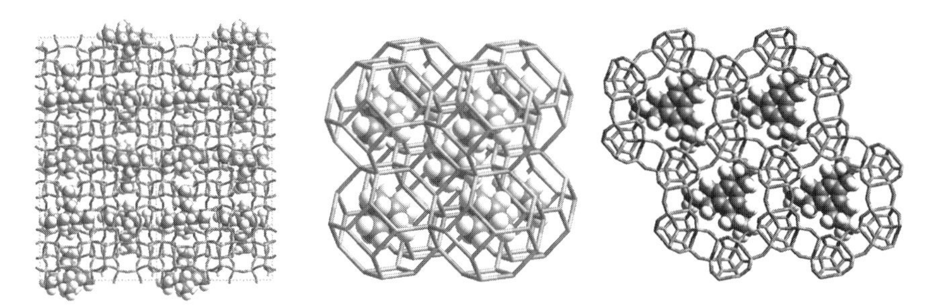

Fig. 2. The relationship between the template molecule and the microporous structure for three zeolitic materials: a schematic of tetrapropylammonium siting at the intersection of the channels in ZSM-5 (left), tetramethylammonium in the cages of sodalite (centre) and a triquaternaryamine in the structure of ZSM-18 (right). In each case the structure is that obtained from Monte Carlo docking calculations which are in agreement with experimentally determined template positions.

template the formation of different microporous structures, and that few examples exist where a single (or a number of closely related) molecule form a specific structure. Thus, whilst the use of SDAs has brought about such a dramatic increase in the number of structures formed, the templating effect is far from being specific.

The remainder of this chapter focuses on two aspects of modelling of templates in microporous solids. Firstly, the determination of template–host interactions will be considered. Can we use host–guest interactions to locate the most stable configurations of templates within microporous structures? Can 'templating' be quantified? Can such quantification lead to a better understanding of any templating effect and thus perhaps to improved templates or new materials? Secondly, the development of methods for the design of new templates will be discussed.

2. Locating templates in microporous structures

In determining the likely position of template molecules within the cavities of microporous framework, we must explore the internal conformational space of the template and also its location within the pore structure. For many cases, manual (by eye) docking (or positioning) of the template with the framework, using any convenient visualisation software, may be a reasonable start. For example, there are few alternative locations for the template in ZSM-18 other than that shown (Fig. 2). The system may then be energy minimised to quantify the interaction energy. However, in many cases, more than one position may be available for the template molecule. To ensure that all these positions, and any likely conformational change within the template, are sampled, the Monte Carlo docking method developed by Freeman et al. [7] has been utilised in many studies. The method is implemented as part of the Cerius^2 modelling package from Accelrys Inc.

The approach is based on a combination of molecular dynamics, Monte Carlo and energy minimisation methods. In a typical calculation, a molecular dynamics (MD) trajectory for the template molecule in the gas phase is used to generate a library of sorbate conformations, each of which is then inserted randomly into the framework using a Monte Carlo — in the sense that the molecule is inserted at a random position and orientation — procedure. Low-energy configurations of the template in the microporous structure are retained for subsequent energy minimisation. The MD ensures adequate sampling of the conformational space of the template molecule. However, it is apparent from the literature, that the vast majority of templates remain in (or close to) their lowest-energy conformation and this stage may be neglected in many cases. Energy minimisation of these crudely docked structures yields representative low-energy binding sites for the molecule within the host structure. The method can be applied either to finite clusters, representative of the zeolite structure, or with periodic boundary conditions. The latter is obviously preferred although the former are often computationally less expensive. Through careful choice of unit cell and the use of periodic boundary conditions the packing of templates can also be ascertained [8,9]. However, care must be taken to ensure that the template concentration is reasonable and that any constraints resulting from the framework symmetry are considered.

Cox and co-workers [9] developed an extension of this method, where local minima are avoided through the use of a simulated annealing procedure. Here, rather than performing simple energy minimisation, a series of MD runs are performed at a range of decreasing temperatures, thus allowing the system to find global minima more effectively. Such an approach is particularly effective when multiple templates are present in each unit cell, which often leads to many potential local minimum energy configurations.

Figure 2 shows representative results obtained using the docking procedure. Where available, comparison to experimental structures is found to be excellent. For example, the position of tetrapropylammonium in ZSM-5 and 1-aminoadamantane in NU-3 are within 1% of the experimental positions [10,11].

Thus, it is evident from the body of work described above, and from other examples such as the work of Cox et al. [8,12], that docking calculations can be used to identify low-energy configurations of templates within microporous structures.

3. Quantifying template–host interactions for template selection

It is clear from experimental observation that a degree of correlation exists between the van der Waals shape of an organic template and the structure that it forms [2–5]. Initial modelling investigations therefore attempted to provide not only a method by which templates could be located within microporous structures (see above), but also to provide a rationalisation in energetic terms for the templating effect. Our work [10,11] attempted to provide a measure of likely templating efficacy by considering the interaction energy between the template and the microporous host.

Using the docking methodology described above, the interaction (or binding) energy between the template and the framework can be determined:

$$E_{\mathrm{inter}} = E_{\mathrm{host+template}} - (E_{\mathrm{host}} + E_{\mathrm{template}}),$$

where $E_{\mathrm{host+template}}$ is the energy of the template in the zeolites, E_{host} is the energy of the empty zeolite and E_{template} is the energy of the template when energy minimised in the gas phase. By considering a large number of experimental template–zeolite, a correlation between the size and shape of the template and the binding energy was obtained (Fig. 3) giving a quantification of the 'good fit' noted experimentally. However, what is not obtained is a general discriminator between these successful and combinations of template and zeolite that do not form experimentally. For example, the binding energy of tetramethylammonium in ZSM-5 is actually higher than in sodalite.

A further problem is that, strictly, energies obtained from molecular mechanics for different molecules cannot be compared, since all the energy terms have different zero points. However, if we consider a homologous series, only the number of different atom types, bonding terms, etc. increase. Thus, here we are able at least to draw firmer correlations. For example, the trend observed for the tetraalkylammonium cations is clearly better than the overall trend in Fig. 3. A progression therefore, is to ask whether this type of calculation

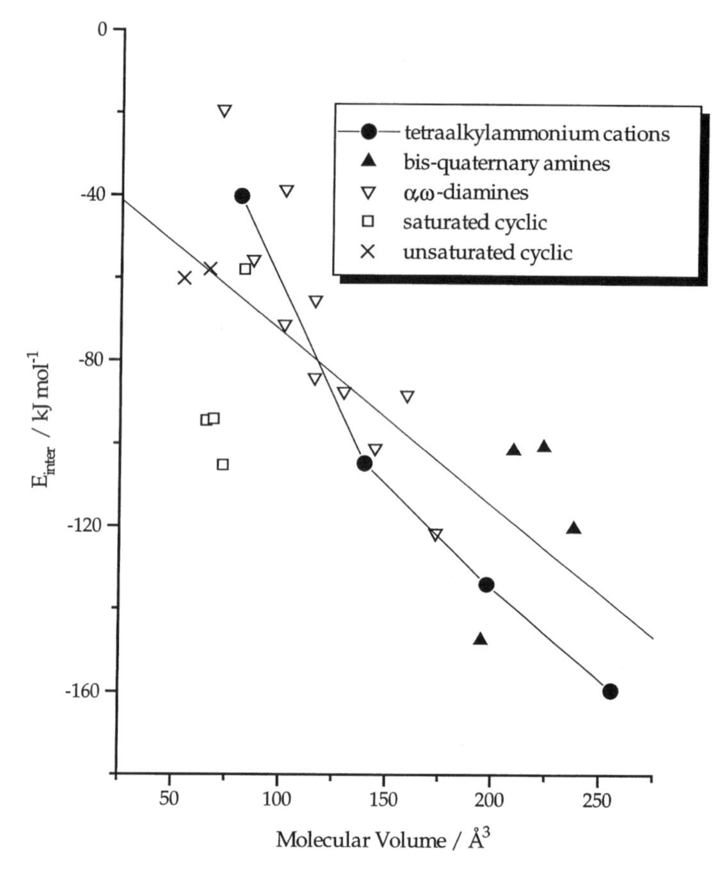

Fig. 3. Interaction energy of experimental framework/template combinations as a function of molecular volume. Straight line is the best fit through all datapoints. The results for the tetraalkylammonium salts are also highlighted with a line joining the datapoints.

could determine which of a homologous series would be the most successful in forming a specific structure.

We were able to demonstrate how the structures formed by the homologous series of tetraalkylammonium cations — from tetramethylammonium to tetrabutylammonium — could be rationalised in terms of the binding of the template within the host. Table 1 shows the binding energies of isolated molecules within each structure known to form from these templates. It is apparent that the structure formed is, in general, the structure in which the template is most strongly bound. Thus, TEA forms zeolite β, whilst TBA forms ZSM-11. However, the periodic packing of the template within the zeolite must also be considered; in these early calculations, the initial interaction energy is

Table 1
Interaction energies and packing energies for the tetraalkylammonium cations

Template	ZSM-5		ZSM-11		Zeolite β
	E_{inter} (kJ mol^{-1})	E_{pack} (kJ mol^{-1})	E_{inter} (kJ mol^{-1})	E_{pack} (kJ mol^{-1})	E_{inter} (kJ mol^{-1})
TMA	−51.7		−38.7		−43.1
TEA	−92.1		−73.0		−104.7
TPA	−133.9	−29.7	−119.9	−18.3	−83.4
TBA	−165.5	+14.9	−159.5	−8.5	−56.7

TMA, tetramethylammonium; TEA, tetraethylammonium; TPA, tetrapropylammonium; TBA, tetrabutylammonium. E_{inter} is defined as the stabilisation energy on placing a single template molecule within the zeolite pores. E_{pack} is defined as the additional stabilisation on placing two templates at adjacent sites.

determined for a single template. As we might expect, a single TBA molecule binds more strongly in ZSM-5 than does TPA, simply because it can fit and it has more interactions with the framework. However, for templating to occur, templates must also pack periodically. By using either periodic boundary conditions, or simply repeating the molecule a number of times, we find that TBA will not template ZSM-5 because the template is too large to pack at adjacent sites. The importance of packing is illustrated for ZSM-5 in Fig. 2. In a similar fashion, it was possible to identify templating trends amongst the homologous series of di-bisquaternary amines (from trimethonium to octamethonium) [11].

Thus, for a series of molecules, such calculations can suggest as to which would be the more potent template for a target structure [10,11]. Others have extended this work, using periodic boundary conditions and the inclusion of charge interactions (see below for a further discussion of the inclusion of Coulombic interactions). For example, ZSM-5 and ZSM-11 often appear as intergrowths and the loading of templates has been found to be critical. For ZSM-11 both tetrapropylammonium and tetrabutylammonium cations are considered as structure-directing agents, and calculations were performed [13] for occlusion of either three or four templates per unit cell of the zeolite. Both estimates of the change in internal energy and Gibbs free energy revealed that the synthesis of ZSM-11 is favoured by the occlusion of three tetrabutylammonium cations per unit cell, consistent with experimental observation. Burchart et al. [14] came to similar conclusions.

More recently, Sabater and Sastre [15] showed how accurate treatment of template–framework interaction could be used to reveal potential preferential substitution of Al within a framework. Once the position of the template in ZSM-18 had been established, they were able to permute large combinations of Al distributions. They found that the most favoured distributions minimised the electrostatic energy and resulted in preferential location of the Al in the three-membered rings. Similarly, Shantz et al. [16] showed how identification of template location from docking calculations, coupled with NMR experiments, could be used to gain insight into potential defects in the as-synthesised zeolite framework. They demonstrated directly a correlation between the charge on the template and the location of such defects.

These calculations reveal how a more detailed study can provide deeper insights into the chemistry of the templating action. However, the simpler calculation discussed above also play a crucial role in providing initial starting points — the template positions in ZSM-18, ZSM-5 and ZSM-11 are unambiguous. They are also far less costly, whilst being accurate enough to determine which template is likely to be the most successful, albeit at the expense of detail. Thus, both types of calculations have their part, depending on the role the calculations play: to understand the mechanisms of templating we need an accurate description, but for identifying templates for synthesis programmes, the simpler calculations provide as much insight.

However, whilst seemingly effective in identifying potential templates, even these simple docking calculations are a relatively time-consuming method for screening a large number of templates and frameworks. An alternative method based on shape analysis has been suggested as a more rapid and cost-effective alternative method of identifying effective templates. Cox and co-workers [12] showed how simple structure–property relationships, such as dipole moment, cluster around certain values for favourable template–framework combinations. Thus, finding template molecules with properties that do not overlap with other templates may suggest that they will form new materials. These calculations suggest that an element of screening of potential templates in known structures may be used to determine if the template is likely to direct the formation of one of these known structures or, if not, to potentially form a new material.

3.1. Template–host interactions and the kinetics of synthesis

Similar calculations were used by Harris and Zones [17] to correlate crystallisation time with the binding energy of the template in the framework. Here kinetic studies and docking calculations were used to help understand the control that organic templates exert in the crystallisation rates during synthesis. A range of template molecules were used to synthesise SSZ-13 (Chabazite structure) and Nonasil. The molecules vary in size and affect rates of zeolite formation. The modelling studies reveal an excellent correlation between the energetics of template 'fit' with the rate of crystallisation. The energy stabilisation mostly arises from increased van der Waals interactions and it was proposed that such favourable interactions help to stabilise nucleation.

3.2. Methods — some thoughts

One major simplification in our initial work, and of many other papers, was the neglect of charge interactions [11,17]. Although we demonstrated that for many systems they are less important than steric factors [11] in determining the location and binding of templates within the host material, there are other systems where they are clearly critical in determining the structure formed, and moreover, in facilitating the crystallisation process [18]. To stabilise the array of cationic templates, we need an anionic framework whose charge is acquired by including aliovalent metal cations in the framework, or by the formation of framework defects.

For example, it is well established that the concentration of such metal cations in the gel can have considerable impact on the nature of the templating effect. Consider the case of how different aluminophosphates are formed from Co-containing gels [19–21]. Using the same template and reaction conditions, simply varying the Co content can produce either the cage-structured Co-AlPO-34 or the channel-structured Co-AlPO-5 structure. Clearly, the Coulombic interactions between the template and the framework are of considerable importance. However, using our docking procedure we again demonstrated that the limiting factor in the templating effect was the steric packing of the templates in the framework [22]. The main role of the Coulombic interaction here is, it seems, to allow the different template packing regimes to be stabilised. Thus, the formation of the cage of Co-AlPO-34 requires a higher Co content to allow the close packing of templates needed to stabilise the structure and allow its crystallisation. Conversely, the templating effect appears less critical for the formation of the AlPO-5 structure which can form even at lower than optimal template concentrations. This effect is reflected in our calculations where the packing of templates has little effect on the overall interaction energy between the templates and the framework. We were therefore, able to explain the role of the Coulombic template–framework interactions in this system without calculating them directly. Indeed, a limiting factor of such calculations has been the availability of high-quality force fields that allow the description of metal ions such as Co within microporous frameworks, or the number of configurations which would need to be considered if a (near) random distribution of framework cations is expected [15]. However, this work has demonstrated that their effect on templating can often be included by careful consideration of the template concentration.

The question remains, should charge interactions be included in studies of the location of templates? The answer as always is 'it depends'. If the aim of the work is to identify likely templates or to identify the possible location of a specific template within the zeolitic lattice, then little will be gained by a careful consideration of the charge distribution in the template. However, if the aim is to consider how, say, the template may direct the localisation of Al in the framework [15], or to determine if similar phases may form intergrowths [13,14], then here, clearly we must consider the role of charge.

If charges are to be included then the number of aliovalent charges must be considered. The effective negative charge on the framework must clearly be balanced by the charge on the template. Hence, the concentration of template is effectively fixed. Care must therefore be taken to establish the likely sites for the templates and the likely concentration prior to fixing the framework charge. The charge on the framework has been considered both explicitly (e.g. by placing Al in the framework [15]) or by smearing the charge over the framework atoms [16].

A related point is that wherever the role of the template is considered to be paramount to the successful synthesis of a specific structure, then in general the concentration of aliovalent cations in the framework is controlled by the concentration of the template. Thus, such calculations can be used as a tool to probe the likely possible *compositional* variations that we may be able to synthesise — examples are given below for DAF-1 and the design of template for DAF-5.

4. Template design

An oft-stated aim of the materials chemist is the desire to design materials for specific applications. In the field of zeolite synthesis, the search for new materials remains to a great extent Edisonian, although the increase in known structures over the past 30 years has demonstrated how considered choice of template is an important feature of new material synthesis. With computational methods now demonstrated to be able to accurately identify template locations and also to provide a semi-quantitative measure of likely templating efficacy, could we not discover new materials by designing new templates? A number of strategies have emerged.

4.1. Design 'by inspection'

If a specific organic species is known to template a given zeolitic host, then modifying that organic species so that it no longer 'fits' that microporous cavity may provide us with a species which will form a new material. Thus, we can envisage taking a known template–host combination and determining the template geometry. We can then modify — through the use of a graphical interface — the organic species (for example, adding side chains etc.) until the binding energy is dramatically reduced. We may now speculate that this modified template will now direct the synthesis of an alternative structure. The difficulty now arises of evaluating whether this 'designed' template will not simply form another already known zeolitic structure. Clearly, we can perform docking calculations using this new template on a representative sample of these structures to ascertain if any of these already known structures is likely to be favoured by our new template. However, without a degree of automation, such a method is time-consuming. Nevertheless, such an approach is likely to provide, if not specific new templates, a deeper understanding of the role of the template, allowing the synthetic chemist to make more informed choice regarding a new synthesis strategy.

For example, when attempting to identify the mode of templating in DAF-1 [23] by the two templates decamethonium and octamethonium (see below), we postulated, on the basis of our packing calculations, that dodecamethonium would not form DAF-1. This, indeed, proved to be the case, although the phase form was the ubiquitous $AlPO_4$-5. However, on then studying the templates in $AlPO_4$-5, the stability and packing of dodecamethonium was more favoured than any of the other templates [24].

4.1.1. Example: Modifying framework compositions using templates

Often, improvements in a catalytic process require, not a new catalyst, but rather modifications to the existing catalyst. For example, changes can be made to the exact composition, morphology, etc. Indeed, it may be argued that this approach is most prevalent in industrial research since the cost of plant will be the largest contribution, and minor modifications, perhaps improving yield by a few percent, will result in large cost benefits. Modifying the concentration of active sites is one route by which the catalytic performance of a microporous solid can be improved. The concentration of the template is proportional to the number of aliovalent cations (and their position to some extent) in the framework and hence to the number of catalytic or ion-exchange sites. Certainly, as discussed above, we can estimate the maximum concentration of template within the pores and use this information as a guide to the expected compositions. But can we use docking calculations to determine the mode of templating and thus allow the synthesis to be modified to allow access to new phases?

Wright et al. [23] demonstrated how an understanding of the templating could be used to design new synthesis conditions for accessing compositions with lower metal concentrations of DAF-1. A magnesoaluminophosphate, DAF-1, is unique in having two different channel systems running parallel to each other. The material is known to catalyse the conversion of a mixture of butenes to iso-butene; an important feedstock for the production of anti-knock agents for unleaded petrol. However, synthesis of DAF-1, regardless of gel composition, always resulted in a phase of composition $Mg_{0.22}Al_{0.78}PO_4$. A consequence of such a high Mg concentration is that there are too many catalytic sites, leading to coking. By determining the lowest-energy sites for the templates (octamethonium and decamethonium) an understanding of why such a high Mg concentration is found was obtained. The calculations revealed that there are two types of template molecule within the pore (see Fig. 4, Table 2). Those within the 10MR channels are structure directing. However, there are many, more weakly bound configurations in the 12MR channel, which have 'supercages' along its length, which have no real structure-directing role; they simply fill the space. However, their presence requires the uptake of further Mg for charge balance. Thus, hypothetically a material could be synthesised with the template being present only in the 10MR (as illustrated in Fig. 4), which has a composition of $Mg_{0.06}Al_{0.94}PO_4$. Wright et al. performed a synthesis using ethylene glycol as a 'co-template', rationalising that it would have no structure-directing role and not require any charge compensation, with the aim that it would spacefill the supercages. New phases of DAF-1 were indeed formed which had a composition of $Mg_{0.05}Al_{0.95}PO_4$ and $Mg_{0.15}Al_{0.85}PO_4$. Whilst it cannot be directly shown that the crystals have a true partition of templates between the two channel systems, subsequent diffraction studies by Muncaster et al. [25] revealed strongly ordered octamethonium in the

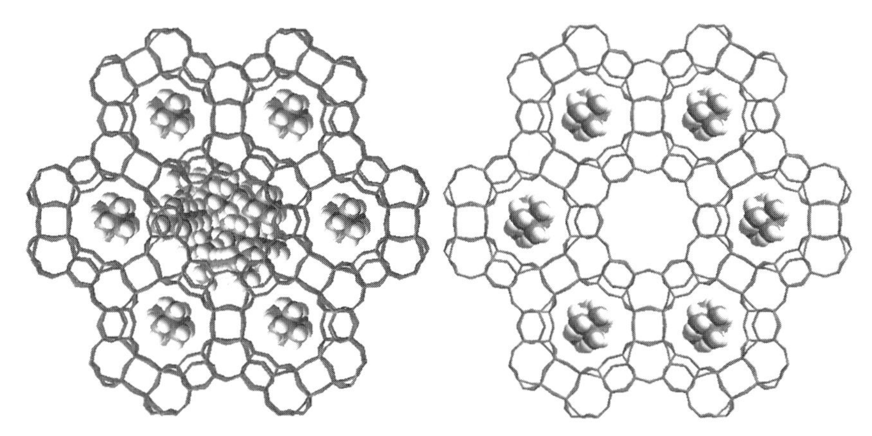

Fig. 4. (Left) Calculated positions of decamethonium in DAF-1. The templates in the 10MR channel are ordered whilst those in the 12MR can occupy many different orientations — here a superposition of all calculcated postions is shown. (Right) Postulated structure with only those decamethonium ions required for structure directing is shown. It was further postulated that the use of ethylene glycol as a 'co-template' would facilitate the formation of such a composition [23].

Table 2
The interaction energies of the experimental and modified templates in various sites within the structure of DAF-1

Template and site	E_{inter} (kJ mol^{-1})
Octamethonium	
10MR channel	−34.0
12MR channel	−23.6
Decamethonium	
10MR channel	−37.5
12MR channel	−27.0

10MR but could not determine any specific ordering in the supercages, which correlates well with the calculations. Thus, whilst no new structure has been formed, a new compositional variation on DAF-1 has been identified and successfully synthesised as a direct result of these calculations.

4.2. De novo design

Taking the above ideas further, we may wish to simply allow the computer to design a new template from the scratch. That is, to take a specific target microporous structure and design a template that has optimal binding. The analogy with drug design methods is clear.

De novo ligand design methods have been applied for some time in the search for biologically active materials [26–30]. These techniques allow a molecule to be 'grown' so as to fit accurately into a targeted space. The molecule is grown, in a computational sense, to promote favourable interactions between the substrate and the target cavity. But owing to the complex, and frequently ill-defined nature of biological active sites, a number of approximations are often made, such as defining the target as a property (CoMFA) map [31] and restricting growth to linking units [26,27].

The template molecules required for the synthesis of microporous inorganic solids are usually much simpler and smaller than those required in a biological context. Furthermore, the description of the target host (the microporous material) is, more often than not, much more clearly defined than those noted in biochemical applications. Firstly, it is crystalline in nature, and can therefore be readily described atomisitically; even hypothetical structures can be defined in this way. Secondly, the frameworks possess reasonable rigidity so that conformational changes in the pore need not be considered. Thirdly, molecular transport is not an issue, since the framework condenses around the template. Finally, as discussed above, the interactions present in the templating of microporous materials can be considered initially as being dominated by van der Waals interactions. Interactions involving charge and hydrogen bonding are less important in determining the final structure formed, although they are significant in the gel chemistry which is an essential precursor of the crystalline product.

Taking these ideas, we developed a de novo molecular design code [32,33], ZEBEDDE (ZEolites By Evolutionary De-novo DEsign) for the application of these techniques to microporous materials. Our aim was to produce a generic code that could be applied to any system, since we found that many of the existing (bioscience) codes were highly specific in their application. We also included many features that are specific to application in solids, particularly the consideration of periodicity and the symmetry of the growing molecule with respect to the host. The technique of de novo ligand design and its implementation in ZEBEDDE is now described (see Fig. 5).

The target space (the *host*) in which a molecule (the *template*) is to be grown is supplied (as atomic coordinates) by the user, along with a library of fragments from which the template is to be constructed. A typical fragment library for growing templates for microporous materials may include, methyl, ethyl, ammonia, benzene, cyclohexane etc.;

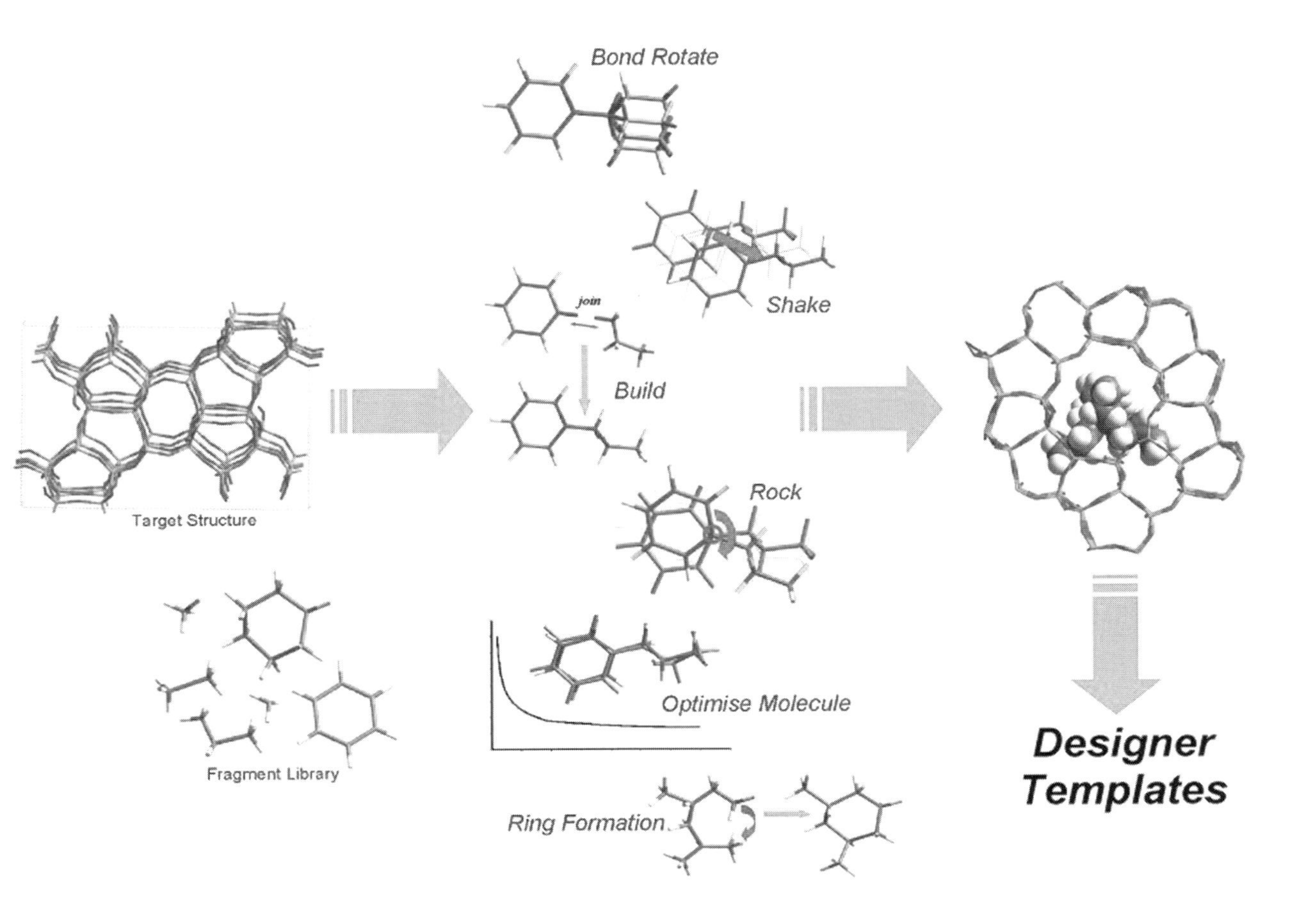

Fig. 5. Overview of the de novo molecular design code ZEBEDDE.

clearly the makeup of this library is application specific. As a starting point a seed molecule is either selected at random from the fragment library or supplied by the user. This seed is either placed at random, at the centre of mass of the host or at defined coordinates. The template then grows by a number of randomly selected actions using the fragment library as the source of new atoms. Actions available are:

(i) *Build*, where a new (randomly selected) fragment is joined onto the existing template. This is achieved by forming a bond between the neighbouring atoms of selected hydrogens, with the selected hydrogens being deleted from the molecule. Addition of a new fragment is controlled by a van der Waals function, which determines if any atom of the template overlaps with others. If the addition of the new fragment violates the overlap function, additional actions, as listed below, are performed to reduce this conflict. If no reduction in the overlap function is possible the new fragment is rejected. To allow better opportunities for growth, van der Waals radii may scaled to allow small overlap.

(ii) *Rotate*, where the last fragment added is rotated about the new bond formed.

(iii) *Shake*, where the template is displaced along a random vector with respect to the host.

(iv) *Rock*, where the template is randomly rotated as a rigid body with respect to the host.

(v) *Random bond twist*, where a randomly selected bond joining fragments in the template is rotated.

(vi) *Ring formation*, where atoms that are nth order neighbours and are within a cutoff distance are joined, forming a ring. Thus, alkyl chains growing in a confined space can convert to cyclic systems. The order n of the neighbours can be specified to form different sized rings; $n = 5$ forms six-membered rings.

(vii) *Template energy minimisation* and

(viii) *Template/host energy minimisation* which allows the template to be energy minimised in the gas phase and with respect to the (fixed) host respectively. Currently the code supports both Discover [37] from Accelrys and MOPAC [38] as external minimisers.

A cost function based on overlap of van der Waals spheres is used to control the growth of the template. When an action other than build is selected, that action is performed repeatedly to maximise the function

$$f_c = \sum_t C(tz),$$

where $C(tz)$ signifies the closest contact between a template atom t and any host atom. During actions (ii)–(v) movements of the template are made as a series of small steps. After each step the move is continued only if f_c has been increased. Thus, although a randomised procedure is used, the template rapidly moves to occupy the largest available space with the optimum possible molecular conformation.

After each action the template may be assessed according to a user-defined probability. If an assessment is required, the cost function is evaluated and the template checked for overlaps with atoms within the molecule or with any symmetry-related images. The new template is only accepted if the cost function has decreased from the last value and there are no van der Waals clashes. Although it may be expected that evaluating the template after every modification may be more beneficial, this would not encourage growth in such confined voids. For example, consider a small template located in the centre of a void. Addition of a large fragment would most likely lead to clashes with the framework and if we evaluate the cost function often, this action would be rejected. However, if the template is allowed to move within the host, by allowing additional actions prior to evaluation of the cost function, such steric hindrances could be removed.

Given the crystalline nature of microporous materials, periodic boundary conditions may also be specified, allowing the option of a unit-cell structure. Typically, these materials have smaller unit cells with higher symmetry in the absence of the template molecules (that is, in their calcined state). For this reason, periodic boundary conditions and the symmetry of the growing molecules may be different from that of the framework. Thus, growing molecules interact not only with the host but also with periodic images of themselves.

The success of the method was initially demonstrated [32] through the successful reproduction of templates for ZSM-5 — for example, TPA — and in the prediction of 2-methylcyclohexylamine for the synthesis of a LEV structured material simultaneously with the successful synthesis of a Co-SAPO-35 (LEV) material by Barrett et al. [34] (see Fig. 6).

4.2.1. Application of de novo template design

Both aluminosilicates and aluminophosphates, with the Chabazite structure (IZA code CHA) act as catalysts for a number of important reactions. For example, SAPO-34 is an effective catalyst for the conversion of methanol to light olefins [35]. However, there are a number of problems in the synthesis of such materials, which arise

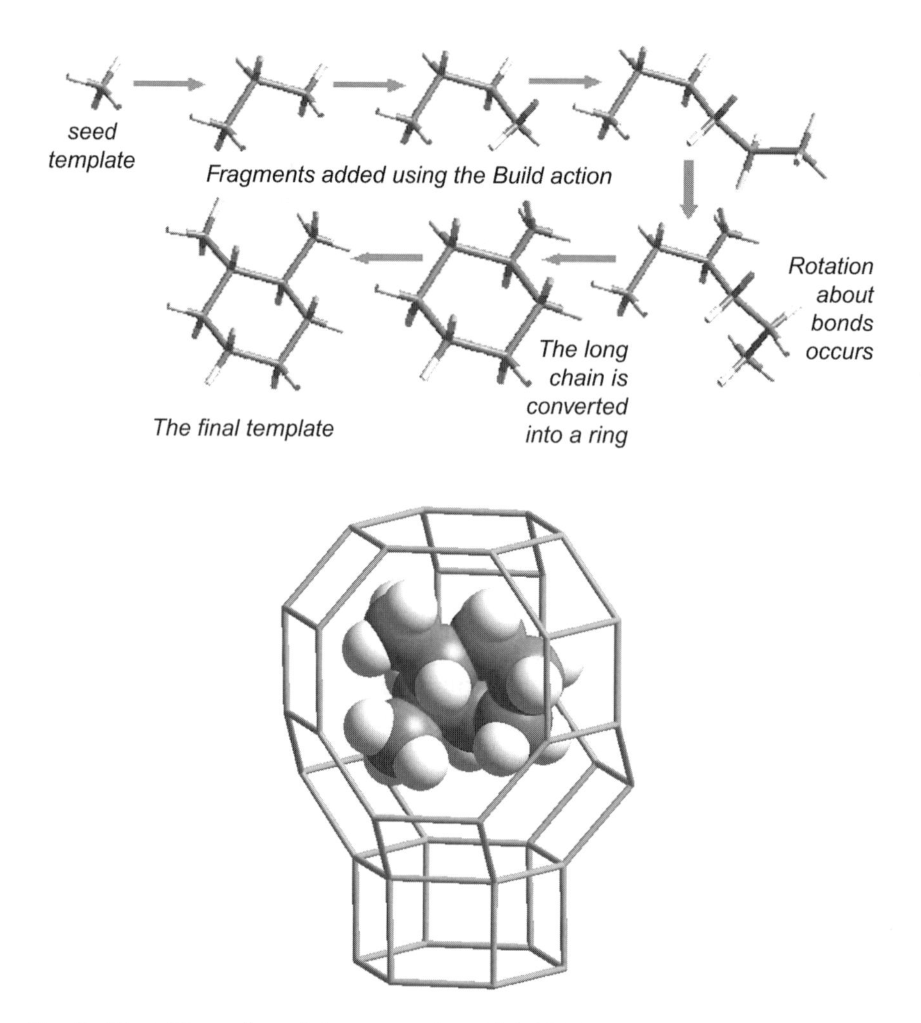

Fig. 6. (Top) Illustration of the process by which ZEBEDDE designed 1,2-dimethyl-cyclohexane as a template for the LEV structure. (Bottom) The final template in the cage of LEV. Barrett et al. [34] used 2-methylcyclohexylamine as a template to synthesise DAF-4, a LEV-structured Co-AlPO.

from the choice of organic species used as the structure-directing agent. Typically, small amines — specificially triethylamine, cyclohexylamine and diethylethanolamine [36] — are used and generally, two such molecules fit within the single CHA cage [37]. Such a high template concentration leads directly to a high Co concentration (since they charge-balance each other). Thus, if the gel concentration of Co is low, $AlPO_4$-5 (AFI) tends to form [19]. Thus, pure CHA phases at low

Co concentrations are difficult to form. Similarly, the kinetics of forming CHA phases is relatively slow. In a joint computational and experimental effort, we [38] designed a new template that would be specific for the CHA structure, with the aim of providing a more efficient route to CoAlPO-34 catalysts. Using ZEBEDDE, we determined that a bicyclic motif (Fig. 7) was optimal for the CHA cage; there is a high binding energy, only one molecule is required per cage, and the molecule is symmetric. We therefore attempted a synthesis using 4-piperidino piperidine as a template. However, prior to synthesis, the new template was docked into $AlPO_4$-5 and a number of other likely competitive

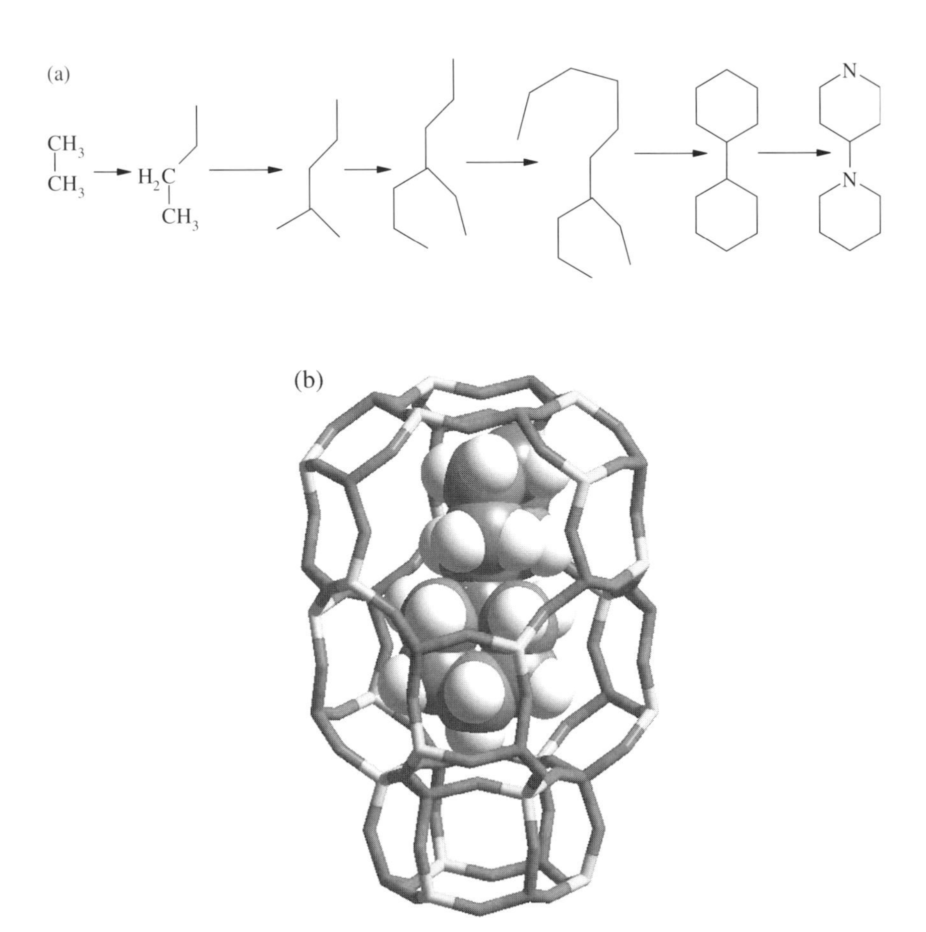

Fig. 7. (a) Growth of 4-piperidino piperidine from ethane seed in a siliceous CHA unit cell using ZEBEDDE. Note that the molecule is orientated to best illustrate the growth process rather than its true position in the CHA cage. The final structure obtained from ZEBEDDE is bicylohexane. Substitution of N into this carbon motif is done manually. (b) Energy minimised location of 4-piperidino piperidine in the CHA cage.

phases. We found that docking of 4-piperidino piperidine has a much lower binding energy in the $AlPO_4$-5 structure than in the CHA structure. Similarly, the binding energy is lower than that of those templates that are found to form $AlPO_4$-5, again supporting the general conclusion that van der Waals interactions provide a good measure of templating efficacy. Not only was a CHA-structured material synthesised (designated DAF-5), but it also formed rapidly and with no competitive formation of $AlPO_4$-5. Furthermore, the synthesis can be performed at higher temperature (with shorter crystallisation time) than could be previously — high temperatures resulted in Co-$AlPO_4$-5 with triethylamine. Subsequent diffraction studies of DAF-5 [39] revealed the template to be intact and in the predicted geometry.

This was, we believe the first, computer de novo designed template that has been successfully deployed in synthesis. Earlier, Schmitt and Kennedy [40] had shown how 'manual' computer-aided design of a new template could be achieved and demonstrated a new template for the synthesis of ZSM-18 in this way. Since ZEBEDDE has been written to be generic we have also been able to use the method to design templates for hypothetical porous transition metal oxides [41,42].

5. Other approaches

Of course the discussion above has been limited to the consideration of template–framework interactions in the *final* crystalline structure. That is, we have attempted to rationalise templating and identify new templates by assuming that the remaining (extensive) experimental parameters can be ignored. Furthermore, these calculations provide no direct insight into *how* the framework is formed around the template. Nevertheless, the success seen suggests that this approach is a useful tool for material design and synthesis. If, however, we are to understand *how* microporous solids form, we must consider the chemistry of the gel, of nucleation and of subsequent crystal growth. A number of approaches have been applied. The role of the template in stabilising small silica fragments has been investigated [18,43] using the same molecular mechanics description used in the docking calculations discussed above. This work highlighted how the presence of the template prevents the collapse of these open fragments, particularly when considered in a hydrated environment. Similarly, the role of charge interactions was shown to be crucial in stabilising fragment-template assemblies, countering the loss of solvation energy. As such they provide evidence to support the view that crystallisation takes place at the interface between the template and the solvent. Clearly there is much more work to be done in understanding the chemistry of the formation of microporous solids, and that modelling plays a critical role, particularly given the difficulties in probing molecular processes in gels experimentally.

6. Conclusions

Over the past few years we have seen how computer modelling can be used to identify how templates are organised in a microporous solid; to provide a quantification of the 'fit' of the template in the structure; and to allow templating efficacy to be determined for a series of templates — in effect allowing us to identify the best template for a structure. Furthermore, methods have been developed to allow the systematic design of new templates. Together, they provide tools that allow the chemist to attempt the synthesis of new compositional variants of existing structures or (the ultimate challenge) the synthesis of wholly novel materials.

New developments such as automated screening methods, the construction of database systems and models that allow the inclusion of other key experimental parameters, will further enhance the role of such modelling in the search for new materials.

A more significant role for such modelling studies should be in elucidating the exact mechanisms of templating. There is a need to model, as accurately as possible, the chemistry of the reactive gel, the formation of likely nucleation species and their growth. In order to do such new methods that consider both reaction pathways, solvation, dynamics and surface chemistry need to be developed.

References

1. Barrer, R.M. and Denny, P.J., *J. Chem. Soc. (London)*, 971 (1961).
2. Lok, B.M., Cannan, T.R. and Messina, C.A., *Zeolites*, **3**, 282 (1983).
3. Davis, M.E. and Lobo, R.F., *Chem. Mater.*, **4**, 759 (1992).
4. Davis, M.E., *Chemistry — A European Journal*, **3**, 1745 (1997).
5. Gies, H. and Marler, B., *Zeolites*, **12**, 42 (1992).
6. Lawton, S.L. and Rohrbaugh, W.J., *Science*, **247**, 1319 (1990).
7. Freeman, C.M., Catlow, C.R.A., Thomas, J.M. and Brode, S., *Chem. Phys. Lett.*, **186**, 137 (1991).
8. Cox, P.A., Stevens, A.P., Banting, L. and Gorman, A.M., *Stud. Surf. Sci. Catal.*, **84**, 2115 (1994).

9. Stevens, A.P., Gorman, A.M., Freeman, C.M. and Cox, P.A., *J. Chem. Soc., Faraday Trans.*, **92**, 2065 (1996).
10. Bell, R.G., Lewis, D.W., Voigt, P., Freeman, C.M., Thomas, J.M. and Catlow, C.R.A., *Stud. Surf. Sci. Catal.*, **84**, 2075 (1994).
11. Lewis, D.W., Freeman, C.M. and Catlow, C.R.A., *J. Phys. Chem.*, **99**, 11194 (1995).
12. Boyett, R.E., Stevens, A.P., Ford, M.G. and Cox, P.A., *Zeolites*, **17**, 508 (1996).
13. Shen, V. and Bell, A.T., *Microp. Mater.*, **7**, 187 (1996).
14. Burchart, E.D., van Koningsveld, H. and van de Graaf, B., *Microp. Mater.*, **8**, 215 (1997).
15. Sabater, M.J. and Sastre, G., *Chem. Mater.*, **12**, 4520 (2001).
16. Shantz, D.F., Fild, C., Koller, H. and Lobo, R.F., *J. Phys. Chem. B*, **103**, 10858 (1999).
17. Harris, T.V. and Zones, S.I., *Stud. Surf. Sci. Catal.*, **84**, 29 (1994).
18. Lewis, D.W., Catlow, C.R.A. and Thomas, J.M., *Faraday Discuss.*, **106**, 451 (1997).
19. Uytterhoeven, M.G. and Schoonheydt, R.A., *Microp. Mater.*, **3**, 265 (1994).
20. Norby, P., Christensen, A.N. and Hanson, J.C., *Stud. Surf. Sci. Catal.*, **84**, 179 (1994).
21. Rey, F., Sankar, G., Thomas, J.M., Barrett, P.A., Lewis, D.W., Catlow, C.R.A., Clark, S.M. and Greaves, G.N., *Chem. Mater.*, **7**, 1435 (1995).
22. Lewis, D.W., Catlow, C.R.A. and Thomas, J.M., *Chem. Mater.*, 8 (1996).
23. Wright, P.A., Sayag, C., Rey, F., Lewis, D.W., Gale, J.D., Natarajan, S. and Thomas, J.M., *J. Chem. Soc., Faraday Trans.*, **91**, 3537 (1995).
24. Lewis, D.W., Bell, R.G., Wright, P.A., Catlow, C.R.A. and Thomas, J.M., *Stud. Surf. Sci. Catal.*, **105C**, 2291 (1996).
25. Muncaster, G., Sankar, G., Catlow, C.R.A., Thomas, J.M., Bell, R.G., Wright, P.A., Coles, S., Teat, S.J., Clegg, W., Reeve, W., *Chem. Mater.*, **11**, 158 (1999).
26. Bohacek, R.S. and McMartin, C.J., *J. Am. Chem. Soc.*, **116**, 5560 (1994).
27. Bohm, H.J., *J. Comput-Aided Mol. Design*, **6**, 61 (1992).
28. Lewis, R.A. and Dean, P.M., *Proc. R. Soc. Lond. (Biol)*, **236**, 125 (1989).
29. Oprea, T.I., Ho, C.M.W. and Marshall, G.R., De novo design: ligand construction and prediction of affinity , In: Reynolds, C.H., Holloway, M.K. and Cox, H.K. (Eds.), *Computer-Aided Molecular Design. Applications in Agrochemicals, Materials and Pharaceuticals.* American Chemical Society, Washington DC, 1995, Vol. 589, p. 65.
30. Gillet, V., Johnson, A.P., Mata, P., Sike, S. and Williams, P.J., *J. Comput-Aided Mol. Design*, **7**, 127 (1993).
31. Cramer III, R., Patterson, D. and Bunce, J., *J. Am. Chem. Soc.*, **110**, 5959 (1988).
32. Lewis, D.W., Willock, D.J., Catlow, C.R.A., Thomas, J.M. and Hutchings, G.J., *Nature*, **382**, 604 (1996).
33. Willock, D.J., Lewis, D.W., Catlow, C.R.A., Hutchings, G.J. and Thomas, J.M., *J. Mol. Catal. A*, **119**, 415 (1997).
34. Barrett, P.A., Jones, R.H., Thomas, J.M., Sankar, G., Shannon, I.J. and Catlow, C.R.A., *Chem. Commun.*, 2001 (1996).
35. Rabo, J.A., Pellet, R.J., Coughlin, P.K. and Shamshoun, E.S., In: Karge, H.G. and Weitkamp, J. (Eds.), *Zeolites as Catalysts, Sorbents and Detergent Builders.* Elseveier, Amsterdam, 1989, Vol. 46, p. 1.

36. Wilson, S.T., Lok, B.M., Messina, C.A., Cannan, T.R. and Flanigen, E.M., *J. Am. Chem. Soc.*, **104**, 1146 (1982).
37. Lewis, D.W., Catlow, C.R.A. and Thomas, J.M., *Chem. Mater.*, **8**, 1112 (1996).
38. Lewis, D.W., Sankar, G., Wyles, J., Thomas, J.M., Catlow, C.R.A. and Willock, D.J., *Angew Chemie*, **36**, 2675 (1997).
39. Sankar, G., Wyles, J., Jones, R.H., Thomas, J.M., Lewis, D.W., Catlow, C.R.A., Clegg, W., Coles, S. and Teat, S., *J. Chem. Soc., Chem. Commun.*, 117 (1998).
40. Schmitt, K.D. and Kennedy, G.J., *Zeolites*, **14**, 635 (1994).
41. Cora, F., Lewis, D.W. and Catlow, C.R.A., *J. Chem. Soc., Chem. Commun.*, 1943 (1998).
42. Cora, F., Catlow, C.R.A. and Lewis, D.W., *J. Mol. Catal.*, **166**, 123 (2001).
43. Catlow, C.R.A., Coombes, D.S., Lewis, D.W. and Pereira, J.C.G., *Chem. Mater.*, **10**, 3249 (1998).

Computer Modelling of Microporous Materials
C.R.A. Catlow, R.A. van Santen and B. Smit (editors)

Chapter 9

The interplay of simulation and experiment in zeolite science

Clive Freeman

Accelrys, 9685 Scranton Road, San Diego, CA 92130, USA

Jörg-Rüdiger Hill

Accelrys, Inselkammerstraße 1, D-82008 Unterhaching, Germany

1. Introduction

Microporous materials are complex with unique properties. Understanding the molecular determinants of these properties has been the main theme of this book: as the detailed understanding of these properties provides the basis for both the rational optimization of materials and the invention of new applications.

In this final chapter the interaction between simulation and experiment in zeolite science is discussed. Four broad areas of application are reviewed: characterization, structural simulation, physisorption, and electronic property simulation. In each of these areas the focus in this chapter is on the use of simulation in providing solutions of practical problems in zeolite science.

2. Characterization

Microporous materials, materials possessing cavities with molecular dimensions, are synthesized under carefully controlled conditions and are generally thermodynamically unstable with respect to more condensed phases. Preparing single crystals, suitable for traditional single-crystal diffraction is often challenging and proves a practical impossibility on many occasions. When only polycrystalline samples are available, the determination of structure must rely on powder diffraction. The intrinsic complexity of zeolitic materials is a product of their often low symmetry and their large unit cells, and may also be compounded at the atomic level by the presence of features such as stacking faults and local disorder. In addition, a range of sample-based difficulties, such as preferred orientation, also affect microporous materials. Powder diffraction studies of zeolite materials remain, therefore, far from routine and must draw upon optimal experimental resources such as synchrotron radiation sources. Additional structural probes are also used, where possible, to augment powder diffraction information. For example, magic angle spinning NMR (MAS-NMR) can provide an account of the number of unique metal atoms in the framework. Transmission electron microscopy (TEM) also reveals information on subunit or layer motifs that underlie a given structure.

2.1. Model building

Model building has long been used to analyze the consequences of experimental observation. Lawrence Bragg considered "...the best way interatomic forces could be satisfied to give stable structures..." in addition to using Fourier methods and isomorphous replacement to solve complex crystal structures (see Ref. [1]). In the 1920s, Bragg's first research student, A.J. Bradley, used models in the computation of powder diffraction patterns to explain the observed powder diffraction of alloys (see Ref. [2]). Since the late 1960s, with the development of Rietveld refinement and the distance least-squares refinement method [3,4], computational techniques have also contributed to the development of structural models of zeolitic materials. Computational model building combines structural library information, with interactive editing and visualization techniques. The ongoing enumeration and accumulation of knowledge on two-dimensional three-connected and three-dimensional four-connected nets provides an additional source of theoretical models [5–8].

A particular advantage of computational methods is the ability to closely couple the calculation of the experimental consequences of the model (powder diffraction patterns, MAS-NMR spectra, and the output of a variety of other possible characterization methods), with the manipulation of the model (see, for example, Ref. [9]). This provides immediate feedback on the degree of similarity between the current structural proposition and observation. For example, Suib and co-workers described the use of such computational techniques in the

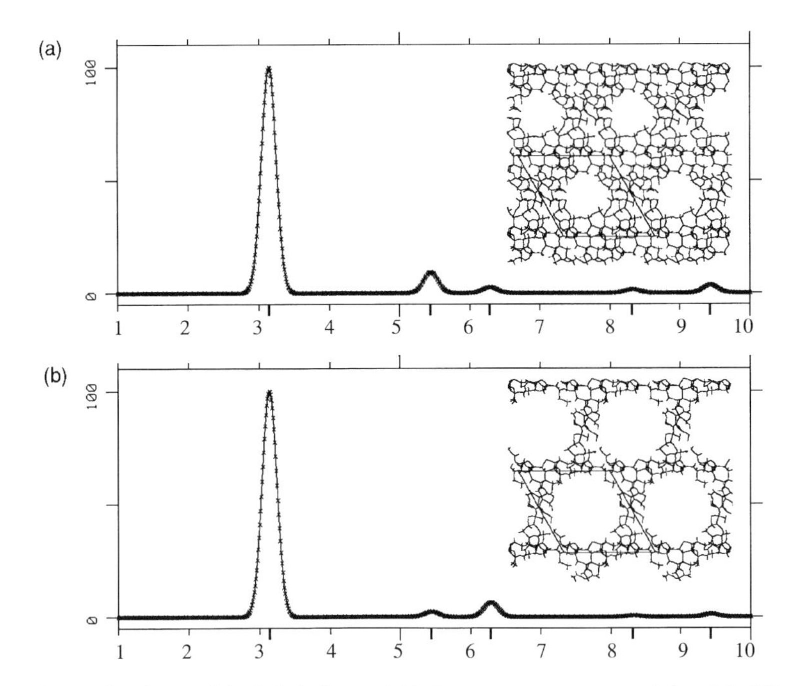

Fig. 1. Two simple models (labeled a and b) for mesoporous materials with differing wall thicknesses. Corresponding simulated powder diffraction patterns for each model are also illustrated. Diffracted intensity is plotted on the y-axis and the diffraction angle, 1–10° 2θ, on the x-axis. The simulated patterns show that wall thickness can be qualitatively estimated on the basis of the relative heights of diffraction peaks around 6° 2θ.

structural characterization of materials with structures based on MnO_6 octahedra [10,11]. Computational model building techniques have been applied to the simulation of the characteristics of materials with high degrees of complexity, such as the mesoporous MCM materials [12,13] (see Fig. 1). While the size of such systems preclude the development of atomic accuracy, such models permit the analysis of averaged properties, such as wall thickness, and surface silanol concentrations. The integrated modeling environments of Cerius^2 and Materials Studio provide the necessary tools to perform such modeling operations [14].

3. Structural simulation

As discussed in Chapter 1, force field-based solid-state computational modeling methods, combined with the calculation of complementary characterization methods, have proved valuable in exploring a range of zeolite structures [15–17]. Such methods were originally designed to examine defects and other situations where crystallographically derived structural information was difficult to obtain [18]. However, the accuracy of such methods, as demonstrated, for example, in the simulation of subtle structural transitions in materials such as silicalite [19,20] (see Fig. 2), has allowed their use in predicting structures based on supplied models. The refined structure of NU-87 was obtained on the basis of a model derived from a high-symmetry idealized structure optimized using a shell model-based energy minimization calculation [21]. Energy minimization calculations played a similar role in the accurate structure description of MeAlPO-36 [22,23]. Energy minimization calculations may also be combined with nondiffraction-based experimental methods. For example, Sankar and co-workers described the synergistic combination of X-ray absorbtion spectroscopy (XAS) and simulation-based models in the analysis of titanium coordination in ETS-10 [24].

Static simulation methods yield both structural and energetic information, resulting in interest in their use in understanding the thermodynamic stability of zeolitic structures [25–27]. Using a harmonic model

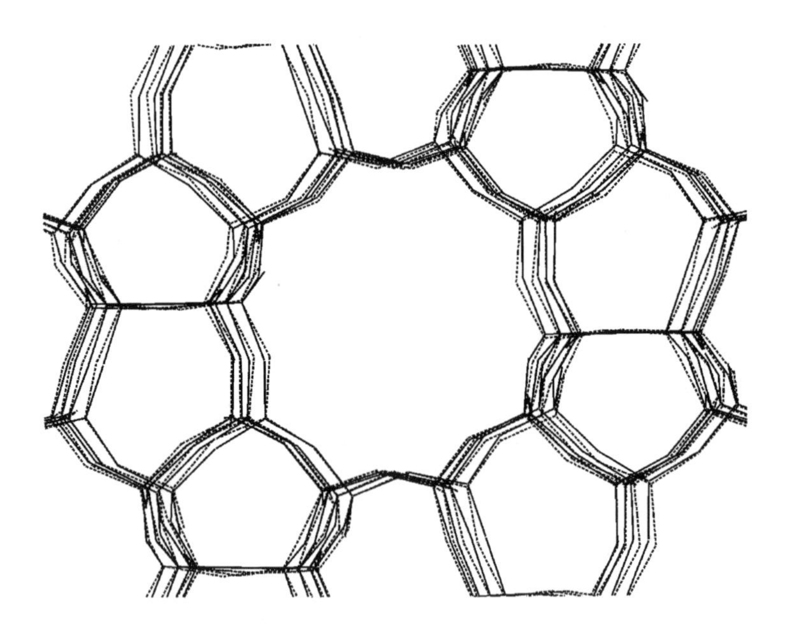

Fig. 2. The experimentally observed (solid line) orthorhombic structure of ZSM-5 and the energy minimized (monoclinically distorted) structure obtained from this starting point [19] (dashed line).

it is possible to use static simulation methods to obtain the free energy of zeolitic systems. For example, such simulations demonstrated that many zeolitic materials possess a negative thermal expansion coefficient [28], as confirmed by subsequent experiments.

A particular concern in the preparation of materials for a variety of practical applications is the incorporation of extra-framework species. In addition to describing framework structure, energy minimization has been employed in understanding extra-framework cation locations [15,29–31]. Here there are large numbers of configurations that must be enumerated if accurate energy calculations are to be attempted for all possible arrangements. A possible simplifying strategy has been suggested based on the precomputation of energy grids (see Ref. [32]). Comparison with experimentally derived site occupancies indicates that the procedure may be used to evaluate extra-framework preferences on the basis of the framework type, the silicon to aluminum ratio, and the number and type of counterions [32,33]. Framework flexibility can be included in the simulations where appropriate (see, for example, Refs. [31,34]). A practical illustration of these approaches is provided by the work of Grillo and Carrazza who employed simulation in the analysis of nonframework Na^+ and K^+ cations in the ETS-10 structure leading to a description of the material with four potential monovalent cation-binding sites with different preferences for each ion [35].

3.1. Automating computational model building

Following the success of potential-based simulation methods and the increasing availability of computational resources, automated techniques for the prediction of crystal structure have been increasingly attempted and reported [36–40]. These methods are based on a variety of global optimization schemes including simulated annealing and genetic algorithms. Applications to microporous materials include the solution of novel structures [41]. An important extension of such methods, however, is the inclusion of a term in the object function to be optimized which measures the degree of agreement expressed by a proposed model with available experimental observation in the form of powder diffraction data [42,43]. Various extensions have been developed including the use of improved sampling schemes [44] and Fourier-based crystallographic methods [45,46]. Figure 3 summarizes the methodology typically employed. A variety of structures have been solved using these methods, a number of which are illustrated in Fig. 4 [43,47–49]. Simulated annealing and related methods for

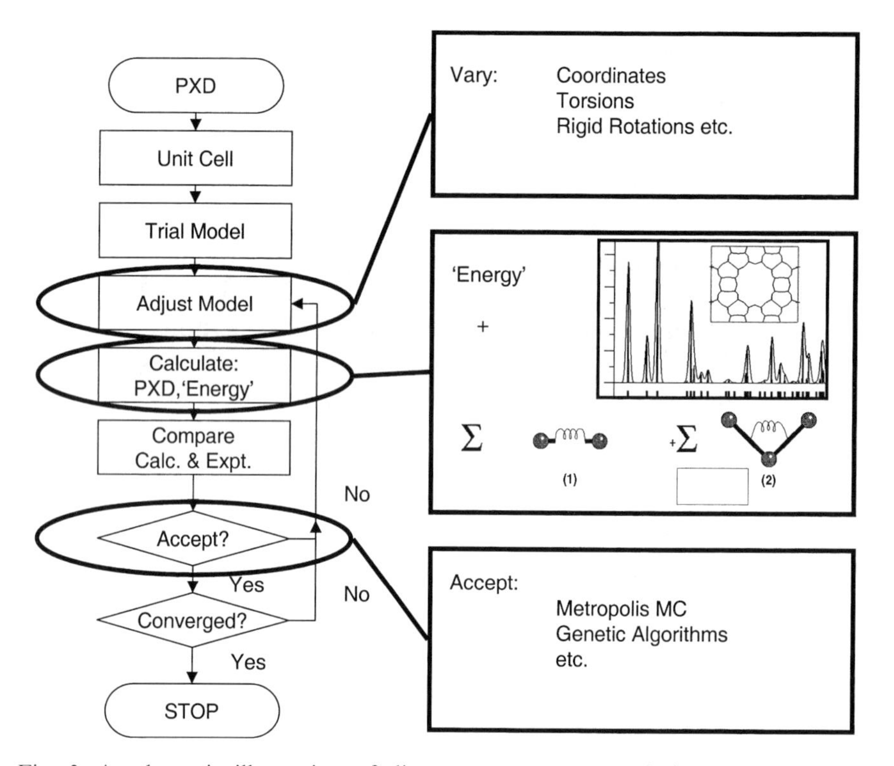

Fig. 3. A schematic illustration of direct space structure solution processes using simulated annealing and related optimization methods. The essential procedure is summarized in the flow diagram on the left; an initial model is iteratively adjusted, its calculated properties compared with observed properties, and appropriate candidate models are retained or rejected. Common choices of structural variables, calculation terms, and adjustment algorithms are summarized on the right of the figure. The scheme illustrated here is for powder X-ray diffraction.

inorganic materials have typically used an atomic representation, although a simplifying focus on only metal centers ('T' atoms) is often employed. A recent extension to such automated model building procedures has been described which uses secondary building units of arbitrary size [50].

Simulation methods using Monte Carlo moves have been applied to a range of experimental observations such as electron diffraction data and extended X-ray absorption fine structure (EXAFS) (see Ref. [52] for a review) and ^{29}Si NMR [53]. In the latter case, simulation provides the means to probe the extent to which Lowenstein and Dempsey's 'rules' (governing the proximal siting of aluminum sites on the zeolite lattice) are discerned in reported ^{29}Si NMR results.

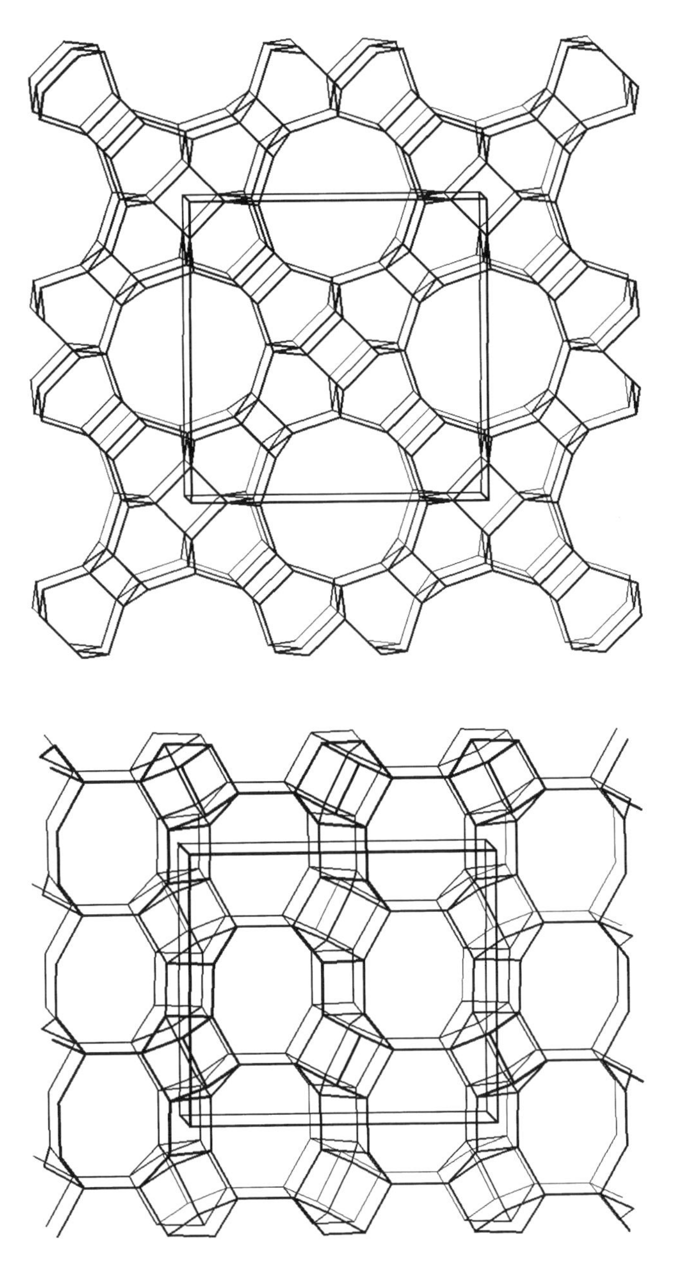

Fig. 4. Three zeolite structures solved using simulated annealing. (a) The OSI framework of UIO-6 [47], (b) the ZON framework of UIO-7 [51], (c) the ESV framework of ERS-7 [48].

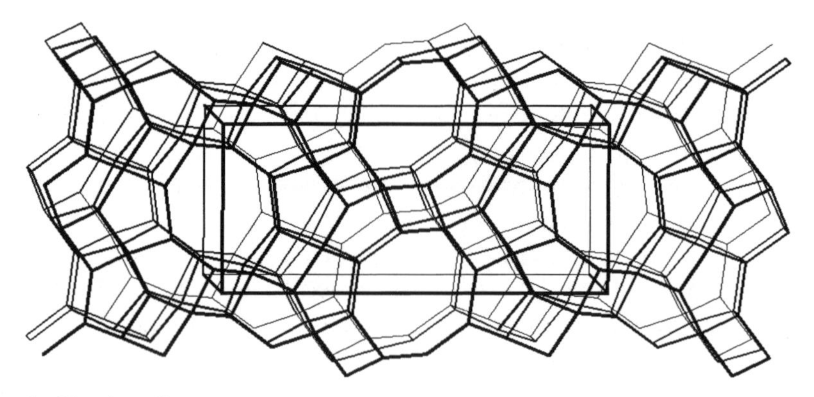

Fig. 4. (Continued)

4. Physisorption

The simulation of the sorption of guest molecules within zeolitic materials continues to receive considerable attention. Early studies by Kiselev and co-workers [54] illustrated the ability of simulation to account for the sorption of gases using Monte Carlo methods. Simulations in this tradition have continued and provide an important means for the explanation of physical observation on the basis of interatomic interactions. For example, Richards and co-workers used Grand Canonical Monte Carlo simulation to predict adsorption isotherms for nitrogen, oxygen, and argon in zeolite Li-X [55]. As in the simulation of polymers, the environment within zeolite materials imposes considerable constraints on the conformational freedom of complex molecules. Making simulation practical in such regimes is taxing, generally requiring important sampling methods [56]. As discussed in Chapter 2, Smit and co-workers showed that, for appropriately sized alkane chains in silicalite, it is possible for a large proportion of the molecules within the zeolite to become frozen in place at appropriate loadings [57]. Experimental adsorption isotherms for alkanes show steps or kinks which resemble their simulated counterparts, providing support for the suggestion that, given appropriate loading and molecular registry with the pore structure of the zeolite, it is possible for a portion of the guest molecules to lose significant mobility. The inclusion of framework flexibility in sorption has received relatively little attention with the exception of the simulation, using Grand Canonical Molecular Dynamics, of the water adsorption isotherm of Na-MAP [58], leading to a qualitative understanding of the form of the experimentally observed isotherm.

In parallel with thermodynamic simulations of sorptive properties, there have been developments in the use of static simulation methods that seek to account for dominant binding locations without tackling the simulation of loading as a function of pressure [59–61]. Such studies have been of value in examining a variety of organic–zeolite interactions and, when combined with appropriate potential functions, in predicting the observed binding locations of organometallic templating agents [62]. An illustration of the interplay between such simulation studies and experimental work is provided by the use of Monte Carlo calculated energies in probing the organo-cation templated nucleation of high-silica zeolites by Zones and co-workers [63].

The interaction of guest molecules and zeolite host structures is dynamic, and experiment and simulation have long sought to provide a description of the kinetics of such systems. Experimentally several techniques are employed, including pulsed-field gradient NMR and frequency response methods. As discussed in Chapter 3, molecular dynamics provides the most straightforward route to simulated dynamical information for zeolite systems, and good agreement is generally obtained between simulated and measured data, for relatively small adsorbate molecules [64–67]. However, as adsorbate size increases, and diffusivities concomitantly decline, simulation times become prohibitive and recourse to transition-state methods is required [68]. Such methods can be applied to the case of the diffusion of rigid entities in microporous materials and rely on the mapping of the energy surface in the region of transition states in key regions where potentially rate-determining steps in migration mechanism are present. As an alternative to such transition-state-based methods qualitative comparison of diffusion under varying adsorbate or framework conditions can be made on the basis of computed energy profiles [69,70]. An illustration of such a calculated energy profile for the diffusion of toluene in the channel of boggsite is provided in Fig. 5.

5. Quantum mechanical methods

The preceding methods yield structural and physisorptive models that describe the structural and energetic properties of zeolitic materials. As discussed elsewhere in this book, the models are based upon potential energy functions and potential parameters. When potential parameters are not available, or information on chemical bonding is sought, a quantum mechanical approach is appropriate and as shown earlier has become increasingly practical, even for comparatively large zeolite

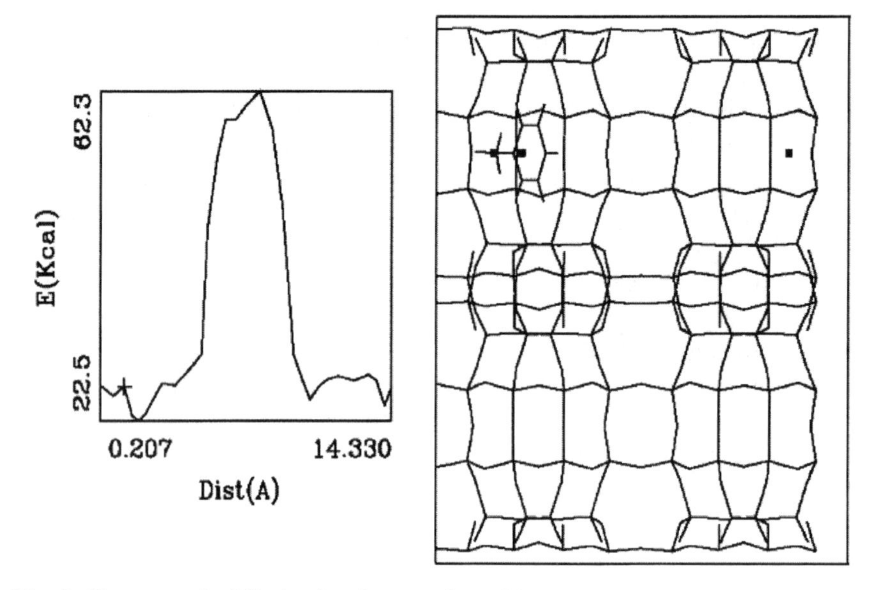

Fig. 5. The energetic diffusion barrier experienced by a toluene molecule in the microporous material boggsite. The simulation model is shown on the right-hand side of the diagram.

systems, through the development of density functional methods and the advance in computational resources.

An example of the growing scope of first principle methods, that is rather different from those discussed earlier, is provided by their use in aiding the Rietveld refinement of Na-MAP [71]. Here, as in the cases of NU-87 and MeAlPO-36 described above, quantum mechanical energy minimization, based on a preliminary starting model, led to a model with improved agreement with the observed diffraction profile.

A particular advantage of quantum mechanical calculations is their ability to provide information on bond strengths. For example, quantum calculations can address the relative acidity of differing acid sites within the zeolite framework and have provided much information to complement, for example, NMR-based studies of reactivity in zeolitic catalytic systems (see, for example, Ref. [72]). As discussed in earlier chapters, quantum mechanical simulations applied to zeolite reactivity have been based both on cluster models and periodic systems. For example, the simulation of the energetics and bond strengths of different possible proton sites in a simplified model of zeolite Y have been reported [73] (see Fig. 6). Full-geometry optimizations were performed using local density functional theory and final energies were determined with gradient corrections. Here, results of low-energy proton-binding

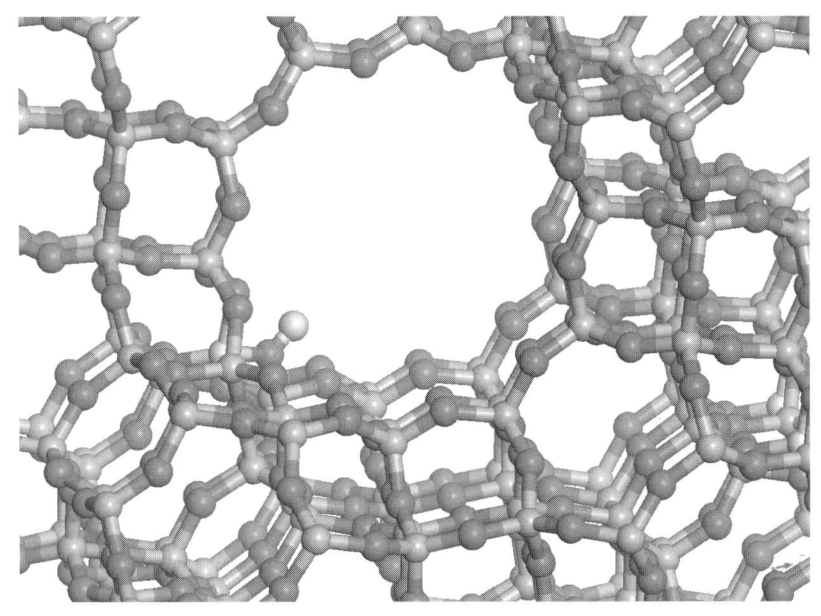

Fig. 6. The structure used in the direct simulation of acid sites in faujasite. The zeolite's framework is composed of SiO_4 tetrahedra linked through bridging oxygen atoms. A single charge-compensating proton is shown bound to a bridging oxygen atom which is attached to a aluminium atom. $DMol^3$ was used in the simulations [74].

configurations are in excellent agreement with experiment and with hybrid quantum mechanical–force field calculations.

An important development is the use in a number of studies of density functional-based molecular dynamics calculations for zeolite systems [75–77]. These simulations, though computationally demanding, are able to include adsorbates and can address proton transfer and other catalytically important events. Overall these and the results described in Chapters 5–7 illustrate the increasingly close coupling of quantum mechanical techniques and real problems in zeolite science.

6. Conclusions

In zeolite science simulation and experiment interact closely. The direct interplay between the two disciplines is important in structural characterization when the ability of a simulation to account for diverse experimental observations may be important in the development of detailed models. Simulation methods may also be of importance in directly generating models which accord with observation, as in simulated annealing studies which reproduce observed diffraction or NMR data.

When structural information is available, computer simulation methods using potential models are able to assess the consequence of available structural information. Such modeling is of value in examining symmetry lowering distortions, for example. Sorption diffusive phenomena are also accessible once structural information is available.

When detailed structural models are available, quantum mechanical methods are of increasing importance. Quantum mechanical methods can now directly simulate reactions relating to the catalytic activity of zeolite materials [76,78,79]. Although such simulations are possible, considerable computational power is required and such calculations are far from routine.

The unique and customizable properties of zeolitic materials will insure their continued use in a variety of industrial and related processes. As a consequence, simulation will continue to be employed to complement and augment experimental probes. There are a variety of additional stimuli, which underlie the evolving close interplay between simulation and experiment in zeolite science. For example, as already noted, computational resources are increasingly powerful and available, and improved algorithms are being developed and applied to zeolite systems. This trend of increasing computational power acts to remove simulation bottlenecks that currently impede the application of quantum mechanical methods to zeolite systems. While simulation costs decline, experimental costs are increasing, and targeting experimental efforts provides a strong incentive for the planning of research programs through the early development of computational models. The availability of computational resources also leads to the increased electronic availability of experimental results in a form suitable for incorporation into simulation studies, again strengthening the interaction between simulation and experiment. It is clear that the interaction of simulation and experiment will continue to be of importance in the development of zeolite science.

References

1. Perutz, M.F., *Proc. R. Inst.*, **62**, 183–198 (1990).
2. Catlow, C.R.A., Thomas, J.M., Freeman, C.M., Wright, P.A. and Bell, R.G., *Proc. R. Soc. Lond. A*, **442**, 85–96 (1993).
3. Rietveld, H.M., *J. Appl. Crystallogr.*, **2**, 65 (1969).
4. Meier, W.M. and Villiger, H., *Z. Kristallogr.*, **129**, 411 (1969).

5. O'Keeffe, M., *Phys. Chem. Miner.*, **22**, 504–506 (1995).
6. Friedrichs, O.D., Dress, A.W.M., Huson, D.H., Kinowski, J. and Mackay, A.L., *Nature*, **400**, 644–647 (1999).
7. Akporiaye, D.E. and Price, G.D., *Zeolites*, **9**, 23–32 (1989).
8. Treacy, M.M.J., Randall, K.H., Rao, S., Perry, J.A. and Chadi, D.J., *Z. Kristallogr.*, **212**, 768–791 (1997).
9. Freeman, C.M., Levine, S.M., Newsam, J.M., Sauer, J., Tomlinson, S.M., Brickmann, J. and Bell, R.G., In: Catlow, C.R.A. (Ed.), *Modelling of Structure and Reactivity in Zeolites*. Academic Press, London, 1992, pp. 133–155.
10. DeGuzman, R.N., Shen, Y.-F., Neth, E.J., Suib, S.L., O'Young, C.-L., Levine, S.M. and Newsam, J.M., *J. Mater. Chem.*, **6**, 815–821 (1994).
11. Shen, Y.F., Zerger, R.P., DeGuzman, R.N., Suib, S.L., McCurdy, L., Potter, D.I. and O'Young, C.L., *Science*, **260**, 511–515 (1993).
12. Feuston, B.P. and Higgins, J.B., *J. Phys. Chem.*, **98**, 4459–4462 (1994).
13. Thomas, J.M., *Nature*, **368**, 289–290 (1994).
14. $Cerius^2$ and Materials Studio. Accelrys, 9685 Scranton Road, San Diego, CA 92121, 2002.
15. Jackson, R.A. and Catlow, C.R.A., *Mol. Simulation*, **1**, 207–224 (1988).
16. Gale, J.D., *J. Chem. Soc., Faraday Trans.*, **93**, 629–637 (1997).
17. Hill, J.-R., Freeman, C.M. and Subramanian, L., *Rev. Comp. Chem.*, **16**, 141–216 (2000).
18. Catlow, C.R.A., Cormack, A.N. and Theobald, F., *Acta Crystallogr. B*, **40**, 195 (1984).
19. Bell, R.G., Jackson, R.A. and Catlow, C.R.A., *J. Chem. Soc. Chem. Commun.*, 782 (1990).
20. de Vos Burchart, E., van Bekkum, H. and van de Graaf, B., *Zeolites*, **13**, 212–215 (1993).
21. Shannon, M.D., Casci, J.L., Cox, P.A. and Andrews, S.J., *Nature*, **353**, 417–420 (1991).
22. Wright, P.A., Natarajan, S., Thomas, J.M., Bell, R.G., Gai-Boyes, P.L., Jones, R.H. and Chen, J., *Angew Chem. Int. Ed. Engl.*, **31**, 1472–1475 (1992).
23. Gale, J.D. and Henson, N.J., *J. Chem. Soc. Faraday Trans.*, **90**, 3175–3179 (1994).
24. Sankar, G., Bell, R.G., Thomas, J.M., Anderson, M.W., Wright, P.A. and Rocha, J., *J. Phys. Chem.*, **100**, 449–452 (1996).
25. de Man, A.J.M., Ueda, S., Annen, M.J., Davis, M.E. and van Santen, R.A., *Zeolite*, **12**, 789–800 (1992).
26. Akporiaye, D.E. and Price, G.D., *Zeolites*, **9**, 321–328 (1989).
27. Henson, N.J., Cheetham, A.K. and Gale, J.D., *Chem. Mater.*, **6**, 1647–1650 (1994).
28. Tschaufeser, P. and Parker, S.C., *J. Phys. Chem.*, **99**, 10609–10615 (1995).
29. Brennan, D., Bell, R.G., Catlow, C.R.A. and Jackson, R.A., *Zeolites*, **14**, 650–659 (1994).
30. Schön, J.C. and Jansen, M., *Angew Chem. Int. Ed. Engl.*, **35**, 1286–1304 (1996).
31. Newsam, J.M., Freeman, C.M., Gorman, A.M. and Vessal, B., *Chem. Commun.*, 1945–1946 (1996).
32. Gorman, A.M., Freeman, C.M., Koelmel, C.M. and Newsam, J.M., *Faraday Discuss.*, **106**, 489–494 (1997).

33. Guliants, V.V., Mullhaupt, J.T., Newsam, J.M., Gorman, A.M. and Freeman, C.M., *Catal. Today*, **50**, 661–668 (1999).
34. Higgins, F.M., Watson, G.W. and Parker, S.C., *J. Phys. Chem. B*, **101**, 9964–9972 (1997).
35. Grillo, M.E. and Carrazza, J., *J. Phys. Chem.*, **100**, 12261–12264 (1996).
36. Pannetier, J., Bassas-Alsina, J., Rodriguez-Carvajal, J. and Caignaert, V., *Nature*, **346**, 343–345 (1990).
37. Freeman, C.M. and Catlow, C.R.A., *J. Chem. Soc., Chem. Commun.*, 89–91 (1992).
38. Boisen, M.B., Gibbs, G.V. and Bukowinski, M.S.T., *Phys. Chem. Miner.*, **21**, 269–284 (1994).
39. Bush, T.S., Catlow, C.R.A. and Battle, P.D., *J. Mater. Chem.*, **5**, 1269–1272 (1995).
40. Newsam, J.M., Freeman, C.M. and Leusen, F.J.J., *Curr. Opin. Solid State Mater. Sci.*, **4**, 515–528 (1999).
41. Deem, M.W. and Newsam, J.M., *Nature*, **342**, 260–262 (1989).
42. McGreevy, R.L. and Pusztai, L., *Mol. Simulation*, **1**, 359–367 (1988).
43. Deem, M.W. and Newsam, J.M., *J. Am. Chem. Soc.*, **114**, 7189–7198 (1992).
44. Falcioni, M.J., *Chem. Phys.*, **110**, 1754–1766 (1999).
45. Grosse-Kunstleve, R.W., McCusker, L.B. and Baerlocher, C., *J. Appl. Crystallogr.*, **30**, 985–995 (1997).
46. Grosse-Kunstleve, R.W., McCusker, L.B. and Baerlocher, C., *J. Appl. Crystallogr.*, **32**, 536–542 (1999).
47. Akporiaye, D.E., Fjellvag, H., Halvorsen, E.N., Haug, T., Karlsson, A. and Lillerud, K.P., *Chem. Commun.*, **13**, 1553–1554 (1996).
48. Campbell, B.J., Bellussi, G., Carluccio, L., Perego, G., Cheetham, A.K., Cox, D.E. and Millini, R., *Chem. Commun.*, **16**, 1725–1726 (1998).
49. Wu, M.G., Deem, M.W., Elomari, S.A., Medrud, R.C., Zones, S.I., Maesen, T., Kibby, C., Chen, C.-Y. and Chan, I.Y., *J. Phys. Chem. B*, **106**, 264–270 (2002).
50. Mellot Draznieks, C., Newsam, J.M., Gorman, A.M., Freeman, C.M. and Férey, G., *Angew Chem. Int. Ed.*, **39**, 2270–2275 (2000).
51. Akporiaye, D.E., Fjellvag, H., Halvorsen, E.N., Hustveit, J., Karlsson, A. and Lillerud, K.P., *J. Phys. Chem.*, **100**, 16641–16646 (1996).
52. McGreevy, R.L., Reverse Monte Carlo methods for structural modelling, In: Catlow, C.R.A. (Ed.), *Computer Modeling in Inorganic Crystallography*. Academic Press, San Diego, 1997, pp. 151–184.
53. Peterson, B.K., *J. Phys. Chem. B*, **103**, 3145–3150 (1999).
54. Bezus, A.G., Kiselev, A.V., Lopatkin, A.A. and Quang Du, P., *J. Chem. Soc., Faraday Trans. 2*, **74**, 367–379 (1976).
55. Richards, A.J., Watanabe, K., Austin, N. and Stapleton, M.R., *J. Porous Mater.*, **2**, 43–49 (1995).
56. Bates, S.P., van Well, W.J.M., van Santen, R.A. and Smit, B., *J. Phys. Chem.*, **100**, 17573–17581 (1996).
57. Smit, B. and Maesen, T.L.M., *Nature*, **374**, 42–44 (1995).
58. Hill, J.-R., Minihan, A.R., Wimmer, E. and Adams, C., *J. Phys. Chem. Chem. Phys.*, **2**, 4255–4264 (2000).
59. Freeman, C.M., Catlow, C.R.A., Thomas, J.M. and Brode, S., *Chem. Phys. Lett.*, **186**, 137–142 (1991).

60. Lewis, D.W., Freeman, C.M. and Catlow, C.R.A., *J. Phys. Chem.*, **99**, 11194–11202 (1995).
61. Millini, R., *Catal. Today*, **41**, 41–51 (1998).
62. Schneider, A.M. and Behrens, P., *Chem. Mater.*, **10**, 679–681 (1998).
63. Zones, S.I., Nakagawa, Y., Yuen, L.T. and Harris, T.V., *J. Am. Chem. Soc.*, **118**, 7558–7567 (1996).
64. June, R.L., Bell, A.T. and Theodorou, D.N., *J. Phys. Chem.*, **94**, 8232–8240 (1990).
65. Catlow, C.R.A., Freeman, C.M., Vessal, B., Tomlinson, S.M. and Leslie, M., *J. Chem. Soc., Faraday Trans.*, **87**, 1947 (1991).
66. Karger, J. and Pfeifer, H., *Zeolites*, **12**, 872–873 (1992).
67. Shen, D. and Ress, L.V.C., *J. Chem. Soc., Faraday Trans.*, **92**, 487 (1996).
68. June, R.L., Bell, A.T. and Theodorou, D.N., *J. Phys. Chem.*, **95**, 8866–8878 (1991).
69. Horsley, J.A., Fellmann, J.D., Derouane, E.G. and Freeman, C.M., *J. Catal.*, **147**, 231–240 (1994).
70. Millini, R., *Catal. Today*, **41**, 41–51 (1998).
71. Hill, J.-R., Stuart, J.A., Minihan, A.R., Wimmer, E. and Adams, C., *J. Phys. Chem. Chem. Phys.*, **2**, 4249–4254 (2002).
72. Haw, J.F., Nicholas, J.B., Xu, T., Beck, L.W. and Ferguson, D.B., *Acc. Chem. Res.*, **29**, 259–267 (1996).
73. Hill, J.R., Freeman, C.M. and Delley, B., *J. Phys. Chem. A*, **103**, 3773–3777 (1999).
74. DMol^3. Accelrys, 9685 Scranton Road, San Diego, CA 92121, 2002.
75. Haase, F., Sauer, J. and Hutter, J., *Chem. Phys. Lett.*, **266**, 397–402 (1997).
76. Stich, I., Gale, J.D., Terkura, K. and Payne, M.C., *Chem. Phys. Lett.*, **283**, 402–408 (1998).
77. Fois, E., Gamba, A. and Tabacchi, G., *Phys. Chem. Chem. Phys.*, **1**, 531–536 (1999).
78. Rozanska, X., Demuth, T., Hutschka, F., Hafner, J. and van Santen, R.A., *J. Phys. Chem. B*, **106**, 3248–3254 (2002).
79. Rozanska, X., van Santen, R.A. and Hutschka, F., *J. Phys. Chem. B*, **106**, 4652–4657 (2002).

Subject Index

X

Z